Topics in Current Chemistry Collections

The series *Topics in Current Chemistry Collections* presents critical reviews from the journal *Topics in Current Chemistry* organized in topical volumes. The scope of coverage is all areas of chemical science including the interfaces with related disciplines such as biology, medicine and materials science. The goal of each thematic volume is to give the non-specialist reader, whether in academia or industry, a comprehensive insight into an area where new research is emerging which is of interest to a larger scientific audience. Each review within the volume critically surveys one aspect of that topic and places it within the context of the volume as a whole. The most significant developments of the last 5 to 10 years are presented using selected examples to illustrate the principles discussed. The coverage is not intended to be an exhaustive summary of the field or include large quantities of data, but should rather be conceptual, concentrating on the methodological thinking that will allow the non-specialist reader to understand the information presented. Contributions also offer an outlook on potential future developments in the field.

Kenneth Kam-Wing Lo • Peter Kam-Keung Leung
Editors

Metal Ligand Chromophores for Bioassays

With contributions from

Shin Aoki • Chandrasekar Balachandran • David Cullinane
Song Gao • Karmel S. Gkika • Li-Jun Guo • Mark J. Hackett
David Hartnell • Yosuke Hisamatsu • Byung Hak Jhun
Guo-Qing Jin • Tia E. Keyes • Massimiliano Massi • Soo Young Park
Peer M. Schenk • Bradley J. Schwehr • Dayoon Song
Yuichi Tamura • Tomohiro Tanaka • Dihua Tian • Miaomiao Wu
Zhi Ping Xu • Kenta Yokoi • Jiaxi Yong • Youngmin You
Jing Zhang • Jun-Long Zhang • Zexi Zhang • Run Zhang

Springer

Editors
Kenneth Kam-Wing Lo
Department of Chemistry
City University of Hong Kong
Hong Kong, China

Peter Kam-Keung Leung
State Key Laboratory
of Terahertz and Millimeter Waves
and Department of Chemistry
City University of Hong Kong
Hong Kong, China

Partly previously published in Topics in Current Chemistry Volume 380 (2022).

ISSN 2367-4067
Topics in Current Chemistry Collections
ISBN 978-3-031-19862-5

This Springer imprint is published by the registered company Springer Nature Switzerland AG
The registered company address is: Gewerbestrasse 11, 6330 Cham, Switzerland

Contents

Preface

Bioassays are broadly defined as analytical methods to determine the concentration or activity of an analyte in biological samples. The nature of analytes varies from small ions, drug molecules, and metabolites, to large protein molecules such as enzymes and antibodies. Due to the advanced development of optical spectroscopic and imaging instruments and their convenience, fast response times, and excellent sensitivity, most bioassays rely on optical detection, such as absorption and fluorescence measurements. Photofunctional d- and f-block metal complexes have started to make extraordinary contributions to the design of sensitive bioassays because they offer intriguing photophysical properties. For example, metal complexes show intense absorption in the UV and visible region, with the emission occurring at much lower energy. Thus, the resulting large Stokes shifts minimize self-quenching of emission, which is commonly observed in fluorescent organic dyes. Also, the emission lifetimes are much longer than organic compounds, allowing more sensitive time-resolved spectroscopic and microscopy techniques. Additionally, many metal complexes are photostable, making it possible to detect and visualize target analytes continuously in real time. Furthermore, the rich redox properties of metal complexes can be exploited in ultrasensitive electrochemiluminescence detection.

The review articles included in this topical collection focus on the contemporary role of photofunctional metal complexes in the design of bioassays.

Near-infrared (NIR) luminescence imaging is an emerging technique that offers low light scattering, minimal cell autofluorescence, deep tissue penetration, and high spatio-temporal resolution. Junlong Zhang summarizes the recent progress of NIR molecular probes based on d-block transition metal and f-block lanthanide complexes. This contribution highlights metal-based NIR-emitting probes focusing on their photophysical properties and applications in bioassays.

Peptides with cell-penetrating, signaling, and receptor-targeting properties have been conjugated to luminescent transition metal complexes to alter their cell permeability and intracellular localization. Tia Keyes describes the progress of peptide conjugation as a strategy to tune the cellular uptake and localization properties of metal complexes. This contribution showcases representative biological applications of these conjugates.

Lipid droplets and phospholipids are important constituents of sub-cellular organelles and are involved in signaling and regulating molecules. Massimiliano Massi summarizes luminescent metal complexes for imaging lipids in biological systems using advanced microscopy. This contribution illustrates the current progress of metal-based complexes in studying lipids and the future challenges in the field.

Youngmin You reviews the biological applications of luminescent iridium(III) complexes as biolabels and sensors. These complexes are utilized in imaging intracellular organelles, such as mitochondria, lysosomes, and cell membranes, and the visualization of tumor spheroids and parasites. Also, sensors for temperature, pH, carbon dioxide, metal ions, anions, biothiols, reactive oxygen species, peptides, and viscosity are introduced. This contribution emphasizes the high application potential of functionally modified iridium(III) complexes.

Shin Aoki presents the broad applications of iridium(III) complexes in organic light-emitting diodes (OLEDs), photocatalysts, bioimaging probes, and anti-cancer reagents. Since these examples involve the modification of thermally-unstable functional moieties such as peptides and sugars, post-complexation functionalization (PCF) is required. This contribution introduces examples of PCF to construct luminescent probes with various emission colors, pH sensitivity, anti-cancer properties, and long emission lifetimes.

Run Zhang reviews recent advances in developing ruthenium(II) complexes as chemosensors for biologically-relevant anions, metal ions, and small biomolecules, including amino acids and reactive nitrogen, oxygen, and carbon species. This contribution focuses on the detection of intracellular analytes through luminescence analysis and imaging.

In summary, the development of photofunctional metal complexes for bioassays is highly multi-disciplinary, involving contributions from inorganic and analytical chemists, photochemists, cell biologists, chemical biologists, and medical scientists. We hope this topical collection will highlight the role of photofunctional metal complexes as biological reagents, which has gained considerable attention in recent years. Finally, we thank all the contributing authors for their significant efforts and contributions that have made this volume possible.

Kenneth Kam-Wing Lo
Department of Chemistry, City University of Hong Kong

Peter Kam-Keung Leung
State Key Laboratory of Terahertz and Millimeter Waves
and Department of Chemistry, City University of Hong Kong

Topics in Current Chemistry (2022) 380:31
https://doi.org/10.1007/s41061-022-00386-6

REVIEW

Luminescent Metal Complexes for Bioassays in the Near-Infrared (NIR) Region

Guo-Qing Jin[1] · Li-Jun Guo[1] · Jing Zhang[2] · Song Gao[1,2,3,4] · Jun-Long Zhang[1,4]

Received: 14 February 2022 / Accepted: 14 May 2022 / Published online: 18 June 2022

Abstract

Near-infrared (NIR, 700–1700 nm) luminescent imaging is an emerging bioimaging technology with low photon scattering, minimal autofluorescence, deep tissue penetration, and high spatiotemporal resolution that has shown fascinating promise for NIR imaging-guided theranostics. In recent progress, NIR luminescent metal complexes have attracted substantially increased research attention owing to their intrinsic merits, including small size, anti-photobleaching, long lifetime, and metal-centered NIR emission. In the past decade, scientists have contributed to the advancement of NIR metal complexes involving efforts to improve photophysical properties, biocompatibility, specificity, pharmacokinetics, in vivo visualization, and attempts to exploit new ligand platforms. Herein, we summarize recent progress and provide future perspectives for NIR metal complexes, including d-block transition metals and f-block lanthanides (Ln) as NIR optical molecular probes for bioassays.

Keywords Near-infrared luminescence · Bioimaging and biosensing · D-block transition metals · Lanthanides

---

Chapter 1 was originally published as Jin, G-Q., Guo, L-J., Zhang, J., Gao, S. & Zhang, J-L. Topics in Current Chemistry (2022) 380: 31. https://doi.org/10.1007/s41061-022-00386-6.

✉ Jing Zhang
Zhangj271@ucas.ac.cn

✉ Jun-Long Zhang
zhangjunlong@pku.edu.cn

1 Beijing National Laboratory for Molecular Sciences, College of Chemistry and Molecular Engineering, Peking University, Beijing 100871, People's Republic of China

2 Center of Materials Science and Optoelectronics Engineering, College of Materials Science and Opto-Electronic Technology, University of Chinese Academy of Sciences, Beijing 100049, People's Republic of China

3 The Institute of Spin Science and Technology, Guangdong-Hong Kong-Macao Joint Laboratory of Optoelectronic and Magnetic Functional Materials, South China University of Technology, Guangzhou 510641, People's Republic of China

4 Chemistry and Chemical Engineering Guangdong Laboratory, Shantou 515031, People's Republic of China

Abbreviations

CT	Computed tomography
MRI	Magnetic resonance imaging
PET	Positron emission tomography
NIR	Near-infrared
SBR	Signal-to-background ratio
PLIM	Photoluminescence lifetime imaging microscopy
ISC	Intersystem crossing
SOC	Spin–orbit coupling
ATP	Adenosine triphosphate
ALP	Alkaline phosphatase
SIM	Structured illumination microscopy
IEDDA	Inverse electron-demand Diels–Alder
BCN	(1R,8S,9s)-Bicyclo[6.1.0]non-4-yne
MRSA	Methicillin-resistant *S. aureus*
STED	Stimulated emission depletion
PeT	Photoinduced electron transfer
MLCT	Metal-to-ligand charge transfer
LMCT	Ligand-to-metal charge transfer
TADF	Thermally assisted delayed fluorescence
DOTA	1,4,7,10-Tetraacetic acid
DTPA	Diethylenetriaminepentaacetic acid
hfac	Hexafluoroacetylacetonate

1 Introduction

Generally, molecular imaging permits noninvasive visualization, characterization, and quantification of biological processes taking place at the cellular and subcellular levels within intact living subjects, providing critical diagnostic information regarding the presence and anatomic location of targets. Noninvasive clinical imaging technologies, including computed tomography (CT) [1], magnetic resonance imaging (MRI) [2], ultrasound (US) [3], and positron emission tomography (PET) [4, 5], have been widely employed in biomedical research and clinical practice for diagnosis, progression monitoring, and prognostic evaluation of diseases over the past decade. Optical imaging is bio-friendly, real-time, and highly sensitive, with the detection concentration of ~ 10^{-9} to 10^{-12} M, and has recently attracted considerable attention [6]. However, biological media such as water, lipids, hemoglobin, and melanin in living systems that usually absorb photons and biological tissues interact with photons, significantly limiting the tissue penetration depth in the visible window (~ 1 mm) [7].

In the past 5 years, the emergence of near-infrared (NIR, 700–1700 nm) luminescent probes has ushered in a thorough-paced revolution for in vivo optical imaging with a deeper tissue penetration depth of 2–8 mm and higher signal-to-background ratio (SBR) owing to substantially reduced tissue absorption and scattering. These advantages have facilitated the development of various NIR imaging systems

[8] and chemical probes, including organic fluorophores, semiconductor quantum dots, carbon nanotubes, and rare-earth-doped nanoparticles [9–13], showing in vivo bioimaging with high temporal-spatial resolution. Among all reported NIR materials, metal complex-based luminophores feature merits [14] including molecular scale, microsecond (μs) to millisecond (ms) lifetime, high brightness, robustness to photobleaching, large Stokes shift, and characteristic metal-centered NIR emission (700–1700 nm), exhibiting great potential in biomedical practice, e.g., immunofluorescence imaging, super-resolution imaging, multi-color imaging, and fluorescent surgical navigation. NIR luminescent metal complexes have been widely investigated as burgeoning choices for biomedical assays in the past decade. The luminescence lifetime is hardly affected by concentration variation of the intracellular luminophore or the excitation laser power. The microsecond to millisecond fluorescence or phosphorescence lifetime of NIR metal complexes allows for easy differentiation from the nanosecond autofluorescence of living systems, enabling photoluminescence lifetime imaging microscopy (PLIM) that has been widely employed as a significant tool to improve accuracy and sensitivity [15]. Distinct from typical organic fluorophores, metal complexes facilitate the rapid intersystem crossing (ISC) of peripheral ligands relying on strong spin–orbit coupling (SOC) induced by metal ions with large SOC constants (ζ) [16], thus populating triplet states to provide diverse photophysical properties that are unattainable from singlet states, for example, phosphorescence, oxygen sensitization, heat generation, and triplet state-metal excited-state energy transfer [17]. Numerous scientific efforts have proved the irreplaceable role of luminescent metal complexes, most of which are indispensible, such as organic light-emitting diodes [18], dye-sensitized solar cells [19], and phototheranostics [20]. Consequently, there are many reasons to search for NIR luminescent metal complexes with longer wavelength, better brightness, higher stability, biocompatibility, and targeting capability for biomedical research and clinical practice.

This review summarizes significant progresses in the molecular design of NIR luminescent metal complexes with good potential for biological applications, mainly in the past 3 years. This review highlights recent advancements in NIR luminescent complexes with different metal centers, including noble metals in the platinum group, earth-abundant metals, particularly the first-row transition metals, and NIR-II (950–1700 nm) emissive lanthanide (Ln) complexes with versatile biomedical applications (Fig. 1). At the end of this review, perspectives focused on promising opportunities and challenges in the development of NIR luminescent metal bioprobes are provided for future scientific research and clinical applications.

2 NIR Luminescent Noble Transition Metal Complexes

Luminescence of platinum group noble complexes such as ruthenium (Ru(II)), iridium (Ir(III)), and platinum (Pt(II)) have exhibited promising photophysical and photochemical properties [21] arising from their intrinsically large ligand field splitting and high spin–orbit coupling (SOC) constants ($\zeta > 1000\ cm^{-1}$), enabling efficient populations of NIR luminescent triplet metal-to-ligand charge transfer (^{3}MLCT)

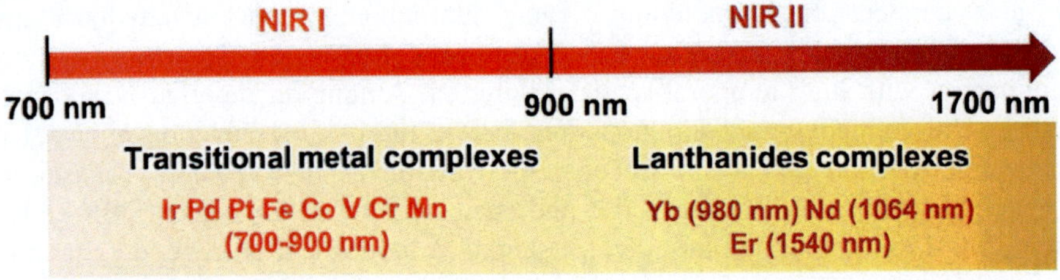

Fig. 1 NIR luminescent metal complexes: transition metal complexes (700–900 nm) and lanthanide (Ln) complexes (900–1700 nm)

states. In past decades, square-planar Pt(II) complexes such as mononuclear, dinuclear, and metallacyclic structures have been extensively investigated due to non-covalent Pt–Pt interactions that enhance NIR luminescence [22–25]. Wong et al. prepared a NIR emissive Pt(II) complex (**1**) consisting of a terpyridyl Pt(II) core, tetraphenylethene (TPE) periphery, and hydrophilic oligo(ethylene glycol) [26]. The terpyridyl Pt(II) core shows strong intermolecular interactions and metal-to-metal–ligand charge transfer (MMLCT) after self-assemblying. The TPE unit was conjugated to the ligand and impaired strong π–π stacking interaction; the hydrophilic oligo(ethylene glycol) unit ensured good biocompatibility. **1** displayed intensive NIR emission and lysosome targeting subcellular localization in NIR cell imaging. Wong and cooperators reported a soft salt [27] of Pt(II) (**2**) with bright NIR emission centered at 750 nm in the solid-state [28]. The single crystal structure was analyzed to display strong MMLCT induced by the Pt–Pt and π–π interactions, facilitating NIR emission of Pt(II). This complex was successfully applied for in vivo NIR bioimaging.

Yam et al. reported a series of Pt(II) terpyridyl complexes with Pt–Pt interactions. Using terpyridyl and dihydroxy phenyl ligands, they prepared a Pt(II) complex (**3**) capable of fluorescently visualizing conformational and microenvironmental changes in DNA aptamers [29]. Specifically, electrostatic contacts between a negatively charged aptamer and positive Pt(II) complex, in either the random-coil 42-mer lysozyme aptamer (LA) or 15-mer thrombin binding aptamer (TBA), rendered them approach to each other and aggregate to generate Pt–Pt and π–π interactions for NIR luminescence. Similarly, polyanionic phosphate derivatives such as adenosine triphosphate (ATP) and phosphopeptide induce supramolecular assembly of cationic alkynylplatinum(II) terpyridine complexes, which aids in fluorescence sensing of phosphorylation relevant enzymes such as alkaline phosphatase (ALP), ATPase, and v-Src kinase [30]. In 2013, Yam et al. reported a pH-sensitive Pt(II) terpyridyl complex (**4**) with NIR emission centered at 795 nm [31]. This complex was found to aggregate in acidic aqueous solutions, showing NIR luminescence induced by strong Pt–Pt and π–π interactions, yet no emission was observed after de-aggregation when pH is >7.4. Consequently, pH-responsive NIR luminescent performance makes it capable of visualizing microenvironmental changes in the acidic organelle, for example, the lysosome. In 2019, the same group developed a sulfonated Pt(II) complex (**5**) to supramolecular self-assemble upon contacting with amyloid and switch on NIR luminescence [32]. This has been successfully demonstrated in sensitive and specific detection of amyloid fibrillation, by introducing positively charged guanidine groups into Pt(II) complexes [33], a Pt(II) probe (**6**) that exhibited switching-on NIR emission after RNA binding has been reported, which was further applied in selective imaging of cell nucleoli in the living cell and the drug screening of RNA inhibitors. In 2021, Tunik et al. reported a series of cyclometalated Pt(II) phosphine chloride complexes, which readily reacted with imidazole or histidine functionalities to give phosphorescent products with a considerably higher brightness than the starting complex of platinum chloride [34]. Furthermore, after conjugation with amyloid-β targeting peptides capable of penetrating the blood–brain barrier (BBB), the resulted Pt(II) complex was successfully conducted in neuroimaging of both wild and amyloid peptide expressing strains of *Drosophila melanogaster* (Fig. 2).

Real-time super-resolution imaging of Pt(II) drugs is significant in elucidating their behavior and distribution in living cells, which is helpful for the development of next-generation Pt(II) drugs. In 2020, Yam et al. designed a NIR phosphorescent dinuclear Pt(II) complex (**7**) with large stokes shift, anti-photobleaching, and super-resolution imaging tolerance [35]. Under light irradiation, one unknown matter transportation route was observed where **7** escaped from the autolysosome to the nucleus. As shown in Fig. 3b, the luminescence intensity in the nucleus gradually increased. In contrast, the signal in autolysosomes decreased in a piecemeal manner, revealing that **7** escaped from autolysosomes to the nucleus and providing a new mechanism to improve the targeting design of Pt(II) drugs.

Chao et al. developed an Ir(III) complex (**8**) capable of super-resolution tracking of mitochondrial dynamics in living cells under structured illumination microscopy (SIM) [36]. It displayed intensive NIR phosphorescence centered at 700 nm and pH-sensitive phosphorescence in PBS between pH 4–10. As shown in Fig. 4, this complex enabled precise observation of mitochondrial cristae with a resolution

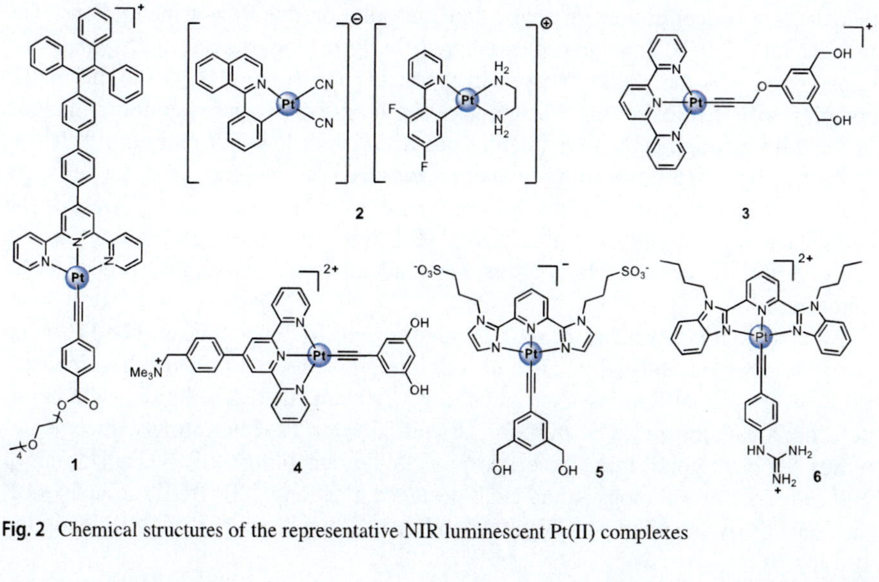

Fig. 2 Chemical structures of the representative NIR luminescent Pt(II) complexes

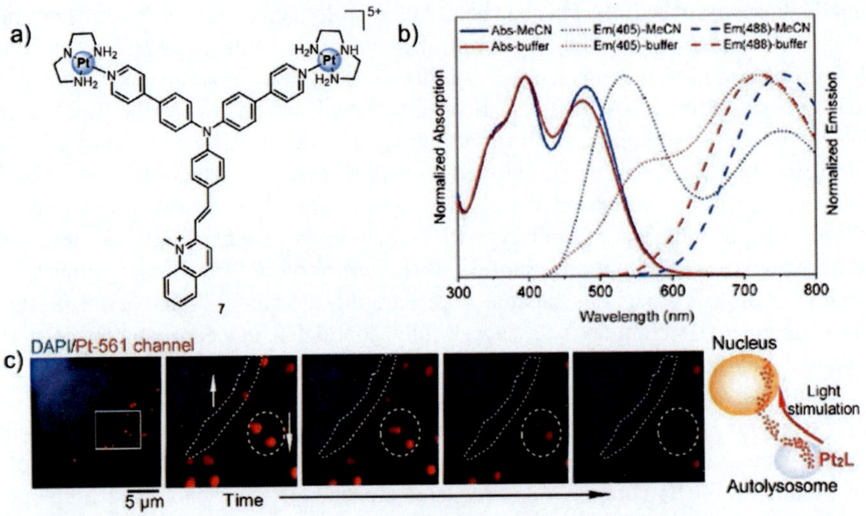

Fig. 3 **a** Chemical structure of the dinuclear Pt(II) complex (**7**). **b** UV/Vis absorption and emission spectra of **7** (10 μM) measured in acetonitrile and Tris–HCl buffer (10 mM, pH 7.4, 100 mM K^+) with excitation at 405 nm and 488 nm. **c** The dynamic photoactivated escape process of Pt(II) from the autolysosomes to the nucleus with the luminescence intensity gradually decreasing in autolysosomes and increasing in the nucleus. Reprinted with permission from Copyright 2020 John Wiley and Sons Ltd [35]

up to 80 nm. It was also used to visualize mitochondrial-lysosome fusion involved in mitophagy and discover new mitochondrial-lysosomes contact under different conditions. Living cell imaging of lysosome localization and movement is critical to study the status of lysosomes and the function of pharmacological interventions. To create a stable lysosomal probe for long-term imaging of living cells, Chao et al. recently synthesized a peptidyl cyclometalated phosphorescent Ir(III) complex (**9**) with a NIR emission tail at ~ 750 nm [37]. It tended to self-assemble into nanoparticles at low concentrations in acidic aqueous solution due to π–π interactions. The self-assembled Ir(III) nanoparticles entered the lysosome via cell endocytosis and transformed into the stable network in the acidic lysosome (pH ~ 5.5). This Ir(III) complex also exhibited high brightness for long-term super-resolution imaging of the 14th-passage cells after treating passage 0 with low-concentration **9**. It was applied to investigate lysosomal dynamics on a starvation assay of HeLa cells. As shown in Fig. 4b, c, the starved lysosomes were redistributed quickly towards the perinuclear region within 30 min. Using SIM, lysosome-mitochondria contact was successfully observed in which lysosomes gradually approach and interact with the mitochondrion.

Xu et al. synthesized two novel NIR luminescent Ir(III) complexes (**10**,**11**) using phenylbenzo[g]quinoline ligands and different N^N ligands [38]. Both complexes exhibited strong NIR emission centered at ~ 750 nm in PBS solution, with luminescence quantum yields of 62% for **10** and 85% for **11**. Interestingly, these complexes showed good mitochondrial targetability, enabling NIR visualization of mitochondrial morphology. Chao et al. designed a cationic NIR Ir(III) complex (**12**) that selectively accumulated in mitochondria and displayed notable NIR emission

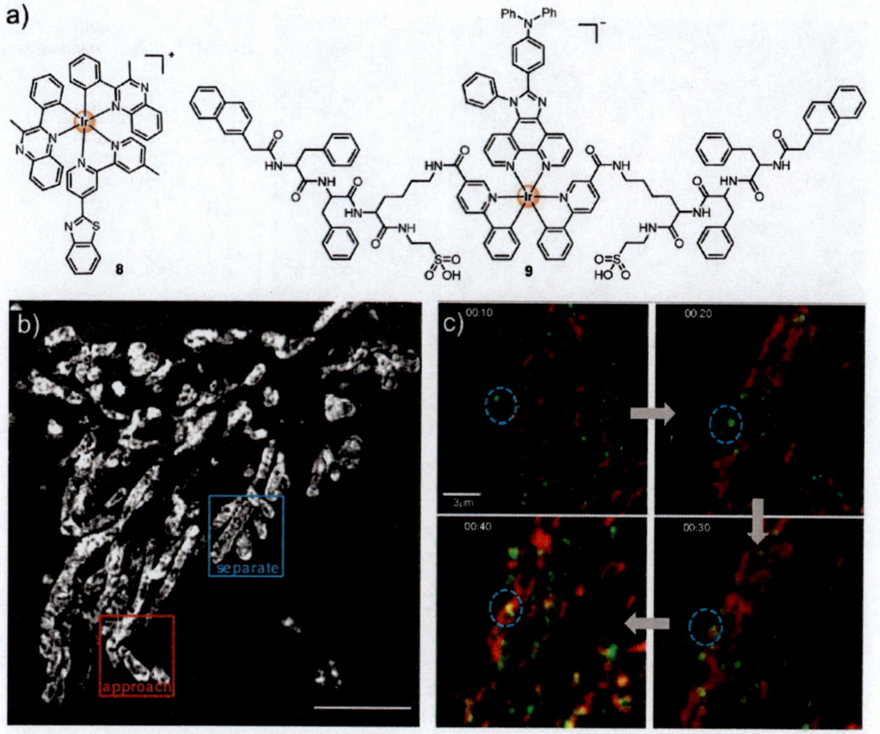

Fig. 4 **a** Chemical structures of Ir(III) complexes **8** and **9**. **b** Iridium(III) complex (**8**) for imaging of mitochondria under SIM. Red rectangles indicate dynamic processes, and blue rectangles indicate dynamic separation processes. Reprinted with permission from Copyright 2021 John Wiley and Sons Ltd [36]. **c** Time-lapse SIM images of HeLa cell treated with **9** co-stained with MitoTracker Deep Red. Reprinted with permission from Copyright 2021 John Wiley and Sons Ltd [37]

under two-photon excitation [39]. **12** was successfully applied for phosphorescent lifetime imaging and time-gated imaging of cells (spheroid) and hippocampal slice. Two-photon excitation microscopy showed deeper tissue penetration, less defocused interference, and superior spatial resolution compared to one-photon excitation microscopy. Using phenanthroline imidazole ligand, Chao et al. developed a highly efficient three-photon phosphorescent Ir(III) complex (**13**) with the maximal two-photon absorption cross-section values of 101 GM and three-photon absorption cross-section value of 2.63×10^{-78} $cm^6s^2photon^2$ when excited at 750 nm and 980 nm, respectively [40]. It was demonstrated in 3D imaging and reconstruction of the brain vasculature of living specimens. As shown in Fig. 5, under three-photon excitation (980 nm), more cerebral blood vessels were observed than under two-photon excitation (750 nm) at the same imaging depth, with substantially improved spatiotemporal resolution. In 2020, a lysosome-targeted Ir(III) complex (**14**) with nearly 100-fold increased NIR phosphorescence (λ_{max} ~ 730 nm) upon stimulus by $ONOO^-$ was reported, enabling accurate detection of $ONOO^-$ with excellent sensitivity (limit of detection, LOD = 49.8 nM) and fast response to $ONOO^-$ (within

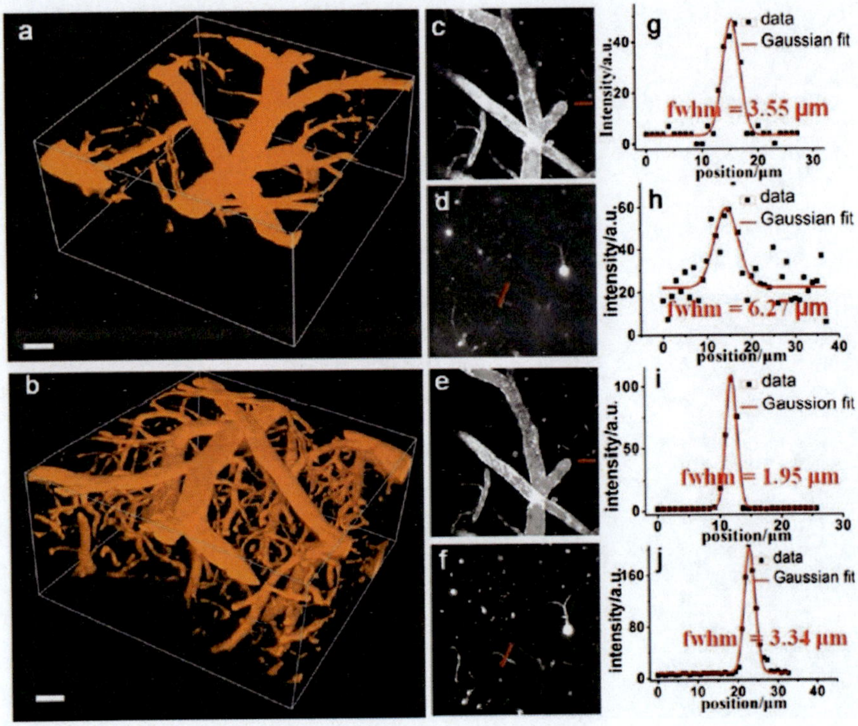

Fig. 5 In vivo 2PE and 3PE images labeled by **13-DSPE-PEG2000**. Three-dimensional reconstruction performed by 2PE (**a**) and 3PE (**b**) imaging, with imaging depth of ca. 200 μm (**a**) and ca. 450 μm (**b**), respectively. **c–f** 2PE (**c**, 60 μ,m and **d** 200 μm) and 3PE (**e** 60 μm and **f** 200 μm) images at different imaging depths. **g–j** FWHM of a blood vessel in the corresponding images. Scale bars: 100 μm. Reprinted with permission from Copyright 2020 John Wiley and Sons Ltd [40]

200 s) [41]. It was successfully applied to explore crosstalk of NO/$O_2^{\cdot -}$ and monitor drug-induced hepatotoxicity in preclinical studies. Independently, Zhao et al. reported a water-soluble triscyclometalated Ir(III) complex (**15**) that self-assembled into nanoparticles due to the introduction of the quaternary ammonium chain, significantly enhancing the phosphorescence intensity [42]. In 2022, Lo et al. developed biorthogonal Ir(III) bis-tetrazine complexes (**16**, **17**). The Ir(III) complexes were non-emissive under photoirradiation but exhibited up to a 3900-fold phosphorescence increase upon the inverse electron-demand Diels–Alder (IEDDA) cycloaddition with (1R,8S,9s)-bicyclo[6.1.0]non-4-yne (BCN) substrates [43]. Considering the fast kinetics (up to 1.5×10^4 $M^{-1}S^{-1}$), **16**, **17** successfully reacted with bi-BCN peptides and a tetra-BCN PEG linker for stapling peptides and synthesizing nanosized hydrogels, respectively (Figs. 6, 7, 8, 9, 10).

Osmium (Os(II)) polypyridyl complexes exhibit outstanding photostability and deep-red to NIR emission in the 700–850 nm spectral region, making them attractive candidates as NIR metal probes. In 2018, Zhang et al. reported two Os(II) complexes (**18**, **19**) with unusual dual emission at 573 nm (green) and 738 nm (NIR)

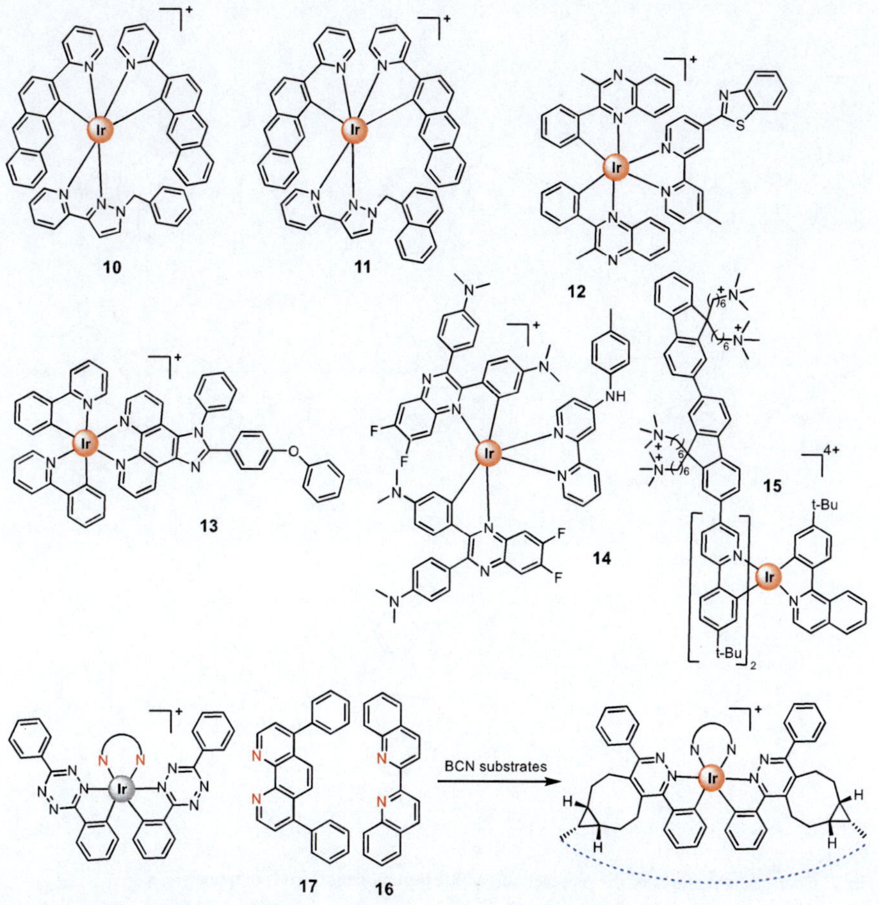

Fig. 6 Chemical structures of the representative NIR luminescent Ir(III) complexes

[44]. It exhibits higher binding activity to RNA than DNA, enabling NIR imaging of RNA in living or fixed cells without autofluorescence as compared to the commercial SYTO RNA-Select. This is the first example of an Os(II) complex capable of NIR RNA-specific labeling, revealing RNA as a new target spot for Os(II)-based prodrugs. In 2020, Elliott et al. reported a set of NIR emissive triazole-based Os(II) complexes (**20**) with DNA binding ability and investigated their antimicrobial activity against methicillin-resistant *S. aureus* (MRSA) [45]. Interestingly, the *fac-/mer*-isomers of **20** exhibited distinct photophysical properties with *mer*-**20** displaying higher antibacterial activity and brighter cell staining properties than *fac*-isomer because of the higher cell uptake of *mer*-counterpart. In 2021, Thomas et al. developed a dinuclear osmium(II) complex (**21**) $[Os(TAP)_{22}tpphz]^{4+}$ (where tpphz = tetrapyrido[3,2-a:2′,3′-c:3″,2″-h:2‴,3‴-j]phenazine and TAP = 1,4,5,8-tetraazaphenanthrene). As **21** showed intensive NIR emission (λ_{max} = 780 nm) with good photostability and no adverse cytotoxicity, super-resolution NIR imaging of

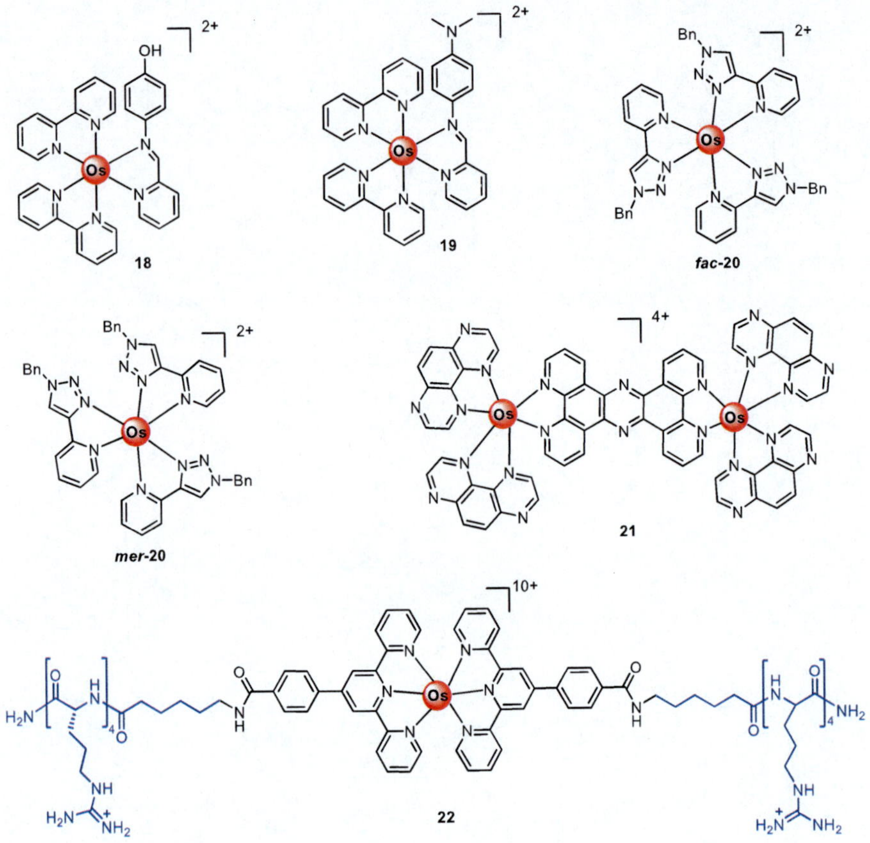

Fig. 7 Chemical structures of the representative NIR luminescent Os(II) complexes

nuclear DNA was performed under stimulated emission depletion (STED) microscopy [46]. In 2021, Keyes et al. prepared an Os(II)-bridged polyarginine conjugates (**22**) with eight positively charged arginine peptide sequence for improved cell permeability in unilaminar cells and three-dimensional (3D) cell spheroid. This conjugate exhibited characteristic NIR fluorescence of Os(II) centered at 754 nm and essentially oxygen-insensitive emission with a lifetime of 89 ns in PBS [47].

The heavy atom effect of transition metals usually leads to the population of triplet states and the accumulation of reactive oxygen species, leading to remarkable phototoxicity, which is unfavorable for biological imaging. Zhang's group developed a set of NIR highly luminescent Pd complexes (quantum yield as high as 14%) (**Pd-3, 25**) by constructing palladium(Pd)-carbon bonds with macrocyclic benzitripyrrin ligands [48]. The presence of the Pd–C bond leads to the destabilization of the d-d excited state, hindering the generation of triplet states and oxygen sensitization. Interestingly, the complex **Pd-2** (**24**) displays enhanced NIR fluorescence under weak acid conditions (pH 4–6) owing to inhibited photoinduced electron transfer (PeT) quenching effect. As a result, it was applied in NIR lysosomal imaging under

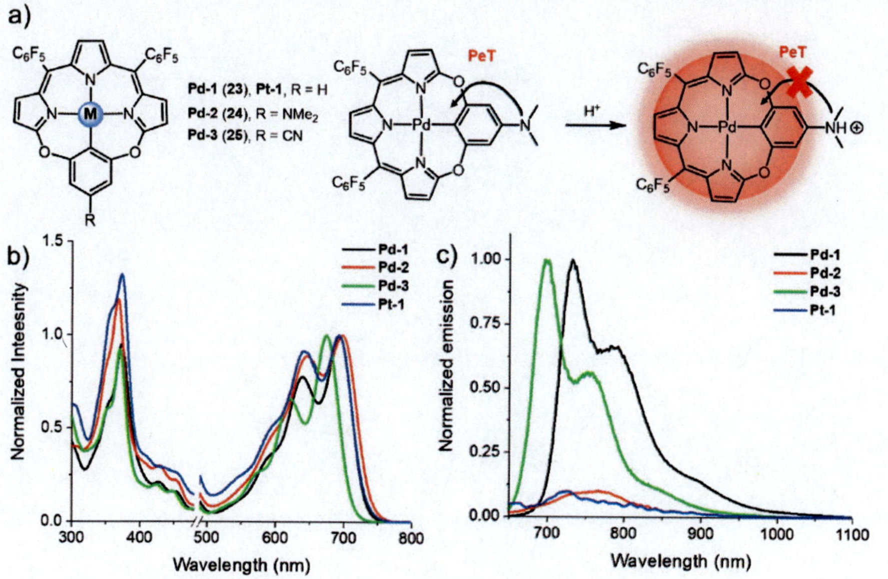

Fig. 8 **a** Chemical structures of NIR luminescent Pd(II) complexes **Pd-1–3** (**23–25**), **Pt-1** (left). Scheme of pH sensitivity of **Pd-2** due to the on/off switching of photoinduced electron transfer (PeT) via deprotonation/protonation. **b** Absorption and **c** emission spectra of **Pd-1–3** and **Pt-1** in CH_2Cl_2. Reprinted with permission from Copyright 2019 Royal Society of Chemistry [48]

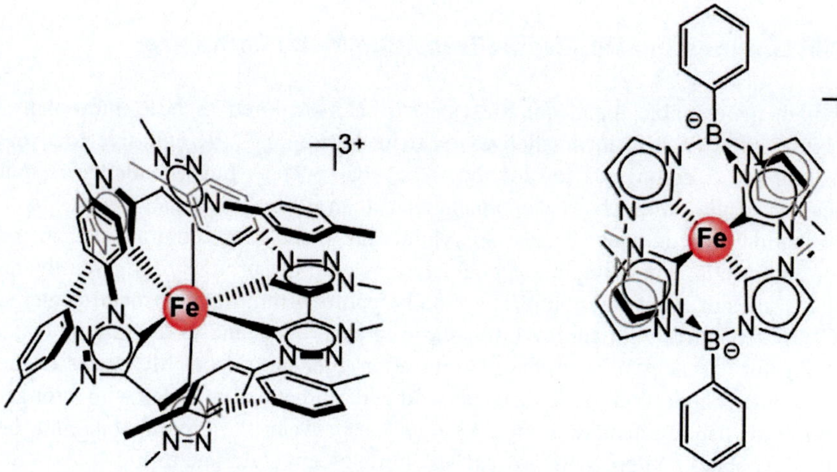

26 **27**

Fig. 9 Chemical structures of NIR luminescent Fe(III) complexes

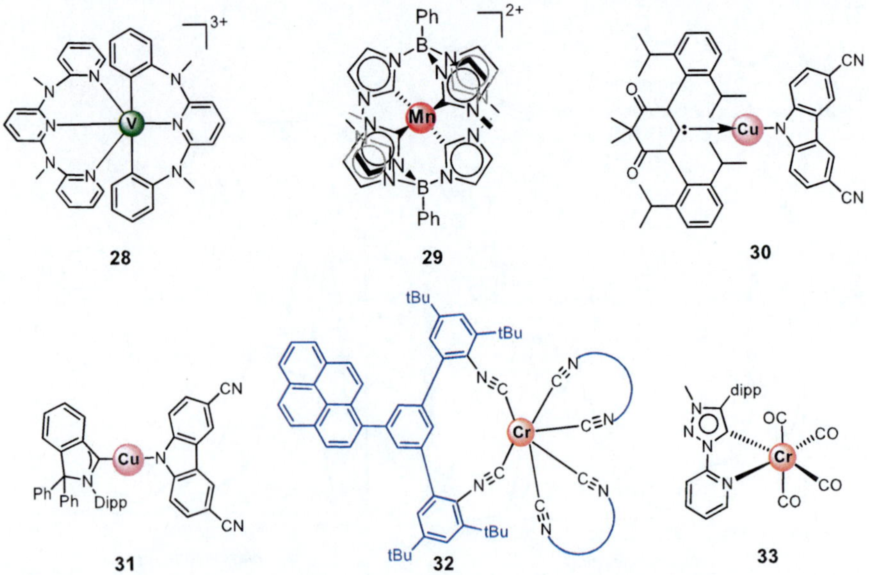

Fig. 10 Chemical structures of NIR luminescent first-row transition metal complexes

different pH values, showing good promise for activated NIR bioimaging of the acidic tumor microenvironment.

3 NIR Luminescent Non-Noble Transition Metal Complexes

Although many noble metal complexes have been explored as NIR fluorophores for bioimaging, significant challenges exist in developing NIR emissive first-row transition metal complexes toward the same goal [49]. A huge challenge is that 3d metals display a much weaker ligand field than 4d or 5d species for the same coordination environment because 3d orbitals are much more contracted than the 4d or 5d orbitals, which impairs spatial overlaps with the orbitals of a given ligand [50]. As a result, 3d metal complexes usually feature a multitude of metal-centered (MC) excited states with energy levels close to each other and to the ground state, allowing for cascades of nonradiative relaxation quenching from MC states to the ground state. The individual MC states of 4d and 5d metal complexes with stronger ligand fields usually display a larger energy gap between the ground state and the lowest MC state, rendering the lowest MC luminescent state due to much less efficient nonradiative relaxation process dissipated by the same number of vibrational quanta [51]. Energetically high-lying MC states allow for the installment of substantially different excited states at lower energies for attractive photophysical properties via metal-to-ligand charge transfer (MLCT) and ligand-to-metal charge transfer (LMCT) commonly governed by the lowest excited state. For instance, thermally assisted delayed fluorescence (TADF)-based Cu(I) complex with > 99% quantum efficiency and microsecond lifetime had been reported for efficient blue-emitting organic light-emitting diode (OLED) [52]. To develop the first-row transition metal complexes with NIR emission, the first task is to select metal ions with potential NIR emission and prepare relevant metal complexes that are NIR luminescent in solution at room temperature.

Iron (Fe(III)) is the most abundant transition metal element in the earth's crust, and the study of its luminescence properties at room temperature has received extensive research attention. However, the photoexcitable charge-transfer states of most Fe(III) complexes are limited by picosecond or subpicosecond deactivation through low-lying MC states, leading to inefficient electron-transfer reactivity and complete lack of photoluminescence. In 2017, Wärnmark et al. reported a Fe(III) complex (**26**) using 3,3′-dimethyl 1,1′-bis(p-tolyl)-4,4′-bis(1,2,3-triazol-5-ylidene) ligand [53]. This ligand showed superior σ-donor and π-acceptor electron properties and stabilized the excited state of low-spin d^5 Fe(III) sufficiently to realize visible to NIR photoluminescence (540–800 nm) at room temperature with a fluorescent quantum yield of 0.03%. The spin-allowed emission of such Fe(III) complex arises from a long-lived doublet LMCT state owing to the absence of intersystem crossing. In 2019, Wärnmark et al. developed an octahedral Fe(III) complex (**27**) using two mono-anionic facial tris-carbene ligands and significantly increased emission quantum yield to 2% [54, 55]. These pioneering works laid the foundation for the development of NIR highly luminescent Fe(III) probes for biological applications.

Heinze et al. reported a vanadium(V(III)) complex *mer*-$[V(ddpd)_2][PF_6]_3$ (**28**) emitting a long-lived phosphorescence around 1100 nm at room temperature in acetonitrile (CH_3CN) with a quantum yield of $1.8 \times 10^{-4}\%$ (ddpd = N,N′-dimethyl-N,N′-dipyridine-2-ylpyridine-2,6-diamine), which is the first earth-abundant first-row transition metal complex emission beyond 1000 nm at room temperature [56, 57]. Telser et al. developed a *bis*-(tris(carbene)borate) manganese(Mn(IV)) complex (**29**), which shows much longer NIR $^2E_g \rightarrow {}^4A_{2g}$ luminescence at 828 nm in the solid state at 85 K than typical Mn(IV)-doped solids owing to a strong σ-donor characteristic of the tris(carbene)borate ligands [58]. In 2021, The same ligand was introduced to establish a cobalt (Co(III)) complex by Persson et al. that exhibited microsecond NIR emission between 700–850 nm in CH_3CN at room temperature [59].

For a $3d^{10}$ electron configurational metal such as Cu(I), no energetically low-lying MC state exists, making it comparatively easy to adopt emissive MLCT and ligand-to-ligand charge transfer (LLCT) excited states. In 2019, Thompson et al. reported a Cu(I) N-heterocyclic carbene complex (**30**) using diamidocarbene and dicyano carbazole as π-electron acceptor and donor, respectively, which leads to a 2% fluorescence quantum yield in 2-MeTHF at room temperature with λ_{max} at 666 nm from an LLCT excited state [60]. Later, by developing Cu(I) cyclic (amino)(aryl)-carbene (CAArC) complexes (**31**) with much more π-extension and electrophilicity, Steffen et al. demonstrated substantially red-shifted emission of Cu(I) to the NIR region (λ_{max} = 738 nm) in THF at room temperature, providing access to NIR Cu(I) luminophores [61]. Recently, a *tert*-butyl isocyanide ligand was developed by Wenger et al. to produce the first luminescent Mn(I) complexes because the strong ligand field of the π-acceptor isocyanide ligands sufficiently elevated the detrimental $^3T_{1g}$ and $^5T_{2g}$ MC states and inhibited nonradiative relaxation [62]. In 2021, the same group conducted a proof-of-concept work for luminescent isoetectronic chromium(0) (**32**).

They utilized an enlarged π-conjugated pyrene diisocyanide to significantly increase the MLCT lifetime at 675 nm (1.57 ns (69%), 5.61 ns (31%)) and a fluorescence quantum yield of **32** up to 0.03% in n-hexane at 20 °C [63]. In the same year, the first Cr(0) complex (**33**) with NIR II emission at 965 nm (380 ± 10 ns), albeit exclusively in the solid state, was achieved by Sarkar et al. using bidentate 1,2,3-triazol-5-ylidene carbene ligands and CO coligands [64] (Fig. 11).

NIR emissive Cr(III) complexes in solution at room temperature achieved significant progress in brightness, stability, and biological applications [65]. Owing to the continuous optical improvement of luminescent Cr(III) complexes, they have gradually been used in sensing [66], photosensitizing [67], upconversion [68] and circularly polarized luminescence (**37**) [69] in the past few years. We summarized representative NIR Cr(III) complexes with high brightness and long lifetime that have been reported in recent years. A breakthrough work on NIR highly luminescent Cr(III) complexes (**34**, **35**) was contributed by Heinze et al. [70]. They prepared a novel tridentate polypyridine ligand (ddpd = *N*,*N*′-dimethyl-*N*,*N*′-dipyridine-2-yl-pyridine-2,6-diamine)) as strong σ-donator and π-acceptor capable of chelating Cr(III) with substantially larger N–Cr–N bite angles than typically used bipyridine (bpy), 1,10-phenanthroline (phen), or 2,2′:6′,2″-terpyridine (tpy), resulting in an unusually large energy gap of 7100 cm^{-1} between the relaxed ^2E and $^4T_2$ excited states. This complex (**34**) is highly water soluble and exhibits emission at 775 nm with a quantum yield of 11.0% and ultra-long lifetime (898 μs) under excitation by 435 nm in de-aerated water. It showed very good stability to aqueous solution (pH

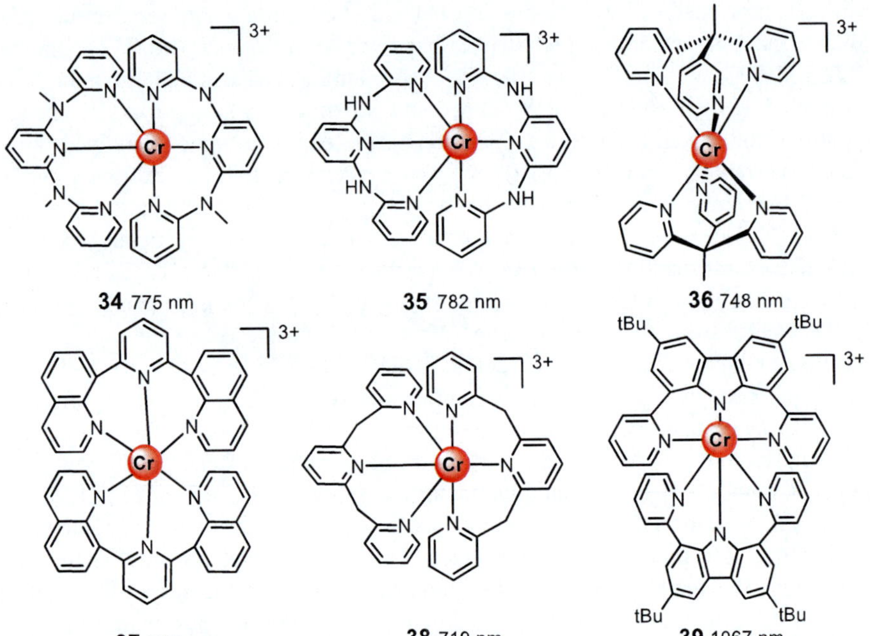

34 775 nm **35** 782 nm **36** 748 nm

37 747 nm **38** 719 nm **39** 1067 nm

Fig. 11 **a** Chemical structures of NIR luminescent Cr(III) complexes

7) as well as in the presence of HCl (pH 2.1) and NaOH (pH1 1.9). Using a partially deuterated polypyridine, a record 30% quantum yield and 2.3 ms in CD_3CN solution at room temperature. was furnished [71, 72]. In 2019, Heinze et al. developed a $[Cr(tpe)_2]^{3+}$ (**36**) (tpe = 1,1,1-tris(pyrid-2-yl)ethane), which exhibited the longest luminescence lifetime to date (2800 μs), with a very high fluorescence quantum yield (5.4%) in fluid solution at room temperature [73]. In 2021, Heinze et al. reported a highly luminescent $[Cr(bpmp)_2]^{3+}$ (bpmp = 2,6-bis(2-pyridylmethyl)pyridine) complex (**38**) with a photoluminescence quantum yield up to 20% and very long lifetime in the millisecond range (1800 μs) in D_2O at room temperature [74]. Interestingly, it displays good potential in optical pH sensing owing to pH-switched NIR luminescence in that it is very luminescent in acidic solutions yet non-luminescent in basic buffer. Different from optimizing ligand strength and optimizing ligand bite angles, in 2021 Wenger et al. devoted to increasing the covalence of metal–ligand bond by designing a new $[Cr(dpc)^2]^+$ complex (**39**) (dpc = 3,6-di-tert-butyl-1,8-di(pyridine-2-yl)-carbazolato) and shifted the emission of Cr(III) from typically 727–778 nm to 1067 nm in acetonitrile at 77 K [75].

4 NIR Luminescent Lanthanide Complexes

Lanthanide (Ln) ions possess unique f-f transitions, and their characteristic luminescence peaks are different from that of transition metals, typically visible luminescence (terbium (Tb(III), green), europium (Eu(III), red), dysprosium (Dy(III), yellow)) and NIR-II luminescence (ytterbium (Yb(III), $^2F_{5/2}\rightarrow{}^2F_{7/2}$, 900–1050 nm), neodymium (Nd(III), $^4F_{3/2}\rightarrow{}^4I_{11/2}$, 1050–1100 nm), and erbium (Er(III), $^4I_{13/2}\rightarrow{}^4I_{15/2}$, 1500–1600 nm) [76–78]. The absorption coefficient of Ln(III) ions is very small (usually lower than 10^{-5} M^{-1} cm^{-1}), making it necessary to introduce additional antenna ligands for efficient sensitization. By way of exception, Nd(III) ion possesses a relatively larger extinction coefficient at 808 nm, it could be directly excited for the purpose of NIR bioimaging. For example, NdDOTA [79] and NdDTPA [80, 81] have been explored for in vivo NIR imaging of tumors and the stomach, respectively. Nevertheless, direct excitation of Nd(III) ions requires huge local concentration of Nd(III) (close to the molar level) that necessitates a high-dose injection of Nd complex, which is not recommended for in vivo applications. In particular, the NIR luminescence of Ln(III) ions is easily quenched by the vibration of high-energy X–H bonds (X = C, H, O) [82]. As a result, it has always been a huge challenge to efficiently sensitize NIR Ln(III) ions and achieve high quantum yields above 5%. In 2020, Maury et al. reported on a cyclam macrocyclic ligand that can sensitize Yb(III) ion (**40**) and the fluorescence quantum yield in methanol or water is too low to be measured [83]. Importantly, **40** can be excited by two-photon light that provides a deeper tissue penetration depth. Benedetti et al. reported a [2.2] paracyclophane-coumarin triazacyclonane macrocyclic ligand capable of sensitizing Yb(III) or oxygen when coordinated with gadolinium (Gd(III)) [84]. Borbas et al. employed chlorine as the antenna to sensitize Yb(III) or Nd(III) and obtain NIR luminescent Ln(III) complexes (**41**,**42**) capable of red-light excitation [85]. As chlorine and Ln ions are not directly coordinated, the fluorescence quantum yields were low. Pecoraro et al. developed a set of gallium(III)/ Ln(III) metallacrowns (Fig. 12b) that sensitized visible to NIR emissive Ln(III) ions including Yb(III), Nd(III), Er(III), thulium(Tm(III)), holmium(Ho(III)), and praseodymium (Pr(III)), providing a wide choice of Ln(III) molecular metallacrowns operating in the NIR region [86, 87].

Zhang's group employed a perfluorinated porphyrin as the antenna to achieve excellent sensitization of Yb(III) (**43**), which represents an unprecedentedly high quantum yield (63%) and long lifetime (714 μs) in CD_2Cl_2 [88]. The highly luminescent Yb(III) probe obtained using this strategy enabled smart design of the Yb(III) probe (**44**) for in vivo intensity imaging for the first time [89]. The key feature of NIR Ln(III) probes that differs from organic fluorophores is long lifetime (μs-ms) and resistance to photobleaching. Consequently, Zhang et al. developed an acidic pH-activated Yb(III) lifetime probe (**44**) and applied it to monitor gastrointestinal pH of a fasting/feeding mouse model via lifetime variations [90]. In 2021, Zhang et al. synthesized a pair of Yb(III) molecular isomers (***trans-/cis-45***) using porphodilactone ligand and observed the *trans*-isomer generate reactive oxygens, while the *cis*-isomer emits only NIR fluorescence due to triplet state energy differences of *trans-/cis*-porphodilactone ligand, providing new ideas for the development of Ln(III) phototheranostics using isomeric chemistry [91]. In the same year, they also reported a NIR Yb(III) carbazole-containing porphyrinoid complex (**46**) with strong NIR absorption in 700–800 nm [92]. Owing to the appropriate energy gap between the triplet state of carbazole porphyrin and the excited state of Yb(III), **46** exhibited balanced deactivation energy dissipation pathways that rendered itself a fascinating NIR phototheranostic agent with combined photoacoustic imaging and photothermal/photodynamic therapy modalities.

In 2021, Tóth et al. incorporated doxorubicin (Dox) and amphiphilic Yb(III) chelates into liposomes via supramolecular self-assembly (Fig. 13a) [93]. The Dox was able to sensitize Yb(III) within the integrated liposomes while Yb(III) was non-emissive once the liposome decomposed. Importantly, in vivo NIR fluorescence imaging was observed after in situ injection of Yb(III)-Dox containing liposomes, which is promising in biomedical applications such as drug delivery monitoring. Also, using supramolecular chemistry, Huang et al. synthesized a series of trinuclear Ln(III) circular helicates $[Ln_3L_3]Cl_3$ in which the unique helical bis-tetradentate 1,2,9-dialdehyde ligand acts as a sensitizing antenna for Nd(III) and Yb(III) ions [94]. Conjugating NIR Ln(III) complexes with functional biomolecules helps bio-specific NIR imaging of disease biomarkers. In 2021, Wong et al. reported a Yb(III)-peptide conjugate (**47**) and successfully applied it for NIR LMP1-targeted imaging in EBV-positive cancer cells [95]. NIR Ln(III) luminescence controlled by photoelectric signals simultaneously has also received research interests. For instance, Rigaut et al. reported a photoelectrically controlled Yb(III) complex (**48**) [96]. By introducing a photochromic dithienylethene unit and electroreductively active Ru(II) complex, NIR luminescence of Yb(III) can be read out by light (**48-closed**) and electrical (**48-oxidized**) control without destruction (Figs. 14, 15).

Upconversion luminescence with molecular and supramolecular Ln(III) complexes recently achieved significant progress [97]. In 2019, Charbonnière et al. constructed heteropolynuclear supramolecular Tb(III)/Yb(III) assembly (**49**) and

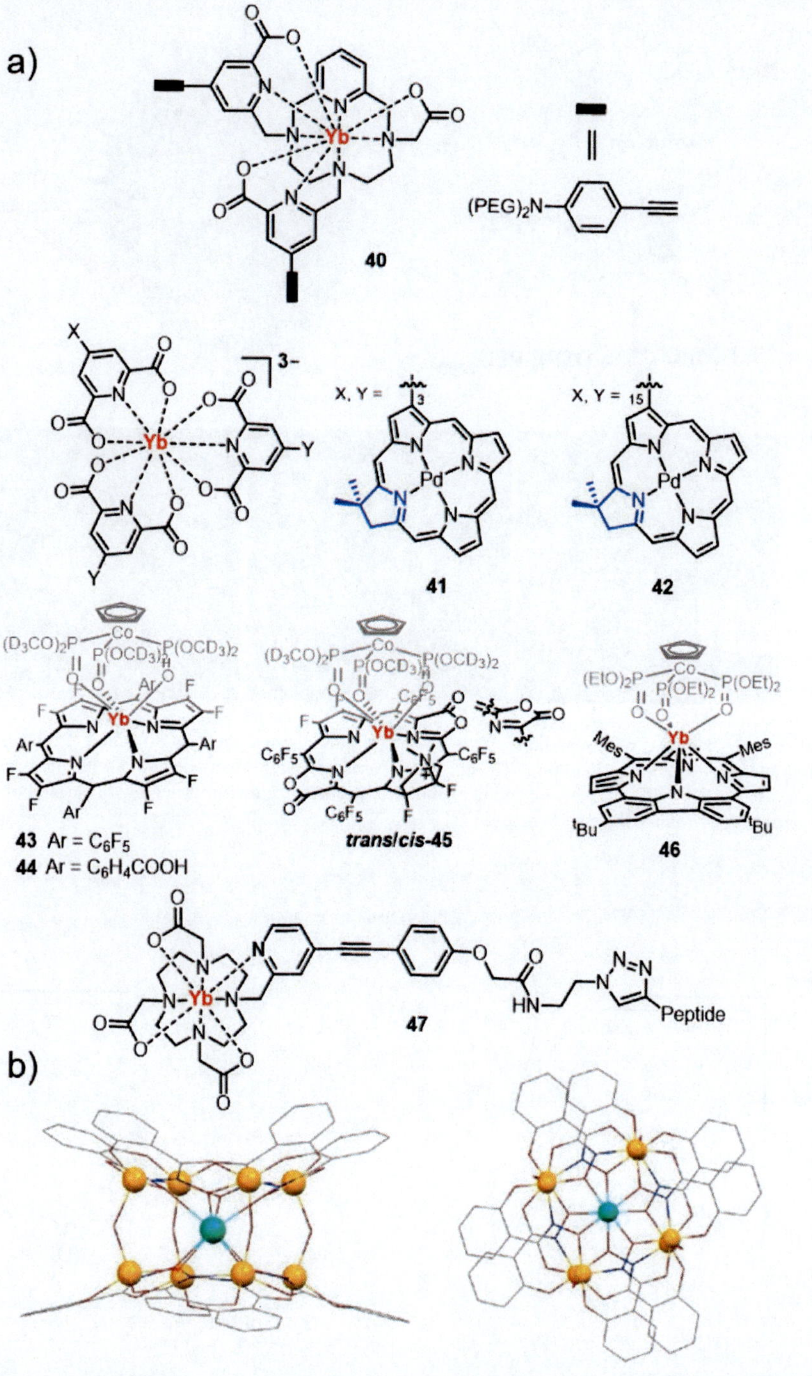

Fig. 12 **a** Representative chemical structures of NIR luminescent Ln(III) complexes. **b** [$NdGa_8(shi)_8(OH)_4$]. Solvents of crystallization, nonintegral counter ions, and hydrogen atoms have been omitted for clarity. Color code: Ga, orange; Na, violet; Ln(III), teal; O, red; N, blue; C, grey. Reprinted with permission from Copyright 2016 American Chemical Society [87]

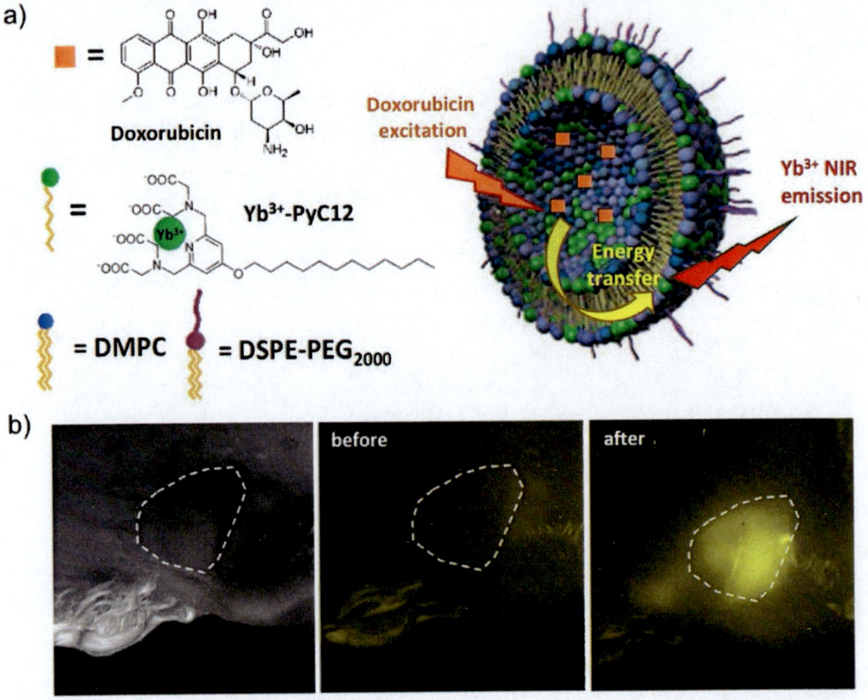

Fig. 13 **a** Schematic representation of the studied liposomes and their sensitization mechanism. **b** Brightfield (left). Yb(III) (long-pass 770-nm filter)-based emission upon excitation with light selected using a bandpass 482/70 nm, before (middle) and after (right) injection. Reprinted with permission from Copyright 2021 John Wiley and Sons Ltd [93]

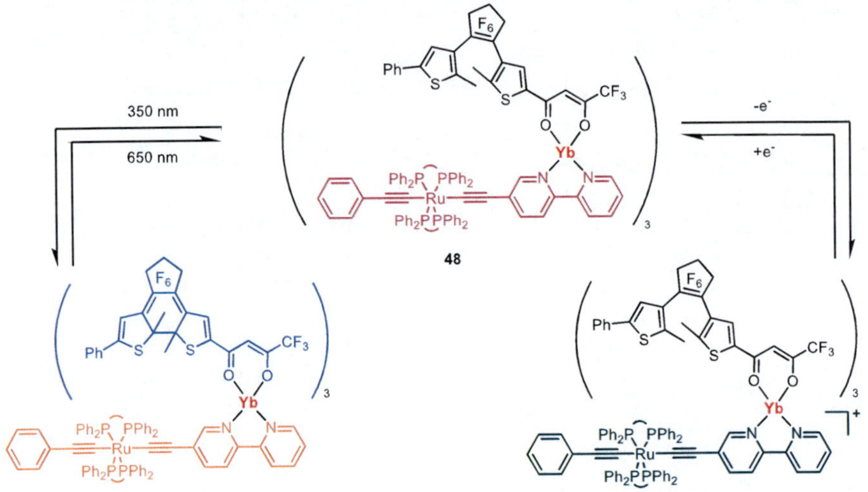

Fig. 14 Synthetic pathway to **48** and redox/optical commutation to **48-oxidized** and **48-closed**. Reprinted with permission from Copyright 2019 American Chemical Society [96]

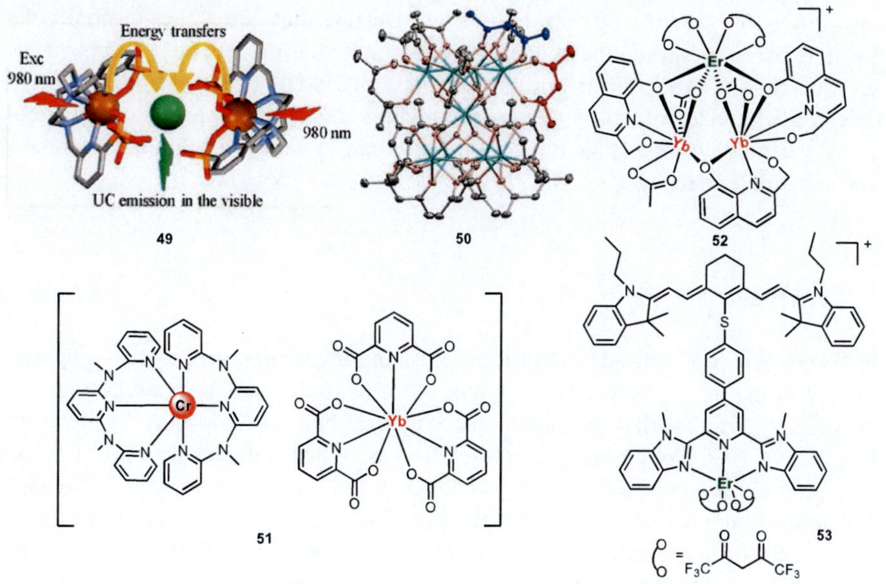

Fig. 15 Single-crystal X-ray structure of **49**. Color code: brown: Yb; Green: Tb. Reprinted with permission from Copyright 2019 American Chemical Society [98]. Single-crystal X-ray structure of **50**. Color code: Green: Yb or Tb. Reprinted with permission from Copyright 2021 John Wiley and Sons Ltd [100]. Chemical structure of Ln-based molecular upconversion system **51–53**

achieved upconversion green emission of Tb(III) in nondeuterated water upon NIR excitation at 980 nm [98, 99]. In 2022, the same group reported a coherent series of homo- and hetero-nonanuclear Yb(III)/Tb(III) complexes and obtained to date the brightest molecular upconversion complex (**50**) with a quantum yield of 2.8×10^{-6} at a low laser power of 2.86 W/cm^2 [100]. Interestingly, Seitz et al. put their efforts into preparing a novel hetero-molecular Cr(III)/Yb(III) complex (**51**) that exhibited laser-power dependent cooperative sensitization of Cr(III)-centered $^2E/^2T_1$ emission at ~ 775 nm on exciting the Yb(III) absorption band at 980 nm in the solid state at room temperature, which represents the first report of Ln(III)-based molecular upconversion example that was NIR-Excitable and NIR-Emissive [68].

In 2021, Zeng et al. reported a discrete heteropolynuclear Yb_2Er^+ assembly (**52**) using 2-methanolquinolin-8-oxide as and hexafluoroacetylacetonate (hfac) ligands [101]. The quinolinoxy O-atom acted as a μ_2 or μ_3 bridge to two Ln(III) centers with the Yb···Er distances around 3.7 Å. **52** exhibited characteristic upconversion emission from Er(III) excited by 980 nm at low excitation power (0.288 W/cm^2) with the introduction of fluoride ions. Also, the upconversion quantum yield was calculated up to 1.88% in CH_2Cl_2 at room temperature under 2 W/cm^2 power densities. As Ln(III)-based molecular upconversion systems possess low-absorption cross section, it restricted their application due to intense incident light. In 2021, Piguet et al. incorporated an NIR organic dye into the coordinating scaffold of a mononuclear Er(III) complex (**53**) that remarkably boosted the upconversion brightness of Er(III) to a measurable range in a dilute solution (5×10^{-4} M) at room temperature using a low-power (0.7 W/cm^2) 801 nm laser [102]. Despite that, researches on molecular Ln(III) upconversion luminescence have been limited at the level of solid power or solution. Nonetheless, the important prospects of Ln(III) upconversion nanoparticles in the biomedical field [103] is encouraging coordination chemists to continuously pursue new molecular Ln(III) upconversion systems with higher brightness and quantum efficiency.

5 Conclusions

In this review, we focused on recent progress in NIR luminescent metal complexes mainly in the past 3 years. We introduced representative structures and designs of NIR highly luminescent transition metal complexes and Ln complexes. The research of different NIR metal-based complexes has witnessed different degrees of scientific progress, including from non-luminescent to luminescent in the NIR region, extending luminescence from the NIR I to NIR II regions using the same metal center, red-shifting excitation wavelength from visible to NIR light, demonstrating applications from basic photophysical studies to NIR in vitro and in vivo imaging and diagnostics. Some challenges for NIR metal complexes should be tackled for further bioapplications. For noble transition metal complexes, although numerous studies have demonstrated their promising applications in bioimaging and biosensing, the emission wavelength of these complexes is in NIR I region that limits the penetration depth of in vivo bioimaging. The penetration depth of transition metal complexes into biological tissues can be enhanced by designing two-photon or NIR excited transition metal complexes, but it also requires sophisticated equipment or smart chemical synthesis. For NIR luminescent non-noble transition metal complexes, most of the research attention has been focused on making them emissive in non-aqueous solvents and then making them brighter through elegant ligand design. With the deepening of research, more research attention is anticipated on improving the luminescence quantum yields of transition NIR metal complexes in aqueous solution and regulating stability, biological toxicity, targeting ability, water solubility, and pharmacokinetics for clinical applications. Perspectives on NIR luminescent metal complexes are vast and exciting. With the emergence of new ligands, we are happy to see richer varieties of NIR metal-based molecular probes with optimized photophysical properties and biomedical performance. By conjugating with clinically important biomolecules including peptides, functional proteins, antibodies, and aptamers, NIR metal complex-based theranostics should see rapid development. In particular, the design of bio-orthogonal NIR metal-based molecular probes and conjugation with targeting antibodies should enable NIR imaging-guided detection of specific disease biomarkers and precision surgery of malignant lesions in vivo with good spatiotemporal resolution.

Acknowledgements Financial support from the National Natural Science Foundation of China (21571007, 21621061, 21778002, and 21861162008), the Chemistry and Chemical Engineering Guangdong Laboratory (1932002) and High-performance Computing Platform of Peking University is acknowledged.

Declarations

Competing interests The authors declare no competing interests.

References

1. Withers PJ, Bouman C, Carmignato S, Cnudde V, Grimaldi D, Hagen CK, Maire E, Manley M, Du Plessis A, Stock SR (2021) X-ray computed tomography. Nat Rev Methods Primers 1:18
2. Wahsner J, Gale EM, Rodríguez-Rodríguez A, Caravan P (2019) Chemistry of MRI contrast agents: current challenges and new frontiers. Chem Rev 119:957–1057
3. Klinkhammer BM, Lammers T, Mottaghy FM, Kiessling F, Floege J, Boor P (2021) Non-invasive molecular imaging of kidney diseases. Nat Rev Nephrol 17:688–703
4. Bodini B, Tonietto M, Airas L, Stankoff B (2021) Positron emission tomography in multiple sclerosis—straight to the target. Nat Rev Neurol 17:663–675
5. Wei W, Rosenkrans ZT, Liu J, Huang G, Luo Q-Y, Cai W (2020) ImmunoPET: concept, design, and applications. Chem Rev 120:3787–3851
6. Zhang Y, Zhang G, Zeng Z, Pu K (2022) Activatable molecular probes for fluorescence-guided surgery, endoscopy and tissue biopsy. Chem Soc Rev 51:566–593
7. Hong G, Antaris AL, Dai H (2017) Near-infrared fluorophores for biomedical imaging. Nat Biomed Eng 1:0010
8. Li C, Chen G, Zhang Y, Wu F, Wang Q (2020) Advanced fluorescence imaging technology in the near-infrared-II window for biomedical applications. J Am Chem Soc 142:14789–14804
9. Liu Y, Li Y, Koo S, Sun Y, Liu Y, Liu X, Pan Y, Zhang Z, Du M, Lu S, Qiao X, Gao J, Wang X, Deng Z, Meng X, Xiao Y, Kim JS, Hong X (2022) Versatile types of inorganic/organic NIR-IIa/IIb fluorophores: from strategic design toward molecular imaging and theranostics. Chem Rev 122:209–268
10. Yang Q, Ma H, Liang Y, Dai H (2021) Rational design of high brightness NIR-II organic dyes with S-D-A-D-S structure. Acc Mater Res 2:170–183
11. Lei Z, Zhang F (2021) Molecular engineering of NIR-II fluorophores for improved biomedical detection. Angew Chem Int Ed 60:16294–16308
12. Antaris AL, Chen H, Cheng K, Sun Y, Hong G, Qu C, Diao S, Deng Z, Hu X, Zhang B, Zhang X, Yaghi OK, Alamparambil ZR, Hong X, Cheng Z, Dai H (2016) A small-molecule dye for NIR-II imaging. Nat Mater 15:235–242
13. Mu J, Xiao M, Shi Y, Geng X, Li H, Yin Y, Chen X (2022) The chemistry of organic contrast agents in the NIR-II window. Angew Chem Int Ed 61:e202114722
14. Ning Y, Jin G-Q, Wang M-X, Gao S, Zhang J-L (2022) Recent progress in metal-based molecular probes for optical bioimaging and biosensing. Curr Opin Chem Biol 66:102097
15. Zhang KY, Yu Q, Wei H, Liu S, Zhao Q, Huang W (2018) Long-lived emissive probes for time-resolved photoluminescence bioimaging and biosensing. Chem Rev 118:1770–1839
16. Montalti MAC, Prodi L, Gandolfi MT (2006) Handbook of photochemistry. CRC Press, Boca Raton
17. Feng G, Zhang G-Q, Ding D (2020) Design of superior phototheranostic agents guided by Jablonski diagrams. Chem Soc Rev 49:8179–8234
18. Li J, Wang L, Zhao Z, Sun B, Zhan G, Liu H, Bian Z, Liu Z (2020) Highly efficient and air-stable Eu(II)-containing azacryptates ready for organic light-emitting diodes. Nat Commun 11:5218
19. Muñoz-García AB, Benesperi I, Boschloo G, Concepcion JJ, Delcamp JH, Gibson EA, Meyer GJ, Pavone M, Pettersson H, Hagfeldt A, Freitag M (2021) Dye-sensitized solar cells strike back. Chem Soc Rev 50:12450–12550
20. Yang Z-S, Yao Y, Sedgwick AC, Li C, Xia Y, Wang Y, Kang L, Su H, Wang B-W, Gao S, Sessler JL, Zhang J-L (2020) Rational design of an "all-in-one" phototheranostic. Chem Sci 11:8204–8213
21. Balzani V, Campagna S (2007) Photochemistry and photophysics of coordination compounds I. Springer, Berlin

22. Wong KM-C, Yam VW-W (2011) Self-assembly of luminescent alkynylplatinum(II) terpyridyl complexes: modulation of photophysical properties through aggregation behavior. Acc Chem Res 44:424–434
23. Yam VW-W, Au VK-M, Leung SY-L (2015) Light-emitting self-assembled materials based on d8 and d10 transition metal complexes. Chem Rev 115:7589–7728
24. Chen W-C, Sukpattanacharoen C, Chan W-H, Huang C-C, Hsu H-F, Shen D, Hung W-Y, Kungwan N, Escudero D, Lee C-S, Chi Y (2020) Modulation of solid-state aggregation of square-planar Pt(II) based emitters: enabling highly efficient deep-red/near infrared electroluminescence. Adv Funct Mater 30:2002494
25. Wang SF, Fu L-W, Wei Y-C, Liu S-H, Lin J-A, Lee G-H, Chou P-T, Huang J-Z, Wu C-I, Yuan Y, Lee C-S, Chi Y (2019) Near-infrared emission induced by shortened Pt–Pt contact: diplatinum(II) complexes with pyridyl pyrimidinato cyclometalates. Inorg Chem 58:13892–13901
26. Wu J, Li Y, Tan C, Wang X, Zhang Y, Song J, Qu J, Wong W-Y (2018) Aggregation-induced near-infrared emitting platinum(II) terpyridyl complex: cellular characterisation and lysosome-specific localisation. Chem Commun 54:11144–11147
27. Wu C, Chen H-F, Wong K-T, Thompson ME (2010) Study of ion-paired iridium complexes (soft salts) and their application in organic light emitting diodes. J Am Chem Soc 132:3133–3139
28. Li J, Ma Y, Liu S, Mao Z, Chi Z, Qian P-C, Wong W-Y (2020) Soft salts based on platinum(II) complexes with high emission quantum efficiencies in the near infrared region for in vivo imaging. Chem Commun 56:11681–11684
29. Yeung MC-L, Wong KM-C, Tsang YKT, Yam VW-W (2010) Aptamer-induced self-assembly of a NIR-emissive platinum(II) terpyridyl complex for label- and immobilization-free detection of lysozyme and thrombin. Chem Commun 46:7709–7711
30. Yeung MC-L, Yam VW-W (2013) Phosphate derivative-induced supramolecular assembly and NIR-emissive behaviour of alkynylplatinum(II) terpyridine complexes for real-time monitoring of enzymatic activities. Chem Sci 4:2928–2935
31. Chung CY-S, Li SP-Y, Louie M-W, Lo KK-W, Yam VW-W (2013) Induced self-assembly and disassembly of water-soluble alkynylplatinum(II) terpyridyl complexes with "switchable" near-infrared (NIR) emission modulated by metal–metal interactions over physiological pH: demonstration of pH-responsive NIR luminescent probes in cell-imaging studies. Chem Sci 4:2453–2462
32. Law AS-Y, Lee LC-C, Yeung MC-L, Lo KK-W, Yam VW-W (2019) Amyloid protein-induced supramolecular self-assembly of water-soluble platinum(II) complexes: a luminescence assay for amyloid fibrillation detection and inhibitor screening. J Am Chem Soc 141:18570–18577
33. Law AS-Y, Lee LC-C, Lo KK-W, Yam VW-W (2021) Aggregation and supramolecular self-assembly of low-energy red luminescent alkynylplatinum(II) complexes for RNA detection, nucleolus imaging, and RNA synthesis inhibitor screening. J Am Chem Soc 143:5396–5405
34. Solomatina AI, Slobodina AD, Ryabova EV, Bolshakova OI, Chelushkin PS, Sarantseva SV, Tunik SP (2020) Blood-brain barrier penetrating luminescent conjugates based on cyclometalated platinum(II) complexes. Bioconj Chem 31:2628–2637
35. Liu L-Y, Fang H, Chen Q, Chan MH-Y, Ng M, Wang K-N, Liu W, Tian Z, Diao J, Mao Z-W, Yam VW-W (2020) Multiple-color platinum complex with super-large stokes shift for super-resolution imaging of autolysosome escape. Angew Chem Int Ed 59:19229–19236
36. Chen Q, Jin C, Shao X, Guan R, Tian Z, Wang C, Liu F, Ling P, Guan J-L, Ji L, Wang F, Chao H, Diao J (2018) Super-resolution tracking of mitochondrial dynamics with an iridium(III) luminophore. Small 14:1802166
37. Jin C, Li G, Wu X, Liu J, Wu W, Chen Y, Sasaki T, Chao H, Zhang Y (2021) Robust packing of a self-assembling iridium complex via endocytic trafficking for long-term lysosome tracking. Angew Chem Int Ed 60:7597–7601
38. Liu Y, Zhang P, Fang X, Wu G, Chen S, Zhang Z, Chao H, Tan W, Xu L (2017) Near-infrared emitting iridium(III) complexes for mitochondrial imaging in living cells. Dalton Trans 46:4777–4785
39. Jin C, Guan R, Wu J, Yuan B, Wang L, Huang J, Wang H, Ji L, Chao H (2017) Rational design of NIR-emitting iridium(III) complexes for multimodal phosphorescence imaging of mitochondria under two-photon excitation. Chem Commun 53:10374–10377
40. Jin C, Liang F, Wang J, Wang L, Liu J, Liao X, Rees TW, Yuan B, Wang H, Shen Y, Pei Z, Ji L, Chao H (2020) Rational design of cyclometalated iridium(III) complexes for three-photon phosphorescence bioimaging. Angew Chem Int Ed 59:15987–15991

41. Wu W, Zhang C, Rees TW, Liao X, Yan X, Chen Y, Ji L, Chao H (2020) Lysosome-targeting iridium(III) Probe with near-infrared emission for the visualization of NO/O2˙⁻ crosstalk via in vivo peroxynitrite imaging. Anal Chem 92:6003–6009
42. Fan Y, Zhao J, Yan Q, Chen PR, Zhao D (2014) Water-soluble triscyclometalated organoiridium complex: phosphorescent nanoparticle formation, nonlinear optics, and application for cell imaging. ACS Appl Mater Interfaces 6:3122–3131
43. Yip AM-H, Lai CK-H, Yiu KS-M, Lo KK-W (2022) Phosphorogenic iridium(III) bis-tetrazine complexes for bioorthogonal peptide stapling, bioimaging, photocytotoxic applications, and the construction of nanosized hydrogels. Angew Chem Int Ed 61:e202116078
44. Ge C, Huang H, Wang Y, Zhao H, Zhang P, Zhang Q (2018) Near-infrared luminescent osmium(II) complexes with an intrinsic RNA-targeting capability for nucleolus imaging in living cells. ACS Appl Bio Mater 1:1587–1593
45. Smitten KL, Scattergood PA, Kiker C, Thomas JA, Elliott PIP (2020) Triazole-based osmium(II) complexes displaying red/near-IR luminescence: antimicrobial activity and super-resolution imaging. Chem Sci 11:8928–8935
46. Dröge F, Noakes FF, Archer SA, Sreedharan S, Raza A, Robertson CC, MacNeil S, Haycock JW, Carson H, Meijer AJHM, Smythe CGW, Bernardino de la Serna J, Dietzek-Ivanšić B, Thomas JA (2021) A dinuclear osmium(II) complex near-infrared nanoscopy probe for nuclear DNA. J Am Chem Soc 143:20442–20453
47. Gkika KS, Noorani S, Walsh N, Keyes TE (2021) Os(II)-bridged polyarginine conjugates: the additive effects of peptides in promoting or preventing permeation in cells and multicellular tumor spheroids. Inorg Chem 60:8123–8134
48. Yao Y, Hou C-L, Yang Z-S, Ran G, Kang L, Li C, Zhang W, Zhang J, Zhang J-L (2019) Unusual near infrared (NIR) fluorescent palladium(II) macrocyclic complexes containing M-C bonds with bioimaging capability. Chem Sci 10:10170–10178
49. Wegeberg C, Wenger OS (2021) Luminescent first-row transition metal complexes. JACS Au 1:1860–1876
50. McCusker James K (2019) Electronic structure in the transition metal block and its implications for light harvesting. Science 363:484–488
51. Caspar JV, Kober EM, Sullivan BP, Meyer TJ (1982) Application of the energy gap law to the decay of charge-transfer excited states. J Am Chem Soc 104:630–632
52. Hamze R, Peltier Jesse L, Sylvinson D, Jung M, Cardenas J, Haiges R, Soleilhavoup M, Jazzar R, Djurovich Peter I, Bertrand G, Thompson Mark E (2019) Eliminating nonradiative decay in Cu(I) emitters: >99% quantum efficiency and microsecond lifetime. Science 363:601–606
53. Chábera P, Liu Y, Prakash O, Thyrhaug E, Nahhas AE, Honarfar A, Essén S, Fredin LA, Harlang TCB, Kjær KS, Handrup K, Ericson F, Tatsuno H, Morgan K, Schnadt J, Häggström L, Ericsson T, Sobkowiak A, Lidin S, Huang P, Styring S, Uhlig J, Bendix J, Lomoth R, Sundström V, Persson P, Wärnmark K (2017) A low-spin Fe(iii) complex with 100-ps ligand-to-metal charge transfer photoluminescence. Nature 543:695–699
54. Kjær Kasper S, Kaul N, Prakash O, Chábera P, Rosemann Nils W, Honarfar A, Gordivska O, Fredin Lisa A, Bergquist K-E, Häggström L, Ericsson T, Lindh L, Yartsev A, Styring S, Huang P, Uhlig J, Bendix J, Strand D, Sundström V, Persson P, Lomoth R, Wärnmark K (2019) Luminescence and reactivity of a charge-transfer excited iron complex with nanosecond lifetime. Science 363:249–253
55. Young Elizabeth R, Oldacre A (2019) Iron hits the mark. Science 363:225–226
56. Dorn M, Kalmbach J, Boden P, Päpcke A, Gómez S, Förster C, Kuczelinis F, Carrella LM, Büldt LA, Bings NH, Rentschler E, Lochbrunner S, González L, Gerhards M, Seitz M, Heinze K (2020) A vanadium(III) complex with blue and NIR-II spin-flip luminescence in solution. J Am Chem Soc 142:7947–7955
57. Dorn M, Kalmbach J, Boden P, Kruse A, Dab C, Reber C, Niedner-Schatteburg G, Lochbrunner S, Gerhards M, Seitz M, Heinze K (2021) Ultrafast and long-time excited state kinetics of an NIR-emissive vanadium(III) complex I: synthesis, spectroscopy and static quantum chemistry. Chem Sci 12:10780–10790
58. Harris JP, Reber C, Colmer HE, Jackson TA, Forshaw AP, Smith JM, Kinney RA, Telser J (2017) Near-infrared 2Eg → 4A2g and visible LMCT luminescence from a molecular bis-(tris(carbene) borate) manganese(IV) complex. Can J Chem 95:547–552
59. Kaufhold S, Rosemann NW, Chábera P, Lindh L, Bolaño Losada I, Uhlig J, Pascher T, Strand D, Wärnmark K, Yartsev A, Persson P (2021) Microsecond photoluminescence and

photoreactivity of a metal-centered excited state in a hexacarbene–Co(III) complex. J Am Chem Soc 143:1307–1312
60. Shi S, Jung MC, Coburn C, Tadle A, Sylvinson MRD, Djurovich PI, Forrest SR, Thompson ME (2019) Highly efficient photo- and electroluminescence from two-coordinate Cu(I) complexes featuring nonconventional N-heterocyclic carbenes. J Am Chem Soc 141:3576–3588
61. Gernert M, Balles-Wolf L, Kerner F, Müller U, Schmiedel A, Holzapfel M, Marian CM, Pflaum J, Lambert C, Steffen A (2020) Cyclic (amino)(aryl)carbenes enter the field of chromophore ligands: expanded π system leads to unusually deep red emitting CuI compounds. J Am Chem Soc 142:8897–8909
62. Herr P, Kerzig C, Larsen CB, Häussinger D, Wenger OS (2021) Manganese(i) complexes with metal-to-ligand charge transfer luminescence and photoreactivity. Nat Chem 13:956–962
63. Wegeberg C, Häussinger D, Wenger OS (2021) Pyrene-decoration of a chromium(0) tris(diisocyanide) enhances excited state delocalization: a strategy to improve the photoluminescence of 3d6 metal complexes. J Am Chem Soc 143:15800–15811
64. Boden P, Di Martino-Fumo P, Bens T, Steiger S, Albold U, Niedner-Schatteburg G, Gerhards M, Sarkar B (2021) NIR-emissive chromium(0), molybdenum(0), and tungsten(0) complexes in the solid state at room temperature. Chem Eur J 27:12959–12964
65. Büldt LA, Wenger OS (2017) Chromium complexes for luminescence, solar cells, photoredox catalysis, upconversion, and phototriggered NO release. Chem Sci 8:7359–7367
66. Otto S, Scholz N, Behnke T, Resch-Genger U, Heinze K (2017) Thermo-chromium: a contactless optical molecular thermometer. Chem Eur J 23:12131–12135
67. Basu U, Otto S, Heinze K, Gasser G (2019) Biological evaluation of the NIR-emissive ruby analogue [Cr(ddpd)2][BF4]3 as a photodynamic therapy photosensitizer. Eur J Inorg Chem 2019:37–41
68. Kalmbach J, Wang C, You Y, Förster C, Schubert H, Heinze K, Resch-Genger U, Seitz M (2020) Near-IR to near-IR upconversion luminescence in molecular chromium ytterbium salts. Angew Chem Int Ed 59:18804–18808
69. Jiménez J-R, Doistau B, Cruz CM, Besnard C, Cuerva JM, Campaña AG, Piguet C (2019) Chiral molecular ruby [Cr(dqp)2]3+ with long-lived circularly polarized luminescence. Chem Eur J 141:13244–13252
70. Otto S, Grabolle M, Förster C, Kreitner C, Resch-Genger U, Heinze K (2015) [Cr(ddpd)2]3+: a molecular, water-soluble, highly NIR-emissive ruby analogue. Angew Chem Int Ed 54:11572–11576
71. Wang C, Otto S, Dorn M, Kreidt E, Lebon J, Sršan L, Di Martino-Fumo P, Gerhards M, Resch-Genger U, Seitz M, Heinze K (2018) Deuterated molecular ruby with record luminescence quantum yield. Angew Chem Int Ed 57:1112–1116
72. Otto S, Förster C, Wang C, Resch-Genger U, Heinze K (2018) A strongly luminescent chromium(III) complex acid. Chem Eur J 24:12555–12563
73. Treiling S, Wang C, Förster C, Reichenauer F, Kalmbach J, Boden P, Harris JP, Carrella LM, Rentschler E, Resch-Genger U, Reber C, Seitz M, Gerhards M, Heinze K (2019) Luminescence and light-driven energy and electron transfer from an exceptionally long-lived excited state of a non-innocent chromium(III) complex. Angew Chem Int Ed 58:18075–18085
74. Reichenauer F, Wang C, Förster C, Boden P, Ugur N, Báez-Cruz R, Kalmbach J, Carrella LM, Rentschler E, Ramanan C, Niedner-Schatteburg G, Gerhards M, Seitz M, Resch-Genger U, Heinze K (2021) Strongly red-emissive molecular ruby [Cr(bpmp)2]3+ surpasses [Ru(bpy)3]2+. J Am Chem Soc 143:11843–11855
75. Sinha N, Jiménez J-R, Pfund B, Prescimone A, Piguet C, Wenger OS (2021) A near-infrared-II emissive chromium(III) complex. Angew Chem Int Ed 60:23722–23728
76. Ning Y, Zhu M, Zhang J-L (2019) Near-infrared (NIR) lanthanide molecular probes for bioimaging and biosensing. Coord Chem Rev 399:213028
77. Jin G-Q, Ning Y, Geng J-X, Jiang Z-F, Wang Y, Zhang J-L (2020) Joining the journey to near infrared (NIR) imaging: the emerging role of lanthanides in the designing of molecular probes. Inorg Chem Front 7:289–299
78. Peng X-X, Zhu X-F, Zhang J-L (2020) Near Infrared (NIR) imaging: exploring biologically relevant chemical space for lanthanide complexes. J Inorg Biochem 209:111118
79. Yang Y, Wang P, Lu L, Fan Y, Sun C, Fan L, Xu C, El-Toni AM, Alhoshan M, Zhang F (2018) Small-molecule lanthanide complexes probe for second near-infrared window bioimaging. Anal Chem 90:7946–7952

80. Li Y, Li X, Xue Z, Jiang M, Zeng S, Hao J (2018) Second near-infrared emissive lanthanide complex for fast renal-clearable in vivo optical bioimaging and tiny tumor detection. Biomaterials 169:35–44
81. Zhuang P, Xiang K, Meng X, Wang G, Li Z, Lu Y, Kan D, Zhang X, Sun S-K (2021) Gram-scale synthesis of a neodymium chelate as a spectral CT and second near-infrared window imaging agent for visualizing the gastrointestinal tract in vivo. J Mater Chem B 9:2285–2294
82. Bünzli J-CG (2015) On the design of highly luminescent lanthanide complexes. Coord Chem Rev 293–294:19–47
83. Hamon N, Roux A, Beyler M, Mulatier J-C, Andraud C, Nguyen C, Maynadier M, Bettache N, Duperray A, Grichine A, Brasselet S, Gary-Bobo M, Maury O, Tripier R (2020) Pyclen-based Ln(III) complexes as highly luminescent bioprobes for in vitro and in vivo one- and two-photon bioimaging applications. J Am Chem Soc 142:10184–10197
84. Wu S, Galán LA, Roux M, Riobé F, Le Guennic B, Guyot Y, Le Bahers T, Micouin L, Maury O, Benedetti E (2021) Tuning excited-state properties of [2.2]paracyclophane-based antennas to ensure efficient sensitization of lanthanide ions or singlet oxygen generation. Inorg Chem 60:16194–16203
85. Xiong R, Mara D, Liu J, Van Deun R, Borbas KE (2018) Excitation- and emission-wavelength-based multiplex spectroscopy using red-absorbing near-infrared-emitting lanthanide complexes. J Am Chem Soc 140:10975–10979
86. Salerno EV, Eliseeva SV, Schneider BL, Kampf JW, Petoud S, Pecoraro VL (2020) Visible, near-infrared, and dual-range luminescence spanning the 4f series sensitized by a gallium(III)/lanthanide(III) metallacrown structure. J Phys Chem A 124:10550–10564
87. Chow CY, Eliseeva SV, Trivedi ER, Nguyen TN, Kampf JW, Petoud S, Pecoraro VL (2016) Ga^{3+}/Ln^{3+} metallacrowns: a promising family of highly luminescent lanthanide complexes that covers visible and near-infrared domains. J Am Chem Soc 138:5100–5109
88. Hu J-Y, Ning Y, Meng Y-S, Zhang J, Wu Z-Y, Gao S, Zhang J-L (2017) Highly near-IR emissive ytterbium(III) complexes with unprecedented quantum yields. Chem Sci 8:2702–2709
89. Ning Y, Chen S, Chen H, Wang J-X, He S, Liu Y-W, Cheng Z, Zhang J-L (2019) A proof-of-concept application of water-soluble ytterbium(III) molecular probes in in vivo NIR-II whole body bioimaging. Inorg Chem Front 6:1962–1967
90. Ning Y, Cheng S, Wang J-X, Liu Y-W, Feng W, Li F, Zhang J-L (2019) Fluorescence lifetime imaging of upper gastrointestinal pH in vivo with a lanthanide based near-infrared τ probe. Chem Sci 10:4227–4235
91. Ning Y, Liu Y-W, Yang Z-S, Yao Y, Kang L, Sessler JL, Zhang J-L (2020) Split and use: structural isomers for diagnosis and therapy. J Am Chem Soc 142:6761–6768
92. Zhu M, Zhang H, Ran G, Mangel DN, Yao Y, Zhang R, Tan J, Zhang W, Song J, Sessler JL, Zhang J-L (2021) Metal modulation: an easy-to-implement tactic for tuning lanthanide phototheranostics. J Am Chem Soc 143:7541–7552
93. Lacerda S, Delalande A, Eliseeva SV, Pallier A, Bonnet CS, Szeremeta F, Même S, Pichon C, Petoud S, Tóth É (2021) Doxorubicin-sensitized luminescence of NIR-emitting ytterbium liposomes: towards direct monitoring of drug release. Angew Chem Int Ed 60:23574–23577
94. Chang F-F, Feng F-D, Geng J, Huang W (2021) Self-assembly and luminescence of trinuclear lanthanide based supramolecular circular helicates. Chem Commun 57:9220–9223
95. Chau H-F, Wu Y, Fok W-Y, Thor W, Cho WC-S, Pa Ma, Lin J, Mak N-K, Bünzli J-CG, Jiang L, Long NJ, Lung HL, Wong K-L (2021) Lanthanide-based peptide-directed visible/near-infrared imaging and inhibition of LMP1. JACS Au 1:1034–1043
96. Al Sabea H, Norel L, Galangau O, Hijazi H, Métivier R, Roisnel T, Maury O, Bucher C, Riobé F, Rigaut S (2019) Dual light and redox control of NIR luminescence with complementary photochromic and organometallic antennae. J Am Chem Soc 141:20026–20030
97. Nonat AM, Charbonnière LJ (2020) Upconversion of light with molecular and supramolecular lanthanide complexes. Coord Chem Rev 409:213192
98. Nonat A, Bahamyirou S, Lecointre A, Przybilla F, Mély Y, Platas-Iglesias C, Camerel F, Jeannin O, Charbonnière LJ (2019) Molecular upconversion in water in heteropolynuclear supramolecular Tb/Yb assemblies. J Am Chem Soc 141:1568–1576
99. Souri N, Tian P, Platas-Iglesias C, Wong K-L, Nonat A, Charbonnière LJ (2017) Upconverted photosensitization of Tb visible emission by NIR Yb excitation in discrete supramolecular heteropolynuclear complexes. J Am Chem Soc 139:1456–1459

100. Knighton RC, Soro LK, Francés-Soriano L, Rodríguez-Rodríguez A, Pilet G, Lenertz M, Platas-Iglesias C, Hildebrandt N, Charbonnière LJ (2022) Cooperative luminescence and cooperative sensitisation upconversion of lanthanide complexes in solution. Angew Chem Int Ed 61:e202113114
101. Wang J, Jiang Y, Liu J-Y, Xu H-B, Zhang Y-X, Peng X, Kurmoo M, Ng SW, Zeng M-H (2021) Discrete heteropolynuclear Yb/Er assemblies: switching on molecular upconversion under mild conditions. Angew Chem Int Ed 60:22368–22375
102. Golesorkhi B, Naseri S, Guénée L, Taarit I, Alves F, Nozary H, Piguet C (2021) Ligand-sensitized near-infrared to visible linear light upconversion in a discrete molecular erbium complex. J Am Chem Soc 143:15326–15334
103. Dong H, Du S-R, Zheng X-Y, Lyu G-M, Sun L-D, Li L-D, Zhang P-Z, Zhang C, Yan C-H (2015) Lanthanide nanoparticles: from design toward bioimaging and therapy. Chem Rev 115:10725–10815

Topics in Current Chemistry (2022) 380:30
https://doi.org/10.1007/s41061-022-00384-8

REVIEW

Metal Peptide Conjugates in Cell and Tissue Imaging and Biosensing

Karmel S. Gkika[1] · David Cullinane[1] · Tia E. Keyes[1]

Received: 20 November 2021 / Accepted: 10 May 2022 / Published online: 15 June 2022

Abstract

Metal complex luminophores have seen dramatic expansion in application as imaging probes over the past decade. This has been enabled by growing understanding of methods to promote their cell permeation and intracellular targeting. Amongst the successful approaches that have been applied in this regard is peptide-facilitated delivery. Cell-permeating or signal peptides can be readily conjugated to metal complex luminophores and have shown excellent response in carrying such cargo through the cell membrane. In this article, we describe the rationale behind applying metal complexes as probes and sensors in cell imaging and outline the advantages to be gained by applying peptides as the carrier for complex luminophores. We describe some of the progress that has been made in applying peptides in metal complex peptide-driven conjugates as a strategy for cell permeation and targeting of transition metal luminophores. Finally, we provide key examples of their application and outline areas for future progress.

Keywords Transition metal luminophores · Photophysics · Imaging · Metal conjugates · Peptide · Imaging · Cell-penetrating peptides · Signal peptides · Octaarginine · MPP

Abbreviations

3D	Three-dimensional
CPP	Cell-penetrating peptide
DAPI	4′, 6-Diamidino-2-phenylindole
DLS	Dynamic light scattering
dpp or dip	4, 7-Diphenyl-1, 10-phenanthroline

Chapter 2 was originally published as Gkika, K. S., Cullinane, D. & Keyes, T. E. Topics in Current Chemistry (2022) 380: 30. https://doi.org/10.1007/s41061-022-00384-8.

✉ Tia E. Keyes
tia.keyes@dcu.ie

[1] School of Chemical Sciences, National Centre for Sensor Research, Dublin City University, Dublin 9, Ireland

Dppz	Dipyrido[3, 2-a:2′, 3′-c]phenazine
FLIM	Fluorescence lifetime imaging
icO_2	Intracellular oxygen
IP	Imidazo[4, 5-f][1, 10]-phenanthroline
MLCT	Metal-to-ligand charge transfer
MMP	Mitochondrial membrane potential
MPP	Mitochondria-penetrating peptide (FrFKFrFK)
phen	1,10-Phenanthroline
PLIM	Phosphorescence lifetime imaging microscopy
STED	Stimulated emission depletion microscopy
TAP	1,4,5,8-tetraazaphenanthrene

1 Introduction

Fluorescence microscopy is one of the most important and ubiquitous tools in the life sciences. Its applications vary from the simple visualisation of fixed samples to quantitative and dynamic determination of biological processes in living cells and tissues. Luminescent metal complexes are emerging as highly useful probes for fluorescence microscopy, competing with more traditional organic fluorophores due to their excellent, tuneable photophysical properties and their amenity to sensing applications.

Indeed, many metal complex luminophores are addressable through multimodal methods and offer prospects for applications in imaging, sensing and theranostics.

Probes with visible to near-infrared (NIR) excitation and emission are required in bioimaging. In particular, emission that coincides with the biological optical window (650–1000 nm) is preferable because NIR light is more isotropically scattered by tissue, and light in this frequency range is not absorbed by biomolecules. It is therefore more penetrative through biological tissue and moreover autofluorescence from endogenous sources upon NIR excitation is minimal. As luminescence from most transition metal complexes is formally phosphorescence, their emission exhibits a large Stokes shift (energy difference between absorption and emission maxima). This is advantageous as it avoids artefactual effects from inner filter effects or self-quenching, which may be more prevalent when the probe is localised at high concentrations. Another rarely considered advantage of the large Stokes shift is that it facilitates dual use of such complexes as probes in tandem luminescence and resonance Raman measurements under resonant excitation, since the Stokes shift enables excitation and detection of the resonance Raman signature away from the overwhelming emission signature [1, 2]. The long-lived and triplet nature of the excited state of many metal complex luminophores, notably those of ruthenium(II) and iridium(III), render them susceptible to quenching by analytes such as molecular oxygen (O_2), reactive redox species or pH. The characteristic luminescence lifetime- or intensity-based response typically reflects the interaction of the metal complex with these species within a cellular or tissue environment.

Luminescence intensity-based sensing can be performed using conventional instrumentation such as a fluorescence microscope or plate reader. An important limitation is that the absolute signal intensity alone, cannot typically be used as a reliable quantitative marker for a single target analyte because intensity *in cellulo* is influenced by many factors. Most notably, it will vary with distribution in the cell which is rarely uniform, and physiochemical issues such as photodamage, probe leaching and interaction with species such as proteins or lipid membranes within the cellular environment can influence emission intensity. Additionally, intensity can be affected by the excitation source or detector drift and sensitivity. A practical approach to facilitate use of emission intensity for sensing is to apply *ratiometric sensing*. The ratiometric approach involves referencing the sensor probe emission signal to a stable emission signal from a dye that does not respond to the analyte or species of interest, but is subject to the same instrumental fluctuations that influence the intensity of the analytical signal.

An alternative way to obtain insight into a particular analyte that is not dependent on dye concentration or instrumental fluctuations is to apply fluorescence lifetime imaging microscopy (FLIM), *or phosphorescence lifetime in the case of most metal luminophores*. It is a quantitative imaging technique that can be used for real-time mapping of the cellular and tissue microenvironment, including cell functions and metabolic changes where the lifetime of a fluorophore is influenced by its local environment. As indicated, unlike intensity-based methods such as confocal fluorescence imaging, the image is independent of luminophore concentration, reflecting only the emission lifetime distribution of the probe.

A challenge that has traditionally impeded the application of metal complexes as bioimaging and biosensing probes is the poor uptake of such materials into cells. This has been widely overcome in recent years with a variety of strategies involving modification of the physicochemical properties of the complex or bioconjugation [3]. Conjugation of complexes to cell-penetrating and signal peptides specifically, has proven to be a particularly attractive and reliable method for achieving efficient cellular uptake without the use of permeabilisation agents. In particular, in the context of metal complex luminophores, this approach has the potential to very specifically drive the probe to target organelles with complex membrane structures such as the mitochondria or nucleus.

This review focuses on the more commonly studied luminescent transition row complexes of Ru(II), Ir(III), Os(II) and Re(I) with some examples from less well studied transition metal complexes such as Pt(II), Pd(II), Rh(III) and Zn(II).

2 Photophysical Profile of an Ideal Chromophore for Bioimaging

Luminescence imaging, including, particularly, confocal fluorescence and luminescence lifetime imaging methods, are widely used techniques in biochemistry and molecular biology as they offer high contrast, sensitivity, good resolution and flexibility in choice of luminophore probe. In addition, with commercialisation of more advanced imaging methods, including super-resolution and multiphoton methods, there is a growing need for probes that meet the demands of these methods, including robust photostability, sensitive environmental responsivity, high membrane permeability and targeted localisation. Indeed, studies to date have demonstrated that metal complexes can be applied in interrogating the cell environment and studying dynamic processes in vivo via a variety of imaging methods and that they have the synthetic versatility to tune to the desirable photophysical properties while maintaining biocompatibility and low cytotoxicity.

2.1 Favourable Properties of Metal Complexes

The ideal photophysical/optical characteristics of a luminescent imaging probe vary depending on the imaging methodology, although a number of characteristics are common to all, including the need for high molecular brightness (product of the molar extinction coefficient and quantum yield) and photostability. A diverse range of probes have been developed for fluorescence/luminescence imaging, including fluorescent proteins, expressed in situ in the cell, or exogenously applied probes, including organic fluorophores, nanoparticles, quantum dots and metal complexes. Organic fluorophores such as rhodamine, cyanine dyes, and the Alexa Fluor and Atto dyes, have been used widely as contrast agents in fluorescence microscopy to date as they exhibit high molecular brightness and in the case of Atto and Alexa Fluor probes, show good photostability. However, intrinsic drawbacks of organic fluorophores include a narrow Stokes shift which leads to inner filter effects and self-quenching at high optical densities, and in many cases, limited photostability. They also frequently show poor solubility in aqueous media, and so, application in cells often requires pre-dissolution in organic solvent that promotes cellular permeation but often through damage to the membrane. Finally, the short emission lifetime of organic fluorophores (usually in the range of 1–5 ns) is typically too short to enable time gating as a method to discriminate probe emission from background autofluorescence, and in sensing applications, limits quenching capability for diffusing species (the dye singlet states limit oxygen sensing also).

Aside from time gating, another approach to avoid autofluorescence interference in cellular or tissue imaging is to use a probe that emits in the red or NIR spectral range. Autofluorescence, excited at short excitation wavelengths, occurs from naturally fluorescent molecules within the cell and tissue environment or medium and usually decays on the nanosecond timescale. Nicotinamide adenine dinucleotide (NAD) and flavin adenine dinucleotide (FAD) are examples of intrinsic biological fluorophores for which several studies on their fluorescent properties have been carried out. Even if, for example, in Stokes-shifted emission, the probe is excited in the blue visible spectral range, which excites autofluorescence, its detection can be avoided if the probe emission is in the red region. In the context of luminescence imaging, but also therapy, a probe absorbing in the low-energy visible or NIR region is also desirable as this allows for deeper light-tissue penetration and avoids biological damage from continuous photo-irradiation into spectral regions where there is absorbance by tissue [4–9]. This is illustrated in Fig. 1 where emission from an Os(II) polypyridyl complex in the 650–800-nm region avoids any significant background signal from biological autofluorescence of a multicellular spheroid [10].

Transition metal complexes have also shown good photostability, which is particularly robust in the case of osmium(II) polypyridyl luminophores, where

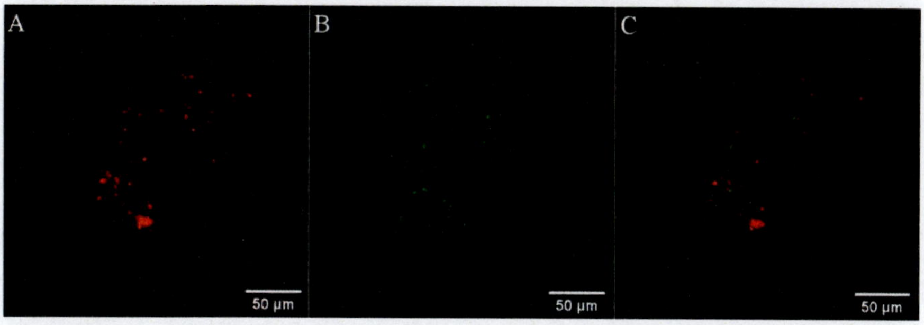

Fig. 1 Confocal imaging of a single live human pancreatic cancer (HPAC) spheroid treated with an Os(II) polyarginine probe, $[Os\text{-}(R_4)_2]^{10+}$ at 100 µM/48 h. Using a 490 nm white light laser for excitation, emission was collected between **A** 650 and 800 nm; Os(II) channel and **B** 500–570 nm; auto-fluorescence window. **C** Os(II)/autofluorescence channel overlay. Reprinted (adapted) with permission from Ref. [10] (https://pubs.acs.org/doi/10.1021/acs.inorgchem.1c00769). Further permissions related to the material excerpted should be directed to the ACS

photodecomposition and photobleaching can be completely avoided. To date, the coordination compounds of the (second and third row) d^6 metals Ru(II), Os(II) or Ir(III) are amongst the most widely studied transition metal imaging probes.

Figure 2 shows examples of luminescent metal complexes discussed in this section, highlighting key ligands used as building blocks for the design and development of metal complex luminophores.

Aside from their favourable photophysical properties, which are highly tunable due to the synthetic versatility of transition metal luminophores, metal complexes can also show good and also tuneable aqueous solubility, cell permeability and uptake and can be driven to subcellular structures through a range of approaches, in particular, as discussed in this review, by bioconjugation to peptides.

2.2 Tuning of Photophysical Properties

The photophysics and photochemistry of the prototype metal complex $[Ru(bpy)_3]^{2+}$ has been very thoroughly studied, and it is often used as an example to describe the photophysical activity of Ru(II) complexes [11–14]. The ultraviolet spectrum of $[Ru(bpy)_3]^{2+}$ is dominated by intense π–π^* ligand bands and the broad metal-to-ligand charge transfer (MLCT) transitions in the visible region. Spin-forbidden transitions are facilitated by spin orbit coupling which can be very large for second and third row transition metals such as Ru(II) and Os(II) complexes. Upon photon absorption, the singlet ^{1}MLCT excited state is populated and undergoes rapid intersystem crossing (k_{ISC}), populating a triplet MLCT (^{3}MLCT) excited state with unity quantum yield. In the case of $[Ru(bpy)_3]^{2+}$, deactivation from the lowest excited MLCT state to the ground state ($^1A_{1g}$ in O_h symmetry) is observed through emission or non-radiative decay via thermally activated (E_a) population of the ^{3}MC state ($^3T_{1g}$ in O_h symmetry). This latter process can lead to ligand dissociation. Indeed, enhanced ligand dissociation following ^{3}MC population is observed for sterically strained complexes or for complexes coordinated to ligands with a weak σ donor

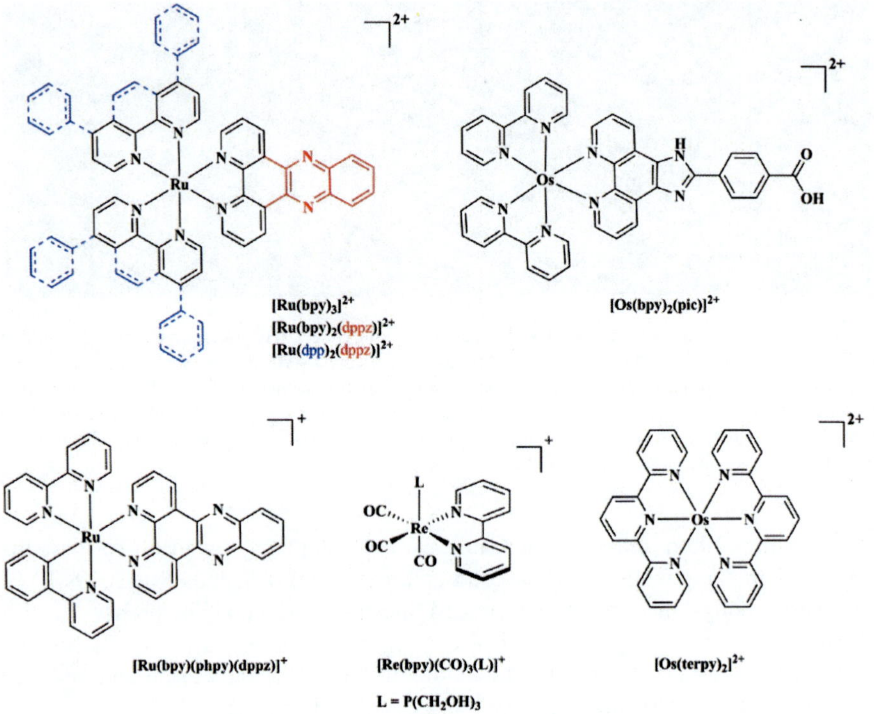

Fig. 2 Chemical structures of selected luminescent metal complexes discussed in this section

such as in Ru(II) 2,2′-biquinoline (biq) complexes [15], where reduced ligand field splitting capacity reduces the energy of the dissociative ^{3}MC, facilitating its thermal population from the ^{3}MLCT state. This is, for example, observed in $[Ru(tpy)_2]^{2+}$, where tpy is terpyridine, which exhibits a weak short-lived emission at room temperature [16]. The rigid tpy ligands cause geometric distortion from the ideal O_h geometry and smaller N–Ru–N trans angles (158.6°) that give rise to a weaker ligand field reducing the energy of the ^{3}MC state, thus facilitating radiationless deactivation [17].

Substitution at the 4′ position of the terpyridine ligands in $[Ru(tpy)_2]^{2+}$ with electron donor or acceptor moieties can enhance the excited state lifetime [18] by destabilising the metal-based highest occupied molecular orbital (HOMO) or stabilising the ligand-based lowest unoccupied molecular orbital (LUMO), respectively. A review by Medlycott and Hanan reports on the various synthetic strategies used to enhance the room-temperature photophysical properties of Ru(II) complexes of tridentate ligands [17].

Emission from Ru(II) complexes typically occurs in the wavelength range of 580–800 nm with $\lambda_{exc} \approx 400$ to 550 nm. Luminescence lifetimes are typically on the order of hundreds of nanoseconds with quantum yields of 1–5% (e.g. $[Ru(bpy)_3]^{2+}$; $\varphi_{air} = 0.04$ in water [19]). Molecular brightness, which is defined as the product of molar extinction coefficient (ε) and quantum yield (φ), is an important photophysical characteristic of an imaging probe, as it can determine the sensitivity and signal-to-noise ratio for luminescence detection. Molar extinction coefficients for Ru(II) complexes are in the range of 5000–20,000 M^{-1} cm^{-1} meaning, their molecular brightness is moderate when compared to organic dyes such as fluorescein. Although less optically tuneable than Ir(III), modification of the σ-donor or π-acceptor properties of Ru and Os complexes can also be used to tune the photophysics of these complexes. For example, coordination of strong π-acceptor ligands, such as 2,2′-biquinoline (biq), decreases ligand field strength and stabilises dπ orbitals, leading to red shifts in absorption and emission of Ru(II) complexes [20]. While, as described above, this can promote population of the ^{3}MC state, simultaneous co-coordination of a strong σ-donor ligand such as pyridyl-1,2,4-triazolate (trz) will promote photostability by raising the energy of the ^{3}MC, thus preventing both thermal population of this state and potential photodecomposition. Thus, strategic co-mixing of ligands can promote red emission whilst impeding photoinstabilty [20, 21].

Alternative strategies to obtain Ru(II) complexes with NIR emission have included widening the bite angle of the coordinated ligand to increase the ligand field. For example, as mentioned, $[Ru(tpy)_2]^{2+}$ complex is short-lived with a weak emission at room temperature [16], whereas $[Ru(terpy)(dgpy)]^{2+}$ (dgpy = 2,6-diguanidylpyridine and terpy = substituted 2,2′:6′,2′-terpyridine) complex exhibits a NIR emission at 900 nm [22]. Widening of the ligand bite angle, such as in the case of $[Ru(bqp)_2]^{2+}$ (where bqp = 2,6-bis(8′-quinolinyl)pyridine), results in red shifted and long-lived luminescence at room temperature ($\tau \approx 3$ μs) [23].

While molecular brightness is modest for most Ru(II) complexes, this is less of an issue for Ir(III) complexes, and they are also inherently more sensitive to the impact of ligand modification due to mixing of ligand and metal states. Factors such as absorbance and emission maxima and molecular brightness are relatively easily modified through ligand modification. The excited states of Ir(III) complexes frequently contain mixed contributions from both ^{3}LC and ^{3}MLCT and permit greater photophysical tuning, leading to complexes with a diverse range of emission properties across the visible to NIR spectrum. The photophysical properties of such complexes can be tuned via a number of strategies via π-extension of the coordinated ligands, or their modification with electron donating/withdrawing substituents to cyclometalated ligands or by introducing an ancillary ligand, e.g. N, N coordinating ligands that have σ-donating or п-accepting properties [24–27]. A recent review on NIR-emitting Ir(III) complexes discusses in detail the different methods that can be utilised to tune the photophysics of Ir(III) complexes [28]. Of note, although Ru(II) complexes are typically weaker emitters and less amenable to photophysical tuning than Ir(III) complexes, an advantage is that they tend to exhibit lower cytotoxicity upon uptake into cells [29, 30].

Shifting the emission maxima toward the NIR region can also be achieved by selecting an alternative metal centre. Os(II) polypyridyl complexes exhibit emission typically centred in the NIR region (> 730 nm), which is advantageous in the context of bioimaging, including cellular and tissue imaging [31–35]. Os(II) complexes share many of the same photophysical properties with their ruthenium analogues, with some key differences; The ^{3}MC state is higher in energy in Os(II) complexes due to increased crystal field splitting that raises the energy of the anti-bonding e_g^* levels, making it thermally inaccessible from the emitting ^{3}MLCT state. Thus, Os(II) complexes are extremely photostable, and their photophysics tend to show weak temperature dependence compared to their ruthenium analogues [16].

However, in comparison to $[Ru(bpy)_3]^{2+}$, the ^{3}MLCT excited state lifetime of Os(II) is much shorter-lived, and quantum yields are lower. This is a feature of the energy gap law which comes into play for red to NIR emission. It predicts that the non-radiative rate decay increases as the energy gap between the excited and ground state decreases [36]. Therefore, the low-energy MLCT in the case of Os complexes leads to efficient non-radiative decay.

The bis-terpyridine $[Os(tpy)_2]^{2+}$ complex, in contrast to $[Ru(tpy)_2]^{2+}$, exhibits an intense long-lived luminescence at room temperature due to the greater ^{3}MC/^{3}MLCT energy gap [16].

Amongst the d^6 complexes, rhenium(I) complexes, *typically rhenium fac tricarbonyl polypyridyls*, also exhibit attractive photophysical properties, including large Stokes shifts, long-lived oxygen sensitive emission, and high photostability. Thus, they have also been applied as bioimaging agents. Photophysical tuning of rhenium complexes is more challenging compared to complexes of Ru(II), Ir(III) and Os(II). In particular, the absorption of such complexes tends to be toward the UV or blue spectral range which limits suitability for imaging applications, especially in tissues. Nonetheless, NIR emission can be achieved by implementing the complex into a D-π-A system [37]. Due to the isostructural relationship between rhenium and technetium-99 m and their characteristic infrared absorption bands, complexes of rhenium(I) have been applied as probes for radio imaging and vibrational imaging, respectively [38, 39]. Furthermore, rhenium(I) tricarbonyl complexes have been developed as agents for photodynamic therapy as they tend to be strong photosensitisers for singlet oxygen generation [40, 41].

Pt(II)/Pt(IV) compounds have historically found application, mainly in therapy as anticancer agents [42, 43], but have also been studied more recently in the context of imaging [44–46]. Luminescence and biocompatibility are prerequisites for use in imaging, and numerous kinetically stable and emissive Pt(II) complexes have been reported. Pt(II) (d^8) luminophores are distinctive from the complexes discussed above because of their square planar geometries, and Pt(II) luminophores have been based mainly on the general structures $[Pt(C\wedge N\wedge N)(L)]^n$ (C^N^N = aryl-substituted N^N ligand, L = monodentate ligand and n = 0 or + 1), cyclometalated tridentate (e.g. [Pt(C^N^C)(Cl)]) derivatives [47]. Tetradentate ligand and π-conjugated porphyrin coordinated Pt(II) luminophores have also been reported [48, 49]. Emission from cyclometalated tridentate complexes is usually attributed to a triplet intra-ligand charge transfer excited state (^{3}ILCT), and so photophysical properties are tuneable through ligand modification [44]. π-Conjugated Pt(II) porphyrin complexes, in particular, can exhibit high quantum yields and NIR emission, but efficient and uniform cellular uptake can be problematic due to the large size of porphyrins. In addition, complexes of platinum(II) exhibit a square planar coordination geometry that can permit self-assembly by non-covalent π–π and/or Pt(II)–Pt(II) interactions and the prospect of triplet metal–metal-to-ligand charge transfer (^{3}MMLCT) excited state emission [50, 51].

Complexes of Rh(III) [52] and Zn(II) [53, 54] have also been applied in bioimaging, but to date, to a lesser extent than the above metals.

2.3 Reducing Toxicity by Ligand Modification

It is important to consider potential cytotoxicity, both dark and photo-induced, when designing a metal complex luminophore for bioimaging and sensing. The metal centre and coordinated ligands dictate the excited state and redox properties of a complex, and these features, along with size, lipophilicity and overall charge, can generally influence cytotoxicity.

Owing to their long-lived triplet excited states, Ir(III), Re(I) and Ru(II) complexes can induce cellular toxicity via a number of photochemical and photophysical routes. For example, incorporation of tap or hat ligands (tap = 1,4,5,8-tetraazaphenanthrene, hat = 1,4, 5, 8, 9,12-hexaazatriphenylene) in complexes of Ru(II), permits efficient proton-coupled electron transfer (PCET) reactions with bio-relevant molecules such as DNA which can lead to cytotoxic effects. For example, the complexes $[(N^\wedge N)_2Ru(tatpp)]^{2+}$ (where N^N = bpy or phen and tatpp = 9,11 ,20,22 -tetraazatetrapyrido[3,2-a:2′,3′-c:3″,2″-1:2‴,3‴-n]-pentacene) were shown to cleave DNA through a redox-mediated mechanism [55]. Sensitisation of reactive oxygen species (ROS), such as singlet oxygen, is another important route to photo-induced toxicity that can be exploited in photodynamic therapy applications [56, 57]

Another strategy to phototherapy and a route to cytotoxicity is photo-induced ligand dissociation or substitution which, as previously mentioned, is observed mainly in Ru(II) complexes upon thermal population of the ^{3}MC states [15, 58]. Turro et al. investigated in detail the factors that affect ligand dissociation and singlet oxygen generation and demonstrated how ligand tuning can be used to promote both reactions. As the ligand dissociation is thermally driven, it can be quite prevalent under cellular imaging conditions and used for application as such complexes as prodrugs [15, 59–62]. For example, Glazer et. al. have described a series of sterically strained Ru(II) complexes that exhibit dramatically increased ligand photorelease which results in a reactive ruthenium-aquo complex that can photo-bind DNA/biomolecules and trigger cellular apoptosis [63]. Conversely, osmium luminophores tend to be very photostable and relatively inert towards ligand substitution, thus reducing cytotoxic effects that occur through ligand dissociation routes [64, 65].

Lipophilicity often determines cellular uptake and influences intracellular accumulation and localisation, thus affecting toxicity. Coordination to cyclometalated ligands has been shown to enhance lipophilicity and improve cellular uptake for iridium and ruthenium complexes [66, 67]. For example, exchanging the bpy ligand in $[Ru(bpy)_2(dppz)]^{2+}$ for the cyclometalating 2-phenylpyridine (phpy) ligand yielded a lipophilic and cell-permeable Ru(II) complex [68]. However, the enhanced lipophilicity with cyclometalation can promote toxicity [69, 70].

While the nature of the ancillary ligands is important, chemical modifications to the ligand itself can also influence the lipophilicity and consequently cellular uptake, localisation and cytotoxicity of the complex [71–73]. Increased lipophilicity and dark cytotoxicity was observed for Ru(II) bis-phen and bis-TAP complexes coordinated to a hydrophobic alkylamide phen ligand [74].

Glazer et al. reported on the uptake of two Ru(II) complexes differing in their charge but coordinated to the highly lipophilic *dip* (or *dpp*) *ligand* [75]. The complexes were successfully internalised by A549 cells where the lipophilic $[Ru(dip)_3]^{2+}$ (log $P = +1.8$) accumulated at the mitochondria and lysosomes, while the anionic and less lipophilic $[Ru((SO_3)_2\text{-dip})_3]^{4-}$ (log $P = -2.2$) localised in the cytosol and was mitochondrial-excluding. Both complexes showed photo-induced toxicity, but interestingly, the mitochondrial accumulating complex also showed dark toxicity with an IC_{50} between 0.62 and 3.75 μM. This study highlights the importance of balancing charge and lipophilicity in order to modulate accumulation and limit cytotoxicity (Table 1).

Recently, Finn et al. reported on functionalised Ru(II) complexes with pendant and lipophilic alkyl-acetylthio chains of varying lengths [80]. The complexes were capable of self-assembling into micelles under aqueous conditions and could traverse the cell membrane.

Polyethylene glycol has been conjugated to metal complexes to increase aqueous solubility and reduce dark cytotoxicity [81]. Reduced cytotoxicity was observed for cell-permeable Ir(III)-poly(ethylene glycol) (PEG) conjugates in comparison to the PEG-free counterparts [82, 83]. The long PEG chains likely protect the Ir(III) complexes from non-specific interactions with proteins, DNA and membranes within the cell.

Table 1 Lipophilicity and cytotoxicity of selected metal complexes upon synthetic modifications

Compound	log $P_{o/w}$ [a]	IC_{50} (μM) [cell line]	References
[Ru(**phen**)$_3$]$^{2+}$	−0.33	268.0 [MCF-7][b]	[76]
[Ru(**bpy**)$_3$]$^{2+}$	−0.41	341.5 [MCF-7][b]	[76]
[Ru(**pic**)$_3$]$^{2+}$	+2.67	66.0 [MCF-7][b]	[76]
[Ru(**bpy**)$_2$(dppz)]$^{2+}$	−2.50	159.9 [HeLa]	[68]
[Ru(**dip**)$_2$(dppz)]$^{2+}$	+1.30	–	[77]
[Ru(**dip**)$_2$(dppz-**NH$_2$**]$^{2+}$	−0.27	> 100 [HeLa][b]	[73]
[Ru(**dip**)$_2$(dppz-**CH$_2$OH**)]$^{2+}$	−0.62	Cell-impermeable [HeLa]	[73]
[Ru(bpy)(**phpy**)(dppz)]$^{1+}$ phpy = 2-phenylpyridine	+1.00	0.6 [HeLa]	[68]
[**Os**(**phen**)$_2$(phpy)]$^{+}$	+2.43	0.4 [A172][b]	[78]
[**Ir**(**phen**)(C^N)$_2$]$^{+}$ *where N^C = 2-(p-tolyl)pyridine*	+0.63	1.68 [HeLa][c]	[79]

[a]Lipophilicity, log $P_{o/w}$, was estimated by the partition coefficient of each compound in octanol/water. Propidium iodide and Hoechst are both commercially available organic nucleic acid markers where the first is permeant only to damaged/dead cells, and the latter is cell-permeable. IC_{50} values for the metal complexes were determined based on incubation periods of 24 h unless stated otherwise where [b]48 h, [c]72 h

3 Rationale for Peptide Conjugation to Transition Row Metal Complexes

Cell membrane permeability is a key barrier to the widespread application of metal complexes in cellular and tissue imaging. One widely used approach to overcome this challenge is the use of organic solvents such as dimethyl sulfoxide (DMSO) or detergents such as Triton-X to permeabilise the cell membrane of mammalian cells. Permeabilising agents act by disrupting the integrity of the membrane bilayer, thus promoting entry of the compound into the cell [84]. This approach is widely used for both organic fluorophores [85–87] and metal complexes [88–91], though it is not always explicitly explained. A key drawback is that above relatively low volume percentages, e.g. for DMSO > 5% vol/vol, solvent permeabilisation can cause irreversible damage to the cell membrane [92], so the approach should be used with care in the study of cultured cells and organic solvent as a permeant is of limited use in tissue or in vivo applications [93, 94].

Other approaches to improving permeability have focused on tuning the lipophilicity, charge and solubility of the complex which in turn can influence cellular uptake and accumulation, as mentioned previously. The use of nanocarriers [95–98], liposomes [99], dendrimers [100], sugars/carbohydrates [101, 102], polyethyleneglycol (PEG) chains [81, 82], vitamins [103–105], antibodies [106], lipophilic moieties such as triphenylphosphonium (TPP) [107], amino acids [108] and cell-penetrating peptides (CPPs) [81, 109–111] has also been shown to increase solubility and improve membrane permeability, facilitating reliable uptake of complexes within cells for a range of applications. Recent reviews describe the preparation and application of ruthenium bioconjugates [112] and vectorisation strategies of metal complex luminophores [3].

Following cellular uptake, subcellular targeting of organelles, such as to the mitochondria or nucleus, is typically of interest in the context of bioimaging/sensing and therapy.

The nuclear envelope is a double membrane comprising of inner and outer nuclear membranes that converge at several sites, generating nuclear pores [113]. Uptake of ions and small molecules is mediated through the nuclear pores through a channel (~ 30 nm in diameter) via passive diffusion. In contrast, uptake of larger molecules is mediated through transport receptors [114]. The mitochondria also feature a double-membrane boundary, though structurally different to the nucleus. The inner mitochondrial membrane is far less permeable than the outer, allowing only very small molecules to cross into the matrix where mitochondrial DNA (mtDNA) and other molecules of analytical interest are contained.

Peptide conjugation has emerged in recent years as a key enabling tool to promote cell uptake, particularly of non-membrane-permeable metal complexes or to enhance uptake and targeting of permeable complexes [115]. The mechanism by which peptides facilitate transport across the cell membrane is often linked to an energy-dependent process such as endocytosis. For example, the recognition of molecules by specific receptors located on the surface of the cell membrane can lead to a receptor-mediated endocytic pathway of uptake. In principle, it is possible that the peptide may lower the uptake efficiency in comparison to the peptide-free complexes—for example, in instances where the complex is highly permeable through the membrane by passive diffusion due to their lipophilic character or in comparison to membrane modification methods such as use of permeabilisation agents. Lower uptake efficiency, however, is often balanced by improved precision in intracellular localisation and decreased cytotoxicity.

Peptide conjugation to metal complexes has been facilitated by a plethora of peptide coupling reactions available to couple peptides to metal complexes, including, but not limited to, amine/carboxyl coupling reactions, "click" chemistry and Sonagashira coupling reactions.

Cell-penetrating and signal peptides specifically are proven reliable vectors for the efficient intracellular delivery of different metal complexes and for targeting organelles with complex membrane structures such as the mitochondria or the nucleus [10, 109, 116–119].

3.1 Peptides

Peptides, short sequences (<50) of amino acids linked by amide bonds, are physiologically important biomolecules that serve in signalling processes and are ligands for many proteins. In the body, they function as hormones, inhibitors, antibiotics and anti-inflammatories, and both natural and synthetic peptides are finding increasing use in therapeutic applications [120]. Peptides have been widely applied in the pharmaceutical industry to promote permeation of drugs across the membrane; in particular, cell-penetrating peptides (CPPs) have been very effective in this regard and have been applied both as appendages to therapeutic molecules and incorporated into nanocarriers [121]. One of the reasons that peptides have become so important in pharmaceuticals as cargo carriers is that they can be readily accessed, including linear and cyclic and branched peptides, through chemical synthesis. The most important route is through solid-phase peptide synthesis (SPPS), and for many peptide sequences, the synthesis can be automated with continuous improvements reported to protocols that lead to gains in speed, purity and yield. Furthermore, functional terminal groups can be readily appended in the synthesis protocol to facilitate conjugation [122]. The success of peptides in the pharmaceutical industry and also their application in driving organic imaging agents into cells has led to their application in recent years as conjugates to metal complexes to promote their cellular access and targeting.

3.1.1 Cell-Penetrating Peptides (CPPs)

The ability of cationic peptide sequences to cross the cell membrane and facilitate uptake of small molecules was first demonstrated in 1965 by Ryser and Hancock with the cationic amino acid-mediated enhanced uptake of albumin followed by studies on conjugation of poly-L-lysine to albumin and horseradish peroxidase [123, 124].

The most studied cell-penetrating peptide is likely the arginine-rich HIV-Tat transduction protein (*RKKRRQRRR*) from the human immunodeficiency virus [125, 126]. Homopolymers of arginine (polyarginines) have shown superior cellular uptake compared to other cationic analogues such as ornithine and histidine [127]. With studies showing no strict requirement for side chain length or backbone chirality (*D-Arg* vs *L-Arg*), it was concluded that the guanidinium head groups of arginine units are the structural features crucial to cellular uptake. Barton et al. first reported peptide-facilitated cellular uptake of rhodium complexes [52]. In order to reduce nonspecific DNA binding owing to the highly charged R8, a shorter peptide sequence, RrRK (where r = D-arginine), was conjugated to a Ru(dppz) complex, achieving cellular uptake and nuclear accumulation above a threshold concentration of 100 μM in complete media [128]. Cargo transduction occurs for arginine sequences of Arg_n or R_n, where $n = 6–11$ residues, with octaarginine (Arg_8) and nonaarginine (Arg_9) being most efficiently transported. Our group has reported the efficient transport of an otherwise cell-impermeable Ru(II) polypyridyl complex, $[Ru(bpy)_2(pic)]^{2+}$, via conjugation to octaarginine [88]. The conjugate was found to passively transport into myeloma cells within 12 min. In addition, studies showed that Arg_5 or lower conjugates are not effective in promoting metal complex permeation. Wender et al. reported similar decrease in uptake efficiency of shorter polyarginines [129]. Polyarginine sequences have been extensively explored for promoting or enhancing cellular uptake of metal complexes with applications ranging from bioimaging to medicinal chemistry [29, 88, 116, 130–133]. More specifically, octaarginine-driven cellular uptake has been reported for a range of otherwise impermeable luminescent complexes differing in their metal centre (e.g. Ru(II), Os(II), Ir(III)) *and* coordinated ligands [10, 88, 116, 117, 131]. The effect of conjugation to peptides on the DNA recognition properties of Ru(II) and Ir(III) complexes has also been explored [30, 116, 134]. Recent studies showed that appending an R8 tail to the Ru(II)-dppz complex increased its affinity for G-quadruplexes and that both the ancillary ligand and the octaarginine tail were key to control the selectivity between quadruplexes [134].

We have also demonstrated that two tetraarginine sequences across a linear osmium(II) complex promote cellular uptake, whereas the analogue containing R8 at each terminal was membrane-impermeable. Our data indicated that a contiguous structure may not be required for octaarginine-facilitated transport and that there is an upper limit to the arginine chain length effective in promoting membrane transport of the metal complex [10].

There have been multiple pathways and mechanisms proposed to explain polyarginine CPP behaviour. Although there are a number of studies that report that polyarginines can promote permeation through a passive mechanism [135] or through local changes at the membrane [136], the key pathway in live cells appears to be ATP-activated endocytosis [137]. Polyarginine interactions with cell surface lipids and formation of neutral complexes that transport across the bilayer have also been reported, as well as surface attachment through interactions with heparan sulfate proteoglycans (HSPG) [138–142].

Penetratin, a cationic peptide sequence (*RQIKIWFQNRRMKWKK*) corresponding to the R-helix of the Antennapedia homeodomain, is capable of crossing lipid

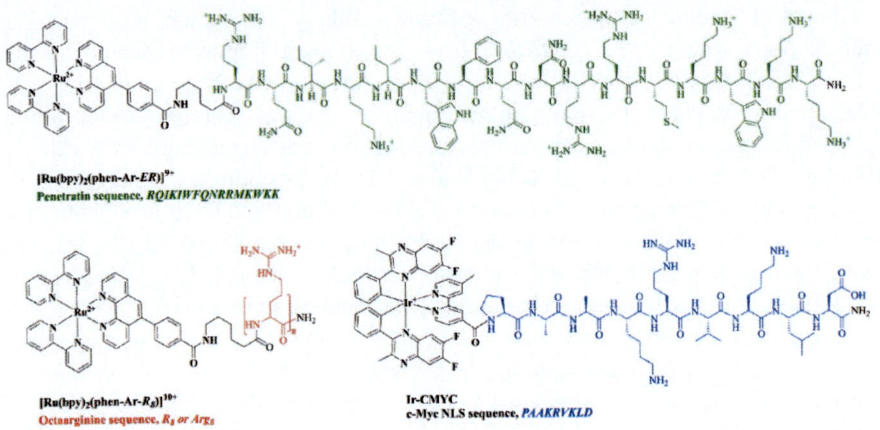

Fig. 3 Chemical structures of peptide conjugates $[Ru(bpy)_2phen\text{-}Ar\text{-}ER)]^{9+}$ cells [117], $Ru(bpy)_2phen\text{-}Ar\text{-}R_8)]^{10+}$ cells [117] and Ir-CMYC [153]

bilayers and is also quite widely applied cell-penetrating peptide [143]. Studies have shown that the uptake mechanism involves direct interaction of the peptide with membrane lipids and does not involve vesicle disruption or pore formation [144, 145]. Peptide conjugation of penetratin to $[Ru(bpy)_2\text{-}phen\text{-}Ar\text{-}COOH]^{2+}$ (Fig. 3) allowed for delivery of the complex to the endoplasmic reticulum in live HeLa cells [117].

3.1.2 Signal Peptides

Although cell-penetrating peptides such as polyarginines can facilitate efficient cell permeability, more targeted subcellular organelle targeting of imaging probes or theranostic agents can be achieved using signal peptides. Natural signal peptides are amino acid sequences appended to the N termini of newly synthesised proteins in the ribosome that direct the protein from the ribosome along its secretory pathway to its destination. Such signal peptides can provide a powerful means of directing exogenous probes to their target within the cell, and naturally derived peptides have been applied in this regard, and designed sequences have been shown to be recognised by proteins in organelle membranes [146].

Nuclear localisation signal (NLS) peptides derived from transcription factors can cross the cellular membrane and enter the nuclear envelope. To date, NLS sequences that have been derived from transcription factors include NFκB, TCF1-α, TFIIE-β, Oct-6 and SV40 [147–149]. Studies by Ragin et al. demonstrated that the NLS peptide, *VQRKRQKLMP*, derived from NFκB, was effective in promoting nuclear penetration [148]. Based on this finding, Keyes's and co-workers exploited NFκB transcription factor bioconjugation for the efficient and selective nuclear uptake of Ru(II) complexes [2, 119, 150]. Gasser et al. reported on functionalisation of Ru(II) complexes with a nuclear localisation signal peptide (*L*-Arg-*D*-Arg-*L*-Arg-*L*-Lys-$CONH_2$) linked via a photolabile protecting group (PLPG); the resulting conjugates showed preferential nuclear accumulation in HeLa and MRC-5 cell lines [151]. A study by the same group described the nuclear delivery of derivatised Re(I) quinolyl complexes using the nuclear localisation sequence, *CRRRK* [152].

In a recent study by Pope et al., an alternative nuclear localisation sequence, *PAAKRVKLD* (Fig. 3), was conjugated to a cyclometalated iridium(III) complex [153]. The c-Myc NLS is derived from the human c-Myc protein and is essential for its nuclear localisation [154]. The Ir-CMYC conjugate was efficiently delivered to the nucleus of human fibroblast cells and was essentially non-toxic, in contrast to the peptide-free parent complex [153].

Cellular uptake of cyclometalated iridium(III) complexes upon conjugation to an endoplasmic reticulum (ER)-targeting sequence (KDEL) and the NLS *PKKKRKV* (derived from SV40 large T antigen) has also been explored [155]. Interestingly, although the ER-targeting conjugate accumulated at the endoplasmic reticulum, the NLS conjugate showed non-specific staining attributed to endosomal trapping upon uptake. Ypsilantis et al. presented a detailed study of the interaction of diruthenium complex peptide conjugates with an oligonucleotide duplex and found that the tethered peptide Gly^1-Gly^2-Gly^3-Lys^1 $CONH_2$ hindered complex binding [156].

Mitochondria-penetrating peptides (MPPs) have been employed for the specific targeting of mitochondria for imaging and therapy. Kelley et al. carried out a detailed iterative study on synthetic peptide sequences relating to signal sequences effective in promoting mitochondrial targeting of fluorescent probes/drug analogues [157]. Amongst the most effective of the sequences studied was an 8-amino acid sequence, *FrFKFrFK*, containing *D*-arginine and hydrophobic residues [157]. Keyes et al. exploited this sequence and the acetyl-blocked sequence, *FrFKFrFK(Ac)*, to effectively and selectively drive mono- and dinuclear Ru(II) complexes to the mitochondria of mammalian cells[118, 119]. As discussed in detail in later sections, such MPP-driven complexes have been applied as bioimaging and sensing tools in live mammalian cells. For example, $[(Ru(bpy)_2phen\text{-}Ar)_2\text{-}MPP]^{7+}$ showed dynamic response to variations in oxygen and ROS levels [118], whereas $[Ru(dppz)(bpy)(bpy\text{-}Ar\text{-}MPP)]^{5+}$ was used as a light switch probe for mitochondrial nucleoid imaging [119]. Bis-conjugation of the MPP sequence to an achiral Os(II) complex generated a NIR probe showing concentration-dependent cell death that could be tracked on the basis of probe localisation using confocal microscopy, offering a potential theranostic probe [158].

3.1.3 Receptor-Targeting Peptides

The peptide sequence Arg-Gly-Asp (RGD) has been applied to mediate specific binding with integrin receptors and has been extensively used in cancer drug research as integrin receptors, such as $\alpha_v\beta_3$, which are overexpressed in certain tumour cells [159–161].

Adamson et al. first reported on RGD-labelled luminescent metal polypyridyl complexes [110]. Complexes of ruthenium(II) were conjugated to a linear RGD peptide with the objective of targeting platelet integrin, $\alpha_{IIb}\beta_3$ to, through emission anisotropy, reflect integrin conformation status. Integrins are adhesion receptors and transmembrane proteins that undergo large conformational changes and clustering on activation that alters their affinity for their receptors, and RGD is a peptide motif recognised by all integrins [162]. The yielded [Ru(N^N)(pic-RGD)]$^{2+}$ (where N^N = bpy or dpp) conjugates showed high binding affinity and specificity for $\alpha_{IIb}\beta_3$, and through alterations in metal complex photophysical behaviour and anisotropy, it was possible to distinguish between different activation states of integrin. A two-step binding was determined for [Ru(dpp)$_2$(pic-RGD)]$^{2+}$ with $K_{d1} = 0.25 \pm 0.29$ μM and $K_{d2} = 4.37 \pm 0.82$ μM. Additionally, confocal imaging revealed that both bpy-RGD and dpp-RGD conjugates selectively bind to CHO cells expressing the resting form of $\alpha_{IIb}\beta_3$.

In spite of their biomedical importance, there are surprisingly few examples of linear- or cyclic-RGD-metal complex conjugates either applied as therapeutic agents [111, 163, 164], luminescent probes [110] or both [165–167].

A zinc phthalocyanine complex conjugated to a cyclic RGD peptide displayed dramatically higher cellular uptake in $\alpha_v\beta_3^+$ U87-MG cells compared with the $\alpha_v\beta_3^-$ MCF-7 cells [168]. A recently reported Ru-cRGD (cyclic RGD) conjugate exhibited strong two-photon luminescence and showed preferential accumulation in malignant cells with promising potential as a theranostic agent [167]. In an alternative system, dual-imaging nanoprobes were prepared by conjugating iridium(III), gadolinium(III) and RGD onto silica nanoparticles [169]. The water-soluble particles permitted in vitro and in vivo studies using confocal luminescence imaging and magnetic resonance imaging.

Certain vectors can be used to target cells that overexpress key receptors such as folate, transferrin and somatostatin at the membrane surface of different disease states [170–173]. For example, enhanced uptake of a somatostatin-targeting Ru(II) conjugate was achieved in A549 cells overexpressing somatostatin receptors [173]. Although the conjugate did not act as a bioimaging probe, it showed excellent photosensitised toxicity with an IC_{50} of 300 μM in the absence of light versus an IC_{50} of 13 μM upon irradiation (PI > 23).

Similarly, a redox active Pt(IV) complex was coordinated to the tumour-penetrating sequence (*TKDNNLLGRFELSG*) that targets the membrane protein *heat shock protein 70 positive* (*memHSP70+*) which is upregulated in colorectal cancer cells but is not usually found in healthy tissues [174]. The Pt(IV) complex is reduced in the cell to Pt(II), releasing the axial ligands and leading to cytotoxicity[175].

This strategy of conjugation to tumour recognition or penetrating sequences can also be exploited in the design of targeted probes for bioimaging. For example, C–X–C chemokine receptor 4 (CXCR4) is overexpressed in over 23 different types of cancer and is more prevalent in malignant cancer tissue [176]. With this consideration, a rhenium(I) tricarbonyl complex was conjugated to a derivative of *T140* (14 amino acid sequence), a known antagonist of CXCR4, and showed potential as an imaging agent for CXCR4 expression that was capable of differentiation between cancerous and healthy tissue [177, 178]. Kuil and co-workers had previously presented an iridium(III)–peptide conjugate for FLIM-based visualisation of CXCR4 expression in cells by conjugating the complex to a series of Ac-TZ14011 peptides [179].

Vallaisamy et al. reported on an iridium(III) complex conjugated to the hexapeptide *MKYMVm, the peptide agonist*, which selectively targeted formyl peptide

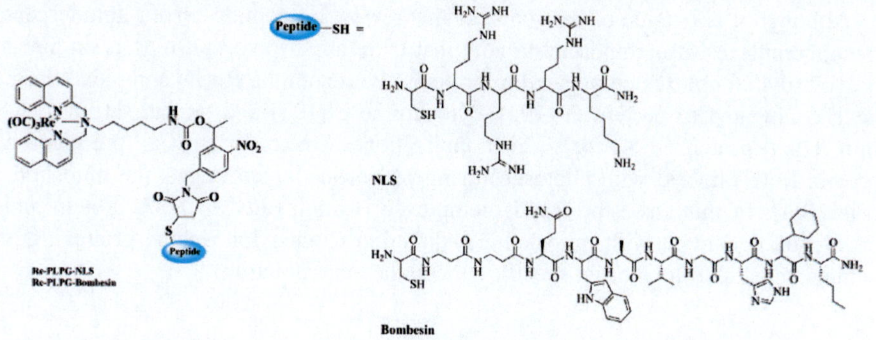

Fig. 4 Structure of Re(I) complex conjugated to a nuclear localisation signalling (NLS) peptide or a bombesin (BBN) derivative peptide sequence used to improve uptake of the complex by cancer cells overexpressing the gastrin-releasing peptide receptor (GRPR) [152]

receptor 2 (FPR2) in live cells [180]. Formyl peptide receptor plays an important role in chemotactic signals and modulation of host defence and inflammation [181].

Gasser et al. have functionalised Ru(II) and Re(I) complexes with the peptide bombesin (BBN, 14 amino acid sequence) in order to enhance uptake and accumulation of the complexes selectively in cancerous cell lines [151, 152]. The peptide BBN, Fig. 4, is structurally similar to the human gastrin-releasing peptide (GRP) and is recognised and internalised by GRP receptors that are overexpressed in some cancer cell lines [182, 183].

A series of zinc phthalocyanine peptide conjugates were synthesised to target gastrin-releasing peptide (GRP) and integrin receptors [184] in order to initiate a targeted therapeutic effect.

Agorastos et al. reported a rhenium tricarbonyl complex functionalised with acridine orange that could selectively stain the cell nucleus of both mouse melanoma (B16-F1) and human prostate adenocarcinoma cell line (PC-3) cells [185]. Conjugation of the complex to the bombesin peptide resulted in cell-specific uptake. The peptide conjugate was membrane-impermeable to B16F1 cells, but readily permeated into PC-3 cells. Cell-specific uptake was achieved by exploiting the lack of gastrin-releasing peptide (GRP) receptors in B16F1 cells, but which are expressed in PC-3 cells. Interestingly, conjugation to the peptide prevented the ability of the complex to enter the nucleus.

A short peptide based on the endogenous opioid pentapeptide ligands was chosen for the preparation of luminescent heterobimetallic Ir(III)/Au(I) conjugates that were found to be membrane-permeable, and localise in the lysosomes of A549 cells [186].

A rather different type of recognition occurs in the case of peptide nucleic acids (PNAs). PNAs are non-natural DNA/RNA analogues that consist of *N*-(2-aminoethyl)glycine units which form a pseudopeptide backbone bearing the four nucleobases. They thus exhibit strong affinity for nucleic acid strands [187]. PNA conjugation has been explored in the development of luminescent rhenium-PNA conjugates for cell imaging and DNA targeting [188–191].

Although it is outside of the scope of this review on peptide-driven luminescent metal complexes, it is important to note that there are also several reports on metal polypyridyl complexes conjugated to proteins. For example, Ru(II) complexes have been conjugated to protein G [192], cytochrome c [193] and human serum albumin (HSA) protein [194] for various applications. Chakrabortty et al. presented a protein-Ru(II) hybrid with photosensitising properties which targets the mitochondria [194]. In this case, the Ru(II) complex was conjugated to HSA protein and covalently decorated with mitochondria-directing triphenylphosphine groups, thus achieving cellular uptake and specific subcellular accumulation.

4 Luminescent Metal Complex Peptide Conjugates Applied in Bioimaging

Early examples of *peptide-conjugated* metal complexes in confocal imaging are described in studies carried out by Barton et al. Conjugation to octaarginine enhanced cellular uptake of rhodium(III) 5,6-chrysenequinone diimine (chrysi) and ruthenium(II) dipyrido-phenazine (dppz) complexes, and interestingly, attachment of a fluorescein moiety, in the case of the Ru(II)-dppz complex, led to nuclear localisation [52, 130]. Our group has focused extensively on the design and development of peptide metal complex conjugates for bioimaging, sensing and theranostics. A series of otherwise cell-impermeable ruthenium(II), iridium(III) and osmium(II) complexes have been conjugated to cell-penetrating and signal peptides and have been studied using confocal microscopy, lifetime imaging and resonance Raman spectroscopy. Recently, a polyarginine Os(II) probe was used in imaging of pancreatic multicellular tumour spheroids, marking the first step towards the application of such luminescent peptide probes in tissue imaging [10].

4.1 Cytoplasm

Octaarginine CPPs drive cellular uptake and internalisation, usually into the cell cytoplasm, resulting in cytoplasmic or non-specific staining. Cell-permeable R8-conjugates have been explored as bioimaging probes, namely $[Ir(dfpp)_2(picCONH)R_8]^{9+}$ [29] and $Ru(bpy)_2(phen\text{-}Ar\text{-}R_8)]^{10+}$, for confocal and high-resolution stimulated emission depletion (STED) imaging [117], and $[Os(bpy)_2(pic\text{-}R_8)^{10+}$ and $[Ru(bpy)_2(pic\text{-}R_8)]^{10+}$ for confocal and phosphorescence lifetime imaging microscopy (PLIM) studies [131], and $[Os\text{-}(R_4)_2]^{10+}$ for confocal/lifetime imaging of two-dimensional (2D) and three-dimensional (3D) cell cultures [10]. Figure 5 illustrates the dye distribution of key examples of octaarginine conjugates of Ir(III), Ru(II) and Os(II). Conjugation to Arg_8 rendered all three complexes membrane-permeable in aqueous solution without the requirement for permeabilisation agent such as detergent or organic solvent. Fei et al. also reported on cytoplasmic and vesicular staining in HeLa cells following incubation with a histidine-targeting Ir(III)-HTat conjugate [195].

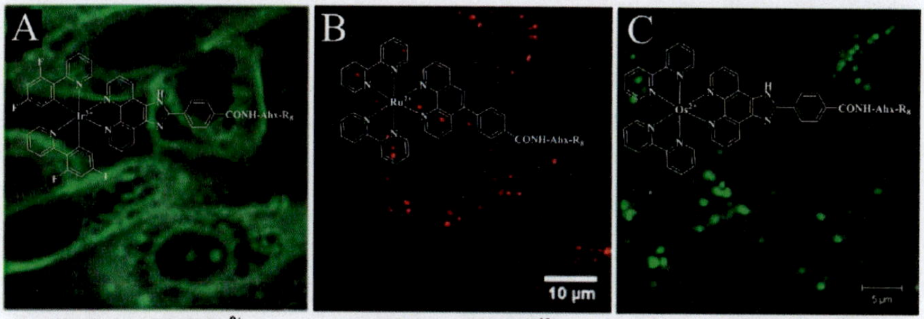

$[Ir(dfpp)_2(picCONH)R_8]^{9+}$ $[Ru(bpy)_2(phen\text{-}Ar\text{-}R_8)]^{10+}$ $[Os(bpy)_2(pic\text{-}R_8)]^{10+}$

Fig. 5 Confocal luminescence imaging of live mammalian cells incubated with octaarginine conjugates of Ir(III), Ru(II) and Os(II). **A** Ir-R8 in CHO cells at 70 μM/15 min incubation [29], **B** Ru-R8 in HeLa cells at 70 μM/4 h incubation (Reproduced from Ref. [117] with permission from the Royal Society of Chemistry) and **C** Os-R8 in CHO cells at 70 μM/2 h incubation in the absence of light at 37 °C (Reproduced from Ref. [131] with permission from the Royal Society of Chemistry)

A fused peptide consisting of a nona-arginine fragment attached to a sequence (*RHVLPKVQA* = Aβ aggregation inhibitor) with anti-amyloid activity was labelled via a histidine residue to a platinum(II) complex [196]. The resulting luminescent conjugate was studied in cells and was shown to stain the cytoplasm of HeLa cells. In vivo studies in *Drosophila melanogaster* showed that the luminescent platinum conjugate could permeate the blood brain barrier of these organisms and evenly distribute in the brain. This work highlighted the use of fused peptides as vectors to penetrate the blood brain barrier while also selectively targeting biorelevant molecules or, in this case, inhibit the formation of amyloids.

Using "click" chemistry, a rhenium(I) tricarbonyl complex was attached to a lipopeptide known to increase cell permeability [197]. The addition of the myristoylated HIV-1 TAT (myr-Tat) peptide to the rhenium complex substantially enhanced uptake in cells compared to the peptide-free complex and showed cytoplasmic accumulation with partial nucleoli staining.

4.2 Nucleus, DNA/RNA

The interaction of metal complex luminophores with nucleic acid materials has been the subject of extensive study since the 1980s. This has led to deep insight into the nature of metal complex–DNA interactions, expanding the prospects for both intracellular sensing and photo therapy by these species. Increased understanding of the factors that can be used to promote metal complex permeation and organelle targeting have led to the application of such complexes to study nucleic acid materials in cells, with several studies now reporting on the nuclear uptake and staining of metal peptide conjugates used for imaging or sensing of DNA within live cells [198–200].

One of the earliest of such studies was reported by Brunner and Barton who utilised functionalised rhodium complexes with octaarginine peptides to study DNA mismatches [52]. The rhodium complexes were capable of specifically binding to DNA mismatches where they can photocleave the DNA adjacent to the mismatch.

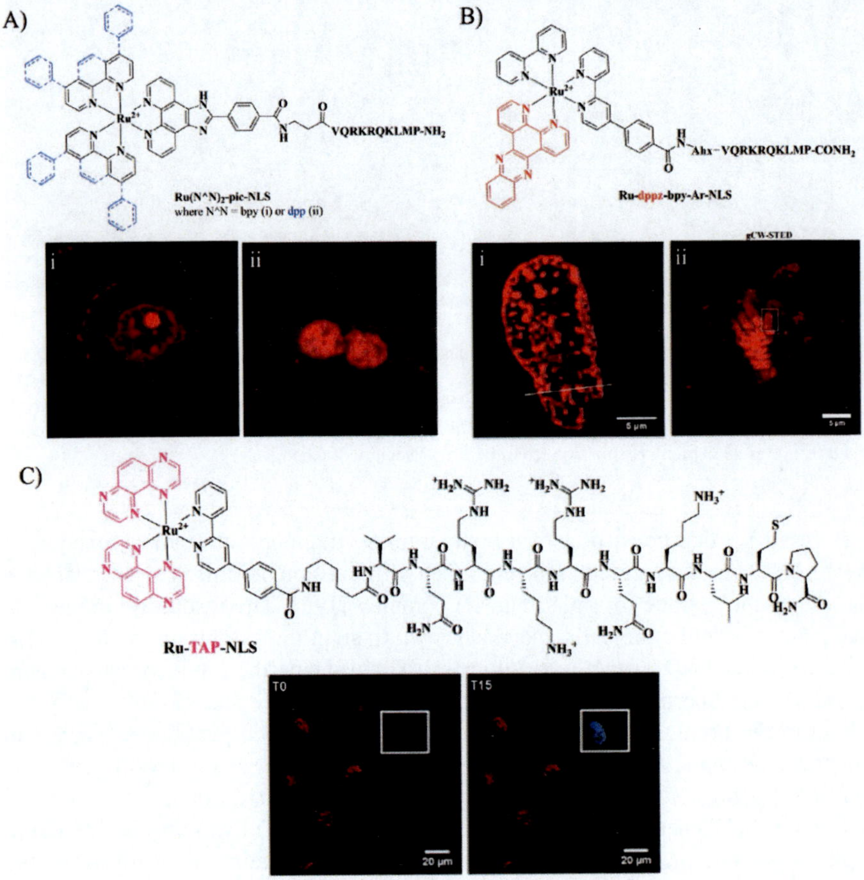

Fig. 6 Chemical structures of NLS-driven nuclear targeting Ru(II) complexes and representative images of their corresponding application in mammalian cells. **A** $[Ru(bpy)_2(pic\text{-}NLS)]^{6+}$(i) and $[Ru(dpp)_2(pic\text{-}NLS)]^{6+}$ (ii) showing nuclei and nucleolus staining, respectively (Reproduced from Ref. [150] with permission from the Royal Society of Chemistry). **B** Confocal (i) and STED (ii) images of $[Ru(dppz)(bpy)(bpy\text{-}Ar\text{-}NLS)]^{6+}$ nuclear and chromosomal DNA staining (Reproduced from Ref. [117] with permission from the Royal Society of Chemistry). **C** Following DNA binding of Ru-TAP-NLS in live HeLa cells, continuous photoirradiation of selected cells resulted in cellular damage as indicated by DRAQ 7 nuclear staining in blue [2] (Reprinted (adapted) with permission from Burke C. S. et. al. [2]. Copyright 2018 American Chemical Society)

The rhodium complex conjugated to an octaarginine peptide appended to fluorescein was rapidly internalised within the cell and localised in the nucleus. It was noted that the presence of the peptide led to binding of matched DNA (electrostatic interaction between peptide and DNA), although the photocleavage only occurred at DNA mismatches as desired.

We reported that specific nuclear targeting could be achieved using the aforementioned transcription factor NFκB and that the localisation of the complex in the nucleus seems to be dictated by the lipophilicity of the complex. As shown in Fig. 6, both $[Ru(bpy)_2(pic\text{-}NLS)]^{6+}$ and $[Ru(dpp)_2(pic\text{-}NLS)]^{6+}$ (where bpy = 2,2-bipyridine, dpp = 4,7-diphenyl-1,10-phenanthroline and pic = 2-(4-carboxyphenyl)-imidazo-[4,5-f][1,10]-phenanthroline) were efficiently transported across the cell and nuclear membrane in Chinese hamsterovary (CHO) cells [150]. However, the more hydrophilic bpy-NLS showed nuclear staining, whereas the dpp-based conjugate showed accumulation in the nucleolus, thus highlighting that the lipophilic character of the metal complex remains important even in intraorganelle distribution.

High-resolution imaging of chromosomal DNA was achieved using a Ru-dppz NLS conjugate, $[Ru(dppz)(bpy)(bpy\text{-}Ar\text{-}NLS)]^{6+}$, which also allowed tracking of the different stages of mitosis in HeLa cells using STED [117]. In a separate study, the $[Ru(tap)_2(bpy\text{-}Ar\text{-}NLS)]^{6+}$ showed nuclear penetration and DNA binding indicated by the extinguished complex emission [2]. In an example of the multimodal addressability of such complexes, they were confirmed to remain present in the nucleus after emission extinction by resonance Raman microscopy. With the aim of extending the application of Ru-NLS conjugates toward theranostics, it was shown that upon in situ photoirradiation, cellular destruction is accomplished, attributed to DNA oxidation by photo-induced electron transfer from a guanine base and the Ru(II) complex, analogous to a mechanism reported for related tap complexes in solution [201].

A rhenium complex conjugated to an NLS peptide was reported to exhibit nucleolar localisation and efficient singlet oxygen generation under light irradiation in polar ($\Phi = 0.25$) or lipophilic ($\Phi = 0.75$) environments [202]. This luminescent probe is attractive for dual application in both imaging and photodynamic therapy as it exhibits low dark toxicity ($IC_{50} = 35\ \mu M$), but enhanced toxicity under UV irradiation. In a separate study, the derivatised and caged Re(I) complex, Re-PLPG, was coupled to an NLS peptide that showed penetration into sub-cellular compartments such as the nucleoli, thus allowing interaction of the complex with nucleic acids [152].

Metal complex peptide nucleic acid (PNA) conjugates are a useful approach to probe different nucleic acid strands due to their ability to hybridise to their complementary oligonucleotide strands with high specificity which is advantageous in sensing and therapy. For example, the Re(I)-PNA conjugate, $[(CO)_3Re(pyridazine\text{-}PNA)(Cl)_2Re(CO)_3]$, suitable for two-photon excitation (λ_{exc} 750 nm), revealed cytoplasmic and nuclear staining in HEK-293 cells attributed to PNA–nucleic acid binding [188]. Notably, small concentrations of DMSO were required for uptake of the conjugate. The emission wavelength was substantially altered depending on sub-cellular localisation and could be used to differentiate between the cytoplasm and the nucleus. The difference in emission energy is attributed to the difference in polarity/rigidity between the different locales. A follow-on study by the Licandro research group on related rhenium complexes conjugated to different PNA sequences revealed difficulties in cell studies, including poor solubility and endosomal entrapment [189]. These issues are frequently encountered in biological studies of metal complexes and can hinder sensing and imaging applications due to a low rate of cell uptake and off-target localisation.

4.3 Mitochondria

The nucleus is the primary location of DNA within the cell, but the mitochondria is also an important repository. Although containing much less DNA, it contains 37 genes in total that encode proteins and RNAs critical for energy transduction.

A histidine-binding Ir(III) complex was bis-conjugated to an HTat sequence and a mitochondrial targeting sequence derived from the mitochondrial protein cytochrome P450 [195]. The conjugate was membrane-permeable and efficiently targeted the mitochondria.

In a recent publication, precision targeting of mitochondrial DNA in live HeLa cells was achieved using an MPP-driven light-switching Ru^{II}-dppz complex [119]. Confocal laser scanning microscopy showed rapid cellular uptake of $[Ru(dppz)(bpy)(bpy\text{-}Ar\text{-}MPP)]^{5+}$ in live HeLa cells, and localisation to mitochondrial sub-structures was confirmed using luminescence lifetime imaging (Fig. 7). Solution titration with ctDNA showed that the DNA binding ability of the parent complex, mediated by dppz intercalation, is retained for the Ru^{II}-dppz MPP conjugate. Additionally, an increased binding constant was reported, which was attributed to electrostatic interactions between the polycationic sequence of MPP and the anionic DNA backbone. The conjugate showed low cytotoxicity in the dark and under imaging conditions, thus facilitating mtDNA visualisation. Photo-induced toxicity was observed only under continuous and intense irradiation, enabling controllable initiation of cell death, making it an interesting prospect for theranostic applications.

Recently, the successful conjugation of an osmium(II) complex to two mitochondrial-penetrating peptides was reported [158]. The bis-MPP conjugate strongly confined to the mitochondria at and below concentrations of 30 μM and leached out of the organelles and into the cytoplasm over time. At increased concentrations, it showed cytoplasmic and even nucleoli staining, leading to cell death. This localisation switch was also reflected by the cell death mechanism, where at 30 μM, loss of the membrane potential was observed, whereas at increased probe concentrations, a moderate effect on depolarisation and a greater caspase activity was observed instead.

4.4 Endoplasmic Reticulum (ER)

The endoplasmic reticulum (ER) in eukaryotic cells is the site of synthesis and processing of many transmembrane and secretory proteins, synthesis of lipids and calcium regulation. Accumulation of unfolded or misfolded proteins trigger an ER stress response which regulates cell functions to either restore ER homeostasis or to induce apoptosis for damaged cells. Complexes that target the ER may be used as imaging tools to study the endoplasmic reticulum and processes, such as ER stress, or as therapeutic tools, as the ER signalling pathways have been linked to various diseases, including cancer.

The ruthenium(II) complex $[Ru(bpy)_2\text{-phen-Ar-COOH}]^{2+}$, exhibiting an emission maximum at 604 nm, was conjugated to the penetratin/ER cell-penetrating peptide, yielding the $[Ru(bpy)_2\text{-phen-Ar-ER}]^{9+}$ bioconjugate [117]. The conjugate was used

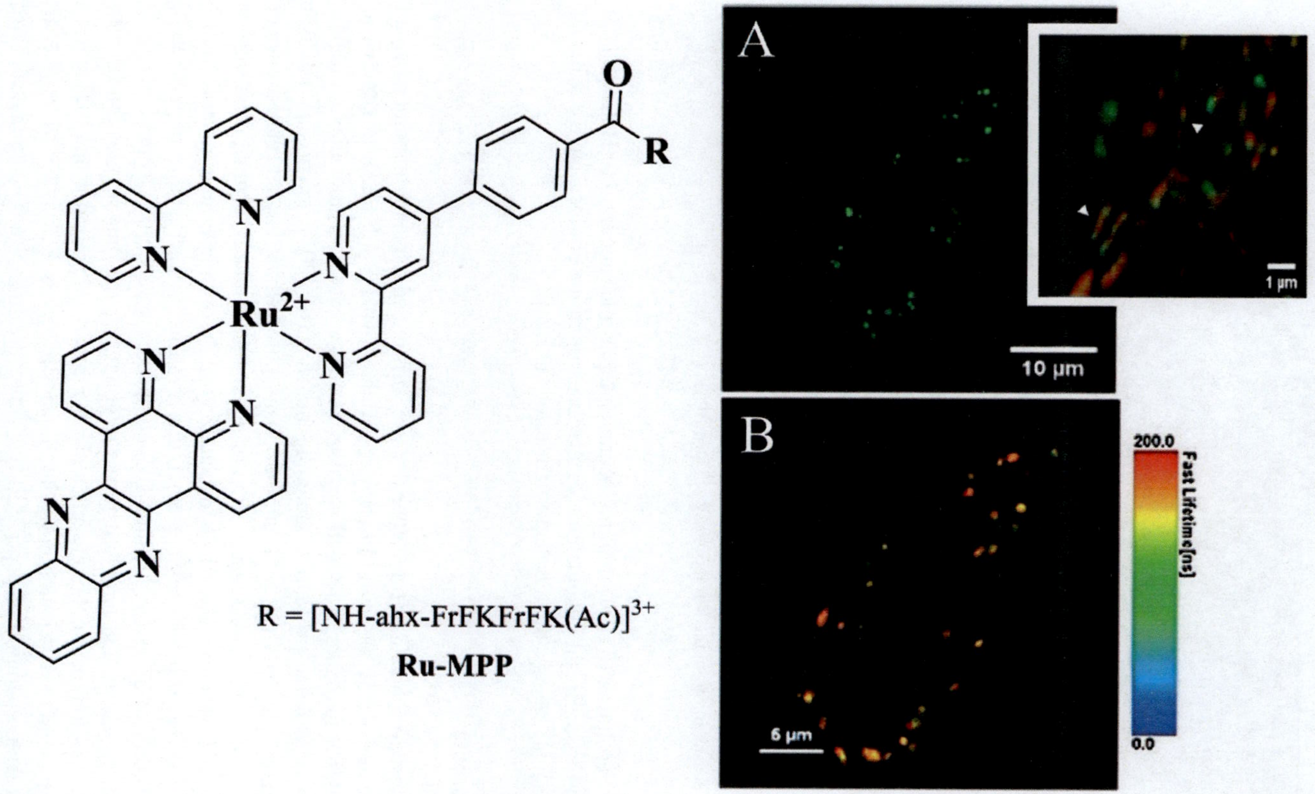

Fig. 7 Precision-targeted [Ru(dppz)(bpy)(bpy-Ar-MPP)]$^{5+}$ (10 µM/2 h) light-switching probe in mitochondria of live HeLa cells. **A** Confocal image with inset showing a close-up of the mitochondria co-stained with MitoTracker Deep Red (red) and **B** luminescence lifetime distribution of the conjugate [119]. Reproduced with permission from Angewandte Chemie International Edition

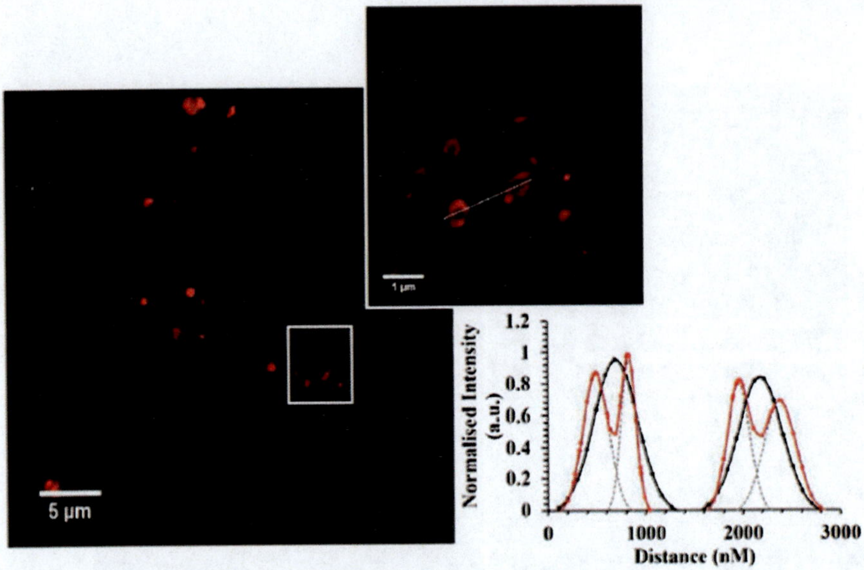

Fig. 8 Confocal (left) and STED (inset) imaging of [Ru(bpy)$_2$-phen-Ar-ER]$^{9+}$ $^-$ (70 µM/4 h)-stained ER in a single fixed HeLa cell. The plot profile compares the full width at half maximum (FWHM) of confocal (*blackcoloured em dash*) versus STED (*red coloured em dash*) resolution. Reproduced from Ref. [117] with permission from the Royal Society of Chemistry

for super-resolution (STED) imaging of the endoplasmic reticulum in HeLa cells as shown in Fig. 8. Bright punctuate spots are observed in the confocal image, whereas STED imaging reveals the tubular structures of the ER.

4.5 Lysosome

Lysosomes are subcellular organelles that are surrounded by a single membrane and are characterised by an acidic interior environment (pH ~ 4.5 to 5), in contrast to mitochondria, for example, which are alkaline (pH ~ 8). Lysosomes play an important role in cellular processes, including homeostasis [203], energy metabolism [204], enzymatic activity [205] and autophagy [206] which is also related but not limited to inflammatory diseases. Monitoring changes in the lysosomal environment including pH variations can aid understanding of lysosomal function and dysfunction. Furthermore, lysosomes are emerging as attractive therapeutic targets [203]. For example, cyclometalated iridium(III) complexes were applied as pH-activatable cell imaging agents and photosensitisers, highlighting the dual application of such probes for photodynamic therapy and real-time therapeutic monitoring [207].

With the aim of developing potential theranostic agents, Fernández-Moreira et al. reported the preparation of peptide-linked bimetallic Ir(III)/Au(I) conjugates [186]. The luminescent properties of the iridium moiety permitted confocal imaging and tracking of the conjugates upon cellular uptake and lysosomal accumulation, and the coordination sphere of the gold appeared to influence cytotoxic activity. The cysteine-containing conjugate showed antiproliferative activity which is thought to be attributed to the readily cleaved Au–S(cysteine) bond.

A Ru(II)-cyclodextrin-RGD nanoassembly reported by Mao et al. was found to accumulate in lysosomes of integrin-rich tumour cells and trigger apoptosis through lysosomal damage, ROS elevation and caspase activation [166]. Uptake of nanoparticles via an endocytic mechanism frequently results in lysosomal accumulation. Endocytosis is generally associated with endosomal entrapment in early or late endosomes which then fuse with lysosomes at a later stage if they are not released to the cytoplasm.

Recently, the bridged octaarginine conjugate, $[Os\text{-}(R_4)_2]^{10+}$ was found to be taken up initially into the cytoplasm of A549 cells prior to accumulating in lysosomal structures at 30 μM/48 h [10]. This permitted both confocal imaging and luminescence lifetime mapping of the intracellular environment, including potential response to redox species as discussed later. Although lysosomal accumulation is desired in the context of therapy or redox and pH sensing, endosomal entrapment can hinder delivery of a luminophore probe or nanoparticle to the desired intracellular destination. The recently reported RuBDP nanoparticles were found to localise in late endosomes and lysosomes of A549 cells [208]. The particles ratiometrically responded to fluctuations in oxygen concentration and, interestingly, exhibited emission enhancement within 4 h following initial uptake, the origin of which is thought to reflect endosomal escape. Endosomal escape is a topic that is particularly important in drug delivery, and approaches addressing this challenge have focused on promoting endosomal membrane fusion and destabilisation or pore formation in the endosomal membrane [209, 210]. In addition, several endosomal escape agents have been identified, such as chemical agents or viral- and bacterial-derived proteins and peptides [211]. Following protocols emerging in this domain, e.g. through modifying the particle composition to achieve pH-induced release [211, 212], efficient endosomal escape and specific organelle targeting may further expand the application of nanoparticles and probes.

5 Sensing Capabilities of Peptide Metal Complex Conjugates

Taking advantage of the excellent targeting capability of peptides, there are several examples of emissive metal complex conjugates that have been applied for sensing of bio-relevant species such as oxygen and molecules including DNA and proteins. The characteristic luminescence lifetime- or intensity-based response typically reflects the interaction of the metal complex with these species within a cellular environment.

Coordination of responsive ligands allows for the design and preparation of complexes with a responsive luminescence. For example, complexes of dipyrido[3,2-a:2′,3′-c]phenazine (dppz) and its derivatives exhibit no luminescence in aqueous solution, but emission is switched on in hydrophobic environments, such as upon DNA binding, leading to the design and development of a range of DNA "light-switch" dppz complexes [213, 214]. Sensing of important molecular structures can be enhanced via coordination of the lipophilic diphenyl phenanthroline (dpp) ligand, for example, which also allows for cellular uptake and targeting of lipid-rich regions [75, 110, 150]. Incorporation of a targeting vector allows for *targeted* sensing. For example, conjugation of a Ru(II) sensor to a mitochondrial-penetrating peptide enables monitoring of *local* oxygen fluctuations in live cells using either emission intensity or lifetime imaging.

5.1 Oxygen

There are several methods applied traditionally for monitoring and measuring dissolved oxygen in biological systems, for example, Clark-type O_2 electrodes [215], electron paramagnetic resonance (EPR) probes [216] and microelectrodes or needle probes [216–220]. However, there is a demand for less invasive techniques for O_2 sensing and particularly for sensing modalities that can be readily followed dynamically intracellularly with as little as possible interference with the cell. For this reason, *quenched phosphorescence*-based O_2 sensors are particularly attractive for intracellular oxygen (icO_2) sensing. Ideal characteristics of an intracellular oxygen sensor include high oxygen responsivity, photostability, cell uptake efficacy, molecular brightness, biocompatibility, cytotoxicity and subcellular targeting ability where desired.

There are numerous examples of emissive probes that have been applied for *oxygen* sensing using lifetime- or intensity-based methods. As mentioned, a key advantage of lifetime sensing is that emission lifetime is largely independent of probe concentration. A drawback though is that phosphorescence lifetime imaging/sensing requires a microscope coupled with a lifetime/FLIM unit, which is rather a specialist technique, not a routine tool in many bio-laboratories. Whereas, intensity-based sensing can be performed using conventional instrumentation such as a fluorescence microscope or plate reader. Intensity-based sensing measurements, as described, can be applied where the probe species is combined with a reference, which overcomes issues of concentration and other artefacts in an intensity-based measurement. The choice of modality for O_2 sensing will also depend on the sample for analysis. For example, using a conventional plate reader, intensity-based measurements permit parallel analysis of monolayer cells exposed to various conditions in a single experiment while also carrying out the measurement in multiplicate. For three-dimensional cell models or tissue samples, PLIM is typically the method of choice as it allows visualisation of spatial O_2 distribution throughout the sample.

Complexes of ruthenium(II), iridium(III) and Pt(II) or Pd(II) porphyrins are amongst the most widely studied oxygen sensors owing to their long-lived triplet excited states which are highly susceptible to oxygen quenching, giving a characteristic lifetime- and intensity- based response to oxygen concentration, as described by the Stern–Volmer equation.

Phosphorescent Pt(II) and Pd(II) porphyrins exhibit phosphorescence lifetimes ranging from 40 to 100 μs and 400 to 1000 μs, respectively [221]. Papkovsky and co-workers have worked extensively on Pt(II) and Pd(II) porphyrin probes which show good photostability and efficient quenching by oxygen [222–225]. The solubility of such probes in water and targeting ability can be improved by conjugation to protein cargos, PEG chains or cell-penetrating peptides [224, 226–229]. Dmitriev et al. presented a Pt(II) coproporphyrin conjugated to a peptide fragment derived from the antimicrobial bactenecin 7 peptide [227]. The conjugate showed efficient cellular uptake across several cell lines, cytoplasmic and mitochondrial accumulation and was used for monitoring intracellular O_2 levels upon exposure to metabolic stimuli reagents. The application of this conjugate and similar Pt(II) coproporphyrins in cells can be hindered by poor photostability and potential photocytotoxic effects.

Although somewhat limited by their photostability, iridium(III) complexes have shown promise for in vitro and in vivo oxygen and hypoxia mapping and low cytotoxicity against 2D cell monolayers [230–235].

Ir(III) dyads such as the iridium–coumarin ratiometric probe, C343-Pro$_4$-BTP reported by Yoshihara et al., have been developed for ratiometric intensity-based O_2 sensing [236]. In this report, the coumarin moiety (C343) is linked to the iridium (BTP) complex through a tetraproline amino acid linker, and upon excitation at 405 nm, energy transfer from C343 to BTP yields emission from both dyad components at 480 nm and > 610 nm, respectively. The phosphorescence emission signal of the iridium is quenched by oxygen, and the ratio of the emission from the dyad moieties exhibits an O_2-dependent response both in solution and in live HeLa cells. In later studies, octaproline [234] and octa- and dodecaarginine [237] linkers were utilised in coumarin–iridium(III) dyads in order to enhance cellular uptake to enable ratiometric imaging of the oxygen gradient in HeLa cells.

Several complexes of ruthenium(II) have demonstrated *in cellulo* oxygen response as molecular probes or part of a dual emissive dyad [238–240]. In the context of peptide conjugates, Keyes' group presented the octaarginine conjugate, $[Ru(bpy)_2(pic\text{-}R_8)]^{10+}$ [1, 88], whose luminescence lifetime, *similar to the parent complex*, was oxygen-sensitive. Confocal imaging revealed rapid uptake of $[Ru(bpy)_2(pic\text{-}R_8)]^{10+}$ in myeloma cells and human blood platelets, and lifetime imaging was used for cellular oxygen mapping where, for example, the probe lifetime was shortest (~400 ns) when the conjugate localised in the cell membrane. This agrees with the increased solubility of oxygen in the cellular membrane. Although the emission lifetime of this complex is strongly oxygen-dependent, the quenching constant by O_2 is largely pH-independent.

The advantage of peptide vectorisation is that it may enable real-time monitoring of *local* oxygen fluctuations at a specific cellular region or organelle, and cross-reactivity may be minimised if other parameters remain unchanged while the analyte of interest is varied. For example, the mitochondrial-targeted Ru(II) conjugate reported by Keyes et al. showed dynamic response to changes in local oxygen concentrations and to elevated levels of reactive oxygen species using luminescence intensity and lifetime imaging [118]. The dinuclear ruthenium(II) probe was bridged across a mitochondrial-penetrating peptide yielding $[(Ru(bpy)_2)_2(phen\text{-}MPP\text{-}phen]^{7+}$. Following exposure of HeLa cells to antimycin A, a mitochondrial uncoupler agent,

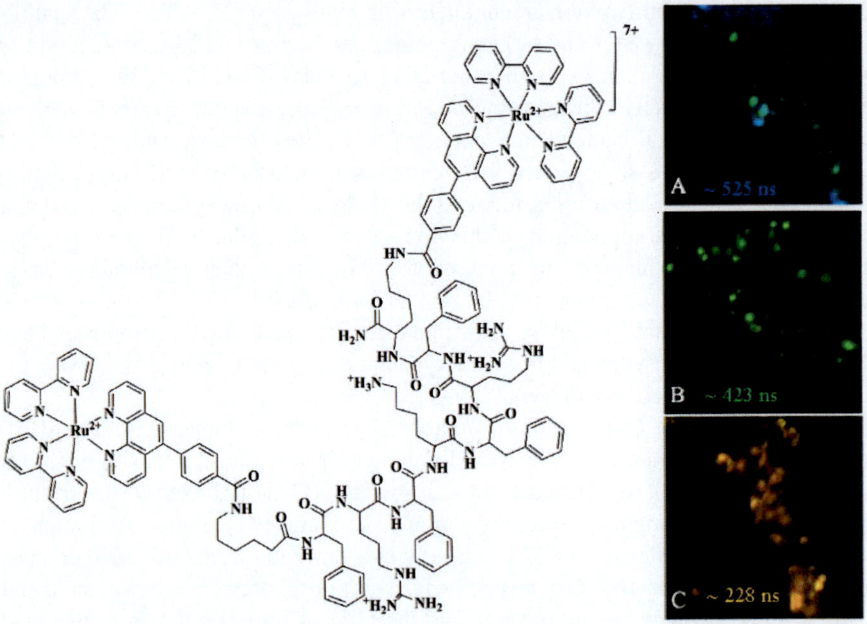

Fig. 9 Luminescence lifetime imaging of $[(Ru(bpy)_2)_2(phen\text{-}MPP\text{-}phen]^{7+}$ in live HeLa cells, in response to antimycin A treatment. PLIM was carried out following **A** incubation with Ru(II) conjugate at 70 µM for 2 h in the absence of light, and treatment with antimycin A (200 µg/ mL) for **B** 10 min and **C** 100 min [118]. Reprinted (adapted) with permission from Martin et al. [118]. Copyright 2014 American Chemical Society

PLIM studies showed that average emission lifetime of $[(Ru(bpy)_2phen\text{-}Ar)_2\text{-}MPP]^{7+}$ in live HeLa cells was quenched from approximately 525–228 ns, as shown in Fig. 9.

Cross-responsiveness with other environmental factors such as pH, proteins/ enzymes and lipidic environments is a key limitation of lifetime imaging for single probes in the cell environment and indeed in molecular probes in general, as it is difficult to decouple or calibrate the analytical probe and/or reference from other potential environmental influences in the complex cell matrix. However, in studies like these, where relative changes to analyte are followed rather than studies of absolute concentration, this may be less of an issue. Nanoparticle probes offer some advantages in this regard.

Nanoparticle (NP)-based systems have inherent advantages in terms of their cell permeability once they are between 50 and 200 nm and have appropriate lipophilicity and surface charge. Several NP O_2 sensor formats have been reported; they comprise of either (1) an O_2 probe alone or (2) a probe reference pair that generates a ratiometric signal through Förster resonance energy transfer (FRET) or (3) a probe reference pair that has independent emission signals that do not require cross-communication but can be excited at a single wavelength [241–250].

Of the latter type, our group has reported a ratiometric core–shell nanosensor, Ru-BODIPY NP, where the BODIPY reference probe was confined to a polystyrene core, offering its protection from environment, and so a stable reference signal and a ruthenium probe conjugated to the poly-lysine shell exterior allowing direct exposure to the environment and O_2 accessibility [241]. In the Ru-BODIPY nanoparticle system, the O_2 indicator and reference dye are spatially separated into a particle core and shell, *thus limiting any potential cross-communication*, and they are simultaneously excited at a single wavelength. The nanoparticles showed good ratiometric response to oxygen in aqueous media with a rate of quenching of 7.52×10^8 M^{-1} s^1. The emission intensity ratiometric data showed moderately good linearity ($R^2 = 0.9525$) over a biologically relevant O_2 range. Following surfactant-mediated uptake of RuBODIPY NPs in CHO cells, lifetime imaging studies showed that the emission lifetime of the BODIPY dye, as expected, was unaffected by the surrounding intracellular environment in contrast to the ruthenium probe, demonstrating the potential of the core–shell approach to designing new ratiometric nanotools.

Aiming to overcome the need for a membrane permeabilising agent, a ratiometric sensor was then developed where the Ru(II) oxygen sensor and reference BODIPY dye are co-encapsulated within the particle core which is permeable to oxygen, and the particle exterior is decorated solely with a poly-L-lysine shell [208]. This approach indeed permitted uptake of the self-referenced O_2 nanoparticles in live mammalian cells, demonstrating the impact of even relatively modest surface modification of the particle on uptake. Importantly, the particles were suitable for both non-invasive hypoxia imaging using confocal microscopy (xyλ scanning) and for quantitative ratiometric intensity-based measurements of oxygen *in cellulo* using a plate reader assay. The isolation of the probe to the particle core protected it from environmental factors other than oxygen, but may impact dynamic response as the oxygen must diffuse through the particle matrix to reach the probe.

5.2 pH

pH is an important regulator of metabolic processes in the cell and is believed to play a role in signalling. pH varies across the cell organelles, and its homeostasis may be a marker of cell health. Therefore, sensing intracellular and organelle pH is an important target analyte that has been the focus of studies in metal luminophore probes. A pH- and oxygen-sensitive iridium(III) complex was prepared by coordinating two cyclometalated ligands [2-(2, 4-difluorophenyl)pyridine; dfpp] to an Ir(III) centre along with the pic(COOH) ligand, 2-(4-carboxylphenyl)imidazo[4, 5-f][1,10]phenanthroline, carrying a terminal carboxyl moiety, thus permitting amide coupling to an octaarginine sequence in order to improve aqueous solubility [29]. The parent complex exhibited a lifetime of approximately 674 ns in degassed organic media which was reduced to 200 ns in degassed aqueous media at pH 6.9. Cytotoxicity studies showed that both the Ir(III) parent complex and conjugate were cytotoxic towards SP2 and CHO cell lines. The cytotoxic character of iridium complexes has been reported in a number of studies [69, 251], and it is likely that increased cytotoxicity compared to other transition metal luminophores is the result of its lipophilic nature inducing rapid uptake and wide distribution of the conjugate within cells.

Chao et al. reported an iridium(III) pH sensor that was coordinated to ligands containing morpholine groups [252]. They observed that morpholine promoted mitochondrial targeting, and the pH dependence of the emission intensity of the probes was explored in HeLa cells where extracellular pH was adjusted and from 6.0 to 8.0 in high-K^+ media; equilibration with the cell interior was achieved by application of nigericin, a membrane-associating antiporter ionophore for K^+ and H^+. The emission intensities from the complexes within the cell were observed to respond to pH in the range of 6.0–8.0. On stimulation of apoptosis in the cells, using mitochondrial uncoupler carbonyl cyanide m-chlorophenyl hydrazine (CCCP), the emission intensity of the probes in the mitochondria was also observed to modulate; however, attributed to pH change, other quenching species may evolve in the mitochondria as a consequence of uncoupling.

As previously described, the octaarginine-driven conjugate $[Ru(bpy)_2(pic-R_8)]^{10+}$ was applied as a probe for oxygen mapping using lifetime imaging [1, 88]. The emission lifetime of this complex is strongly oxygen-dependent, but the quenching constant by O_2 is largely pH-independent, which serves the use of the probe in O_2 mapping. Conversely, resonance Raman spectroscopy, and therefore the Raman signature signal of the probe, is strongly pH-dependent and is insensitive to O_2, thus enabling use of the probe in pH mapping using resonance Raman spectroscopy. The probe permits multi-parameter monitoring and mapping of the intracellular environment using a single probe, single excitation and two imaging techniques enabled by the large Stokes shift of Ru(II) polypyridyl complexes. Ligands such as *pic* in this complex or *dppz* and *bpy* exhibit signature Raman signals when they participate in the MLCT the excitation is resonant with, and in the case of the ionizable pic ligand, its Raman signature grows into resonance depending on the pH of the environment/ionisation of the pic imidazole residue, thus providing a distinctive pH marker.

5.3 Biorelevant Molecules: Receptors, Proteins, Enzymes

Lo et al. presented a series of cyclometalated Ir(III) complexes containing a perfluorobiphenyl (PFBP) moiety and their respective conjugates, afforded through reaction of PFBP with the cysteine moiety in a four amino acid sequence (FCPF, known as "π-clamp") [155]. Following this π-clamp-mediated cysteine conjugation, novel Re(I) conjugates were prepared and applied as imaging agents but also as enzyme sensors [253].

An early study, presented by Stephenson et al., described conjugation of a rhenium complex to the peptide fMLF which is known to deliver to the formyl peptide receptor (FPR) [254]. A qualitative comparison of the cell uptake and distribution between a known FPR-targeting fluorescent probe (fluorescein-labelled fNLFNTK) and the rhenium-fMLF conjugate suggested that the rhenium probe successfully targeted the FPR.

As mentioned earlier, a rhenium(I) tricarbonyl complex was conjugated to a derivative of *T140*, a known antagonist of CXCR4 and a chemokine receptor which is overexpressed in cancer cells [177]. The rhenium conjugate was successful in sensing CXCR4, evident by a strong luminescence signal detected from cells expressing CXCR4, whereas no luminescence was detected from cells lacking the receptor.

Although rhenium peptide conjugates have been exploited as luminescent probes for interrogating different cell receptors, there has been an alternative motivation for synthesising rhenium peptide conjugates as structural analogues for "hot" technetium complex conjugates. However, in several studies, the radioactive technetium peptide conjugate was utilised to study specific cell receptors via radioimaging techniques, for instance, SPECT [255–258].

Although not exploited for its sensing capabilities, a zinc phthalocyanine complex conjugated to a receptor-targeting peptide, LARLLT, was reported to exhibit high selectivity for the epidermal growth factor (EGF) which tends to be overexpressed on the surface of cancerous cells [259]. Conjugation to the peptide increased the photodynamic efficacy and selectivity of complex against cancer cells with different receptor expression levels.

5.4 Reactive Oxygen and Nitrogen Species (ROS/RNS)

Reactive oxygen species and reactive nitrogen species are highly reactive, often radical species, generated as part of metabolic processes within the cell. They are potentially injurious to the cell if not regulated. ROS and RNS are numerous and

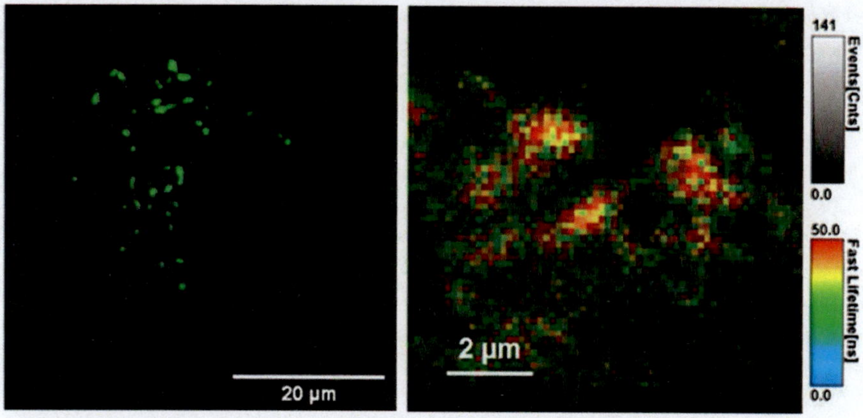

Fig. 10 Confocal fluorescence (left) and luminescence lifetime imaging (right) of $[Os\text{-}(R_4)_2]^{10+}$ confined in lysosomes at 30 μM in live A549 cells. The Os(II) conjugate exhibited lifetimes of 37 ± 1.8 ns (54%) and 9.3 ± 0.6 ns (32%). Reprinted (adapted) with permission from Ref. [10] (https://pubs.acs.org/doi/10.1021/acs.inorgchem.1c00769). Further permissions related to the material excerpted should be directed to the ACS

include species such as superoxide ($O_2^{\cdot -}$), hydroxyl ($^{\cdot}OH$), peroxyl ($RO_2^{\cdot}$) and alkoxyl (RO^-) radicals nitric oxide, peroxynitrite, nitrate, nitrite and nitrogen dioxide. RNS and ROS generation pathways are inter-dependent. For example, peroxynitrite ($ONOO^-$) is the product of a reaction between nitric oxide (NO) free radicals and superoxide and at abnormal levels, can induce oxidative changes in intracellular molecules, including DNA and proteins [260].

Monitoring changes in oxygen, RNS and ROS levels within the cell and particularly within the mitochondria, one of the key cellular sources of such species, is invaluable in understanding both normal physiology and disease, and also in understanding toxicity and therapeutic response.

As mentioned earlier, the mitochondria-targeting Ru(II) probe, $[(Ru(bpy)_2)_2(phen\text{-}MPP\text{-}phen]^{7+}$, is capable of responding to changes in local O_2 concentrations and also to elevated ROS levels [118].

Recent studies have highlighted the potential of Os(II) polypyridyl complexes for detection of oxidative damage and intracellular reactive oxygen species [10, 261, 262]. The absence of oxygen sensitivity in the case of Os(II) complexes but potential redox sensitivity offers an advantage in their application as intracellular sensors over complexes of ruthenium or iridium. For example, phosphorescence lifetime imaging studies revealed that the emission lifetime of the polyarginine Os(II) conjugate, $[Os\text{-}(R_4)_2]^{10+}$, was found to vary with intracellular localisation [10]. When confined to the lysosomes and surrounding cytoplasm, $[Os\text{-}(R_4)_2]^{10+}$ exhibited reduced lifetimes in comparison to when it was initially taken up into the cytoplasm of cells. For example, the dominant amplitude component of the decay was measured as 92.2 ± 2.9 ns upon cytoplasmic uptake and 37 ± 1.8 ns upon lysosomal accumulation. This lifetime quenching is likely due to the presence of redox-active species as the probe luminescence was not sensitive to oxygen or pH changes [10]. Figure 10 shows the confocal and lifetime imaging of $[Os\text{-}(R_4)_2]^{10+}$ upon lysosome localisation at 30 μM/48 h. Lifetime imaging studies were also carried out following uptake and accumulation of the probe within pancreatic 3D multicellular tumour spheroids, thus highlighting the suitability of such probes for monitoring metabolic changes in cells, spheroids or tissues without interference from oxygen.

The mitochondria-localised ruthenium(II) complex–cyanine (Ru-Cy5) scaffold, *although peptide-free*, is a good example of the application of such transition metal probes for *in cellulo* sensing and imaging, in this instance, of peroxynitrite in cells [263]. This energy transfer-based probe constituted a Ru(II) complex as the energy transfer donor and Cy5 as energy transfer acceptor. Following cellular uptake and mitochondrial localisation in HeLa cells, the emission of Cy5 was decreased in the presence of $ONOO^-$ as a result of oxidative cleavage of the polymethine bridge which interrupts the energy transfer between Ru(II) and Cy5. The Ru-Cy5 system showed low cytotoxicity, efficient mitochondrial accumulation and good selectivity for $ONOO^-$ (*over other reactive species*).

Although there are some additional examples of non-peptide metal complexes which exhibit a luminescence response to NO [264, 265] or radical species such as hypochlorite [266], there are to date limited reports of peptide metal conjugates which have been applied for monitoring intracellular redox species.

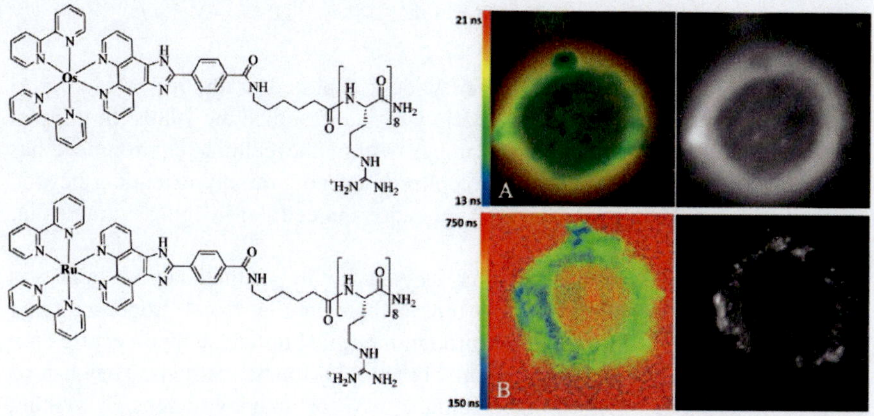

Fig. 11 Phosphorescence lifetime imaging of **A** $[Os(bpy)_2(pic\text{-}R8)]^{10+}$ (70 µM/24 h) and **B** $[Ru(bpy)_2(pic\text{-}R8)]^{10+}$ (350 µM/15 min) in live SP2 cells. The false colour images and intensity mapping highlight the differences in emission lifetime of the luminophore depending on its localisation in the cell. The emission lifetime of the osmium(II) conjugate was 14.5 ± 0.5 ns in the cytoplasm and 18.8 ± 0.6 ns at the cell membrane. Reproduced from Ref. [131] with permission from the Royal Society of Chemistry

5.5 Cell Membrane Markers

Again reflecting the versatility of Ru(II) complexes as multiparameter and multimodal imaging tools, $[Ru(dppz)_2(pic\text{-}Arg_8)]^{10+}$ was used in confocal luminescence and resonance Raman imaging [109]. Owing to the dppz ligands, the complex behaved as a molecular light switch where the luminescence of the complex is extinguished in aqueous solution, *due to hydrogen bonding to the phenazine nitrogens*, but switched on in lipid vesicles. Resonance Raman intensity mapping revealed that the octaarginine conjugate crossed the membrane and distributed throughout the cell, whereas the parent complex accumulated in the cell outer membrane.

PLIM mapping of a related osmium octaarginine conjugate, $[Os(bpy)_2(pic\text{-}Arg_8)]^{10+}$, revealed that the emission lifetime of the complex changed in response to the intracellular environment [131]. For example, the average lifetime was found to be 11.6 ± 0.4 ns in the cytoplasm of CHO cells and 14.5 ± 0.5 ns in SP2 cells. Additionally, a lifetime of 13 ± 1.5 ns and 18.8 ± 0.6 ns was observed for the membrane of CHO and SP2 cells, respectively. As the lifetime of the complex is oxygen-independent, this response may be due to differences in the lipid packing of the cell membrane of each cell line. As mentioned earlier, a variation in lifetime was observed for the Ru(II) analogue, $[Ru(bpy)_2(pic\text{-}Arg_8)]^{10+}$, conjugate which was attributed to the increased solubility of O_2 in the cellular membrane, thus reflecting cell membrane oxygenation [88]. Figure 11 illustrates the PLIM mapping of $[Os(bpy)_2(pic\text{-}Arg_8)]^{10+}$ and $[Ru(bpy)_2(pic\text{-}Arg_8)]^{10+}$ in SP2 cells.

6 Conclusions

Metal complex luminophores have been widely explored as probes of biological molecules, particularly of DNA in its variety of motifs, since the 1980s. In the past decade, their putative application as probes within the cellular environment has been realised, and metal complex luminophores are now rapidly extending beyond the scope of in vitro studies to diverse applications in cellular imaging, intracellular sensing and theranostics.

The beauty of metal complexes is that they are highly synthetically versatile and can be tailored to their application by tuning their photophysical properties (e.g., *molecular brightness, NIR emission*), through modification of the metal centre and/or the coordinated ligands, through relatively facile synthetic methods. They can be tailored to meet the demands of the technique used for imaging or sensing whether that is phosphorescence lifetime imaging microscopy, luminescence intensity-based sensing using a plate reader or vibrational imaging such as Raman mapping.

Critically, metal complex luminophores can be readily conjugated to vectorising functionalities to facilitate cellular uptake and intracellular targeting. Peptides are a valuable means of promoting such permeation and targeting within cells, and their efficacy in this regard has been known for many years in drug delivery. As signal molecules, they can drive their cargo very precisely within the cell. Peptides can confer improved solubility and lower cytotoxicity on their cargo and can be synthesised readily via high throughput and often automated solid-state synthesisers. Conjugation of metal complexes to cell-penetrating and signal peptides has been used for the efficient delivery of complexes in cell monolayers, but also recently in multicellular spheroids. Peptides, to date, have been very effective in driving complex cargo to the cell and targeting, but the mechanism is not fully elucidated; for example, the role of the counter-anions accompanying the charged conjugate seem to have a strong effect on membrane permeability, but remains to be fully explored and exploited. The possibilities are extensive, and metal complex luminophore peptide conjugates are likely to find increasing application in theranostic applications.

Although peptides can promote fairly rapid cellular uptake (i.e. < 1–24 h) required for imaging, the application of the metallo-peptide conjugates in bioassay experiments has been less explored. In terms of the metal complex photophysics, key future challenges lie in maximising probe brightness, as inorganic luminophores do not compete well with organic fluorophores in this regard, but maximising the absorbance cross section and tuning analytically relevant responsive luminescence will ensure that even if they do not match organic probes in terms of emission quantum yield, they offer versatile alternatives.

Another challenge is the *in cellulo* limit of detection and limit of quantification of the species that the probe is sensing. The influence of the incredibly complex cytosol composition or organelle environment complicates photophysical effects and sensing capability.

Metal complex luminophores translate well to confocal luminescence imaging techniques that permit focused interrogation of a specific cell region to yield site-specific biorelevant information with good optical sectioning and indeed look likely to provide important solutions in super-resolution methods too. In particular, lifetime imaging is the method of choice for visualising spatial oxygen distribution in cells and particularly in multicellular spheroid samples. However, metal complex luminophores are less widely tested in the more common and conventional methods used in biological labs. In particular, as probes for plate reader-based bioassays in 96-well format which are a widely used and high-throughput approach where emission, lifetime or absorbance is collected from multiple cells across a well format. Highly selective probe targeting or switchable probes would be required with limited contribution from emission of the complex outside the region of interest for such applications. The targeting ability of the conjugate dye would also have to be extremely precise. Few of the dyes described above meet this criterion of highly specific localisation that would be required for a well format assay yet. Nonetheless, they have proven to be effective imaging probes for investigating cellular membrane dynamics, sensing receptor expression and monitoring redox species in mitochondria, pH fluctuations and DNA interactions under imaging conditions *in cellulo* and look likely with further advances in precision targeting to emerge as powerful tools for diverse bioanalysis.

Acknowledgements The authors gratefully acknowledge the Science Foundation Ireland for funding under grants [19/FFP/6428] and [12/RC/2276_P2].

Funding Open Access funding provided by the IReL Consortium.

Declarations

Conflict of interest The authors declare no conflict of interest.

References

1. Neugebauer U, Cosgrave L, Pellegrin Y, et al (2012) Membrane permeable luminescent metal complexes for cellular imaging. In: Popp J, Drexler W, Tuchin VV, Matthews DL (eds). Brussels, Belgium, p 84270C
2. Burke CS, Byrne A, TiaE K (2018) Targeting photoinduced DNA destruction by Ru(II) tetraazaphenanthrene in live cells by signal peptide. J Am Chem Soc 140:6945–6955. https://doi.org/10.1021/jacs.8b02711
3. Holden L, Burke CS, Cullinane D, Keyes TE (2021) Strategies to promote permeation and vectorization, and reduce cytotoxicity of metal complex luminophores for bioimaging and intracellular sensing. RSC Chem Biol 2:1021–1049. https://doi.org/10.1039/D1CB00049G

4. Drezek R, Brookner C, Pavlova et al (2001) Autofluorescence microscopy of fresh cervical-tissue sections reveals alterations in tissue biochemistry with dysplasia. Photochem Photobiol 73:636–641. https://doi.org/10.1562/0031-8655(2001)073%3c0636:AMOFCT%3e2.0.CO;2
5. Lakowicz JR, Szmacinski H, Nowaczyk K, Johnson ML (1992) Fluorescence lifetime imaging of free and protein-bound NADH. Proc Natl Acad Sci 89:1271–1275. https://doi.org/10.1073/pnas.89.4.1271
6. Skala MC, Riching KM, Gendron-Fitzpatrick A et al (2007) In vivo multiphoton microscopy of NADH and FAD redox states, fluorescence lifetimes, and cellular morphology in precancerous epithelia. PNAS 104:19494–19499. https://doi.org/10.1073/pnas.0708425104
7. Chance B, Schoener B, Oshino R et al (1979) Oxidation-reduction ratio studies of mitochondria in freeze-trapped samples. NADH and flavoprotein fluorescence signals. J Biol Chem 254:4764–4771
8. Nakashima N, Yoshihara K, Tanaka F, Yagi K (1980) Picosecond fluorescence lifetime of the coenzyme of D-amino acid oxidase. J Biol Chem 255:5261–5263. https://doi.org/10.1016/S0021-9258(19)70779-0
9. Chance B (1976) Pyridine nucleotide as an indicator of the oxygen requirements for energy-linked functions of mitochondria. Circ Res 38:I31-38
10. Gkika KS, Noorani S, Walsh N, Keyes TE (2021) Os(II)-bridged polyarginine conjugates: the additive effects of peptides in promoting or preventing permeation in cells and multicellular tumor spheroids. Inorg Chem 60:8123–8134. https://doi.org/10.1021/acs.inorgchem.1c00769
11. Kober EM, Meyer TJ (1982) Concerning the absorption spectra of the ions M(bpy)32+ (M = Fe, Ru, Os; bpy = 2,2′-bipyridine). Inorg Chem 21:3967–3977. https://doi.org/10.1021/ic00141a021
12. Felix F, Ferguson J, Güdel HU, Ludi A (1979) Electronic spectra of M(bipy)2+3 complexions (M = Fe, Ru and Os). Chem Phys Lett 62:153–157. https://doi.org/10.1016/0009-2614(79)80432-7
13. Ferguson J, Herren F, Krausz ER et al (1985) Electronic spectroscopy of M(bpy)2+3 (M = Fe, Ru, Os), Cr(bpy)3+3 and related compounds. Coord Chem Rev 64:21–39. https://doi.org/10.1016/0010-8545(85)80039-4
14. Yersin H, Gallhuber E, Hensler G (1987) Highly resolved polarized absorption spectra of single-crystal [Ru(bpy)3](PF6)2. Chem Phys Lett 134:497–501. https://doi.org/10.1016/0009-2614(87)87180-4
15. Wachter E, Heidary DK, Howerton BS et al (2012) Light-activated ruthenium complexes photobind DNA and are cytotoxic in the photodynamic therapy window. Chem Commun 48:9649. https://doi.org/10.1039/c2cc33359g
16. Sauvage JP, Collin JP, Chambron JC et al (1994) Ruthenium(II) and Osmium(II) Bis(terpyridine) Complexes in covalently-linked multicomponent systems: synthesis, electrochemical behavior, absorption spectra, and photochemical and photophysical properties. Chem Rev 94:993–1019. https://doi.org/10.1021/cr00028a006
17. Medlycott EA, Hanan GS (2005) Designing tridentate ligands for ruthenium(ii) complexes with prolonged room temperature luminescence lifetimes. Chem Soc Rev 34:133. https://doi.org/10.1039/b316486c
18. Heinze K, Hempel K, Tschierlei S et al (2009) Resonance Raman studies of bis(terpyridine) ruthenium(II) amino acid esters and diesters. Eur J Inorg Chem 2009:3119–3126. https://doi.org/10.1002/ejic.200900309
19. Suzuki K, Kobayashi A, Kaneko S et al (2009) Reevaluation of absolute luminescence quantum yields of standard solutions using a spectrometer with an integrating sphere and a back-thinned CCD detector. Phys Chem Chem Phys 11:9850. https://doi.org/10.1039/b912178a
20. Keyes TE, Vos JG, Kolnaar JA et al (1996) Tuning of the photostability of bis(2,2′-biquinoline) ruthenium(II) complexes containing pyridyltriazole ligands by variation of pH. Inorg Chim Acta 245:237–242. https://doi.org/10.1016/0020-1693(95)04817-0
21. Duati M, Fanni S, Vos GJ (2000) A new luminescent Ru(terpy) complex incorporating a 1,2,4-triazole based σ-donor ligand. Inorg Chem Commun 3:68–70. https://doi.org/10.1016/S1387-7003(00)00007-1
22. Pal AK, Serroni S, Zaccheroni N et al (2014) Near infra-red emitting Ru(ii) complexes of tridentate ligands: electrochemical and photophysical consequences of a strong donor ligand with large bite angles. Chem Sci 5:4800–4811. https://doi.org/10.1039/C4SC01604A
23. Abrahamsson M, Jäger M, Österman T et al (2006) A 3.0 μs room temperature excited state lifetime of a bistridentate RuII-polypyridine complex for rod-like molecular arrays. J Am Chem Soc 128:12616–12617. https://doi.org/10.1021/ja064262y

24. Zhao Q, Liu S, Shi M et al (2006) Series of new cationic iridium(III) complexes with tunable emission wavelength and excited state properties: structures, theoretical calculations, and photophysical and electrochemical properties. Inorg Chem 45:6152–6160. https://doi.org/10.1021/ic052034j
25. Hasan K, Bansal AK, Samuel IDW et al (2015) Tuning the emission of cationic iridium (III) complexes towards the red through methoxy substitution of the cyclometalating ligand. Sci Rep 5:12325. https://doi.org/10.1038/srep12325
26. Skórka Ł, Filapek M, Zur L et al (2016) Highly phosphorescent cyclometalated iridium(III) complexes for optoelectronic applications: fine tuning of the emission wavelength through ancillary ligands. J Phys Chem C 120:7284–7294. https://doi.org/10.1021/acs.jpcc.6b01663
27. Stonelake TM, Phillips KA, Otaif HY et al (2020) Spectroscopic and theoretical investigation of color tuning in deep-red luminescent iridium(III) complexes. Inorg Chem 59:2266–2277. https://doi.org/10.1021/acs.inorgchem.9b02991
28. Zhang Y, Qiao J (2021) Near-infrared emitting iridium complexes: molecular design, photophysical properties, and related applications. iScience 24:102858. https://doi.org/10.1016/j.isci.2021.102858
29. Dolan C, Moriarty RD, Lestini E et al (2013) Cell uptake and cytotoxicity of a novel cyclometalated iridium(III) complex and its octaarginine peptide conjugate. J Inorg Biochem 119:65–74. https://doi.org/10.1016/j.jinorgbio.2012.11.001
30. GambaSalvadó II, Brissos RF et al (2016) High-affinity sequence-selective DNA binding by iridium(III) polypyridyl organometallopeptides. Chem Commun 52:1234–1237. https://doi.org/10.1039/C5CC07619F
31. Moucheron C, Kirsch-De Mesmaeker A, Kelly JM (1997) Photoreactions of ruthenium(II) and osmium(II) complexes with deoxyribonucleic acid (DNA). J Photochem Photobiol B Biol 40:91–106
32. Pankuch BJ, Lacky DE, Crosby GA (1980) Charge-transfer excited states of osmium(II) complexes. 1. Assignment of the visible absorption bands. J Phys Chem 84:2061–2067. https://doi.org/10.1021/j100453a013
33. Crosby GA, Demas JN (1970) Quantum efficiencies of transition-metal complexes. I. d-d Luminescence. J Am Chem Soc 92:7262–7270. https://doi.org/10.1021/ja00728a006
34. Braun D, Gallhuber E, Hensler G, Yersin H (1989) Site selective spectra of the lowest excited states of [Os(bpy)3] 2+ in [Ru1-x Os x (bpy)3]X 2(X = PF6, AsF6, SbF6). Mol Phys 67:417–430. https://doi.org/10.1080/00268978900101171
35. Decurtins S, Felix F, Ferguson J et al (1980) The electronic spectrum of tris(2,2′-bipyridine) iron(2+) and tris (2,2′-bipyridine)osmium(2+). J Am Chem Soc 102:4102–4106. https://doi.org/10.1021/ja00532a020
36. Cummings SD, Eisenberg R (1996) Tuning the excited-state properties of platinum(II) diimine dithiolate complexes. J Am Chem Soc 118:1949–1960. https://doi.org/10.1021/ja951345y
37. Yu T, Tsang DP-K, Au VK-M et al (2013) Deep red to near-infrared emitting rhenium(I) complexes: synthesis, characterization, electrochemistry, photophysics, and electroluminescence studies. Chem Eur J 19:13418–13427. https://doi.org/10.1002/chem.201301841
38. North AJ, Hayne DJ, Schieber C et al (2015) Toward hypoxia-selective rhenium and technetium tricarbonyl complexes. Inorg Chem 54:9594–9610. https://doi.org/10.1021/acs.inorgchem.5b01691
39. Clède S, Lambert F, Sandt C et al (2013) Detection of an estrogen derivative in two breast cancer cell lines using a single core multimodal probe for imaging (SCoMPI) imaged by a panel of luminescent and vibrational techniques. Analyst 138:5627. https://doi.org/10.1039/c3an00807j
40. Pierri AE, Pallaoro A, Wu G, Ford PC (2012) A luminescent and biocompatible PhotoCORM. J Am Chem Soc 134:18197–18200. https://doi.org/10.1021/ja3084434
41. Yin Zhang K, Ka-Shun Tso K, Louie M-W et al (2013) A phosphorescent rhenium(I) tricarbonyl polypyridine complex appended with a fructose pendant that exhibits photocytotoxicity and enhanced uptake by breast cancer cells. Organometallics 32:5098–5102. https://doi.org/10.1021/om400612f
42. Bonetti A (2009) Platinum and other heavy metal compounds in cancer chemotherapy: molecular mechanisms and clinical applications. Humana Press, New York
43. Johnstone TC, Suntharalingam K, Lippard SJ (2016) The next generation of platinum drugs: targeted Pt(II) agents, nanoparticle delivery, and Pt(IV) prodrugs. Chem Rev 116:3436–3486. https://doi.org/10.1021/acs.chemrev.5b00597

44. Baggaley E, Sazanovich, Williams JAG et al (2014) Two-photon phosphorescence lifetime imaging of cells and tissues using a long-lived cyclometallated N $_{pyridyl}$ ^C $_{phenyl}$ ^N $_{pyridyl}$ Pt(II) complex. RSC Adv 4:35003–35008. https://doi.org/10.1039/C4RA04489D
45. Wu J, Li Y, Tan C et al (2018) Aggregation-induced near-infrared emitting platinum(ii) terpyridyl complex: cellular characterisation and lysosome-specific localisation. Chem Commun 54:11144–11147. https://doi.org/10.1039/C8CC06839A
46. Baggaley E, Botchway SW, Haycock JW et al (2014) Long-lived metal complexes open up microsecond lifetime imaging microscopy under multiphoton excitation: from FLIM to PLIM and beyond. Chem Sci 5:879–886. https://doi.org/10.1039/C3SC51875B
47. Farley SJ, Rochester DL, Thompson AL et al (2005) Controlling emission energy, self-quenching, and excimer formation in highly luminescent N∧C∧N-coordinated platinum(II) complexes. Inorg Chem 44:9690–9703. https://doi.org/10.1021/ic051049e
48. Sommer JR, Shelton AH, Parthasarathy A et al (2011) Photophysical properties of near-infrared phosphorescent π-extended platinum porphyrins. Chem Mater 23:5296–5304. https://doi.org/10.1021/cm202241e
49. Turner E, Bakken N, Li J (2013) Cyclometalated platinum complexes with luminescent quantum yields approaching 100%. Inorg Chem 52:7344–7351. https://doi.org/10.1021/ic302490c
50. Carrara S, Aliprandi A, Hogan CF, De Cola L (2017) Aggregation-induced electrochemiluminescence of platinum(II) complexes. J Am Chem Soc 139:14605–14610. https://doi.org/10.1021/jacs.7b07710
51. Tsai JL-L, Zou T, Liu J et al (2015) Luminescent platinum(ii) complexes with self-assembly and anti-cancer properties: hydrogel, pH dependent emission color and sustained-release properties under physiological conditions. Chem Sci 6:3823–3830. https://doi.org/10.1039/C4SC03635B
52. Brunner J, Barton JK (2006) Targeting DNA mismatches with rhodium intercalators functionalized with a cell-penetrating peptide. Biochemistry 45:12295–12302. https://doi.org/10.1021/bi061198o
53. Shen Y, Shao T, Fang B et al (2018) Visualization of mitochondrial DNA in living cells with super-resolution microscopy using thiophene-based terpyridine Zn(II) complexes. Chem Commun 54:11288–11291. https://doi.org/10.1039/C8CC06276E
54. Tian X, Zhang Q, Zhang M et al (2017) Probe for simultaneous membrane and nucleus labeling in living cells and in vivo bioimaging using a two-photon absorption water-soluble Zn(II) terpyridine complex with a reduced π-conjugation system. Chem Sci 8:142–149. https://doi.org/10.1039/C6SC02342H
55. Griffith C, Dayoub AS, Jaranatne T et al (2017) Cellular and cell-free studies of catalytic DNA cleavage by ruthenium polypyridyl complexes containing redox-active intercalating ligands. Chem Sci 8:3726–3740. https://doi.org/10.1039/C6SC04094B
56. Ramos LD, da Cruz HM, Morelli Frin KP (2017) Photophysical properties of rhenium(I) complexes and photosensitized generation of singlet oxygen. Photochem Photobiol Sci 16:459–466. https://doi.org/10.1039/C6PP00364H
57. Heinemann F, Karges J, Gasser G (2017) Critical overview of the use of Ru(II) polypyridyl complexes as photosensitizers in one-photon and two-photon photodynamic therapy. Acc Chem Res 50:2727–2736. https://doi.org/10.1021/acs.accounts.7b00180
58. White JK, Schmehl RH, Turro C (2017) An overview of photosubstitution reactions of Ru(II) imine complexes and their application in photobiology and photodynamic therapy. Inorg Chim Acta 454:7–20. https://doi.org/10.1016/j.ica.2016.06.007
59. Albani BA, Peña B, Leed NA et al (2014) Marked improvement in photoinduced cell death by a new tris-heteroleptic complex with dual action: singlet oxygen sensitization and ligand dissociation. J Am Chem Soc 136:17095–17101. https://doi.org/10.1021/ja508272h
60. Knoll JD, Turro C (2015) Control and utilization of ruthenium and rhodium metal complex excited states for photoactivated cancer therapy. Coord Chem Rev 282–283:110–126. https://doi.org/10.1016/j.ccr.2014.05.018
61. Cuello-Garibo J-A, Pérez-Gallent E, van der Boon L et al (2017) Influence of the steric bulk and solvent on the photoreactivity of ruthenium polypyridyl complexes coordinated to L-proline. Inorg Chem 56:4818–4828. https://doi.org/10.1021/acs.inorgchem.6b02794
62. Knoll JD, Albani BA, Turro C (2015) New Ru(II) complexes for dual photoreactivity: ligand exchange and $^1O_2$ generation. Acc Chem Res 48:2280–2287. https://doi.org/10.1021/acs.accounts.5b00227

63. Howerton BS, Heidary DK, Glazer EC (2012) Strained ruthenium complexes are potent light-activated anticancer agents. J Am Chem Soc 134:8324–8327. https://doi.org/10.1021/ja3009677
64. Cebrián-Losantos B, Krokhin AA, Stepanenko IN et al (2007) Osmium NAMI-A analogues: synthesis, structural and spectroscopic characterization, and antiproliferative properties. Inorg Chem 46:5023–5033. https://doi.org/10.1021/ic700405y
65. Büchel GE, Stepanenko IN, Hejl M et al (2012) Osmium(IV) complexes with 1H- and 2H-indazoles: tautomer identity versus spectroscopic properties and antiproliferative activity. J Inorg Biochem 113:47–54. https://doi.org/10.1016/j.jinorgbio.2012.04.001
66. Huang H, Zhang P, Chen H et al (2015) Comparison between polypyridyl and cyclometalated ruthenium(II) complexes: anticancer activities against 2D and 3D cancer models. Chem Eur J 21:715–725. https://doi.org/10.1002/chem.201404922
67. Ma D-L, Zhong H-J, Fu W-C et al (2013) Phosphorescent imaging of living cells using a cyclometalated iridium(III) complex. PLoS One 8:e55751. https://doi.org/10.1371/journal.pone.0055751
68. Huang H, Zhang P, Yu B et al (2014) Targeting nucleus DNA with a cyclometalated dipyridophenazineruthenium(II) complex. J Med Chem 57:8971–8983. https://doi.org/10.1021/jm501095r
69. Lo KK-W, Zhang KY (2012) Iridium(iii) complexes as therapeutic and bioimaging reagents for cellular applications. RSC Adv 2:12069. https://doi.org/10.1039/c2ra20967e
70. Wang F-X, Chen M-H, Hu X-Y et al (2016) Ester-modified cyclometalated iridium(III) complexes as mitochondria-targeting anticancer agents. Sci Rep 6:38954. https://doi.org/10.1038/srep38954
71. Xie Y-Y, Lin G-J, Jiang G-B et al (2013) Cytotoxicity, cellular uptake, cell cycle arrest, apoptosis, reactive oxygen species and DNA-binding studies of ruthenium(II) complexes. Trans Met Chem 38:563–571. https://doi.org/10.1007/s11243-013-9724-6
72. Jiang G-B, Xie Y-Y, Lin G-J et al (2013) Synthesis, characterization, DNA interaction, antioxidant and anticancer activity studies of ruthenium(II) polypyridyl complexes. J Photochem Photobiol B 129:48–56. https://doi.org/10.1016/j.jphotobiol.2013.09.009
73. Mari C, Pierroz V, Rubbiani R et al (2014) DNA intercalating RuII polypyridyl complexes as effective photosensitizers in photodynamic therapy. Chem Eur J 20:14421–14436. https://doi.org/10.1002/chem.201402796
74. Estalayo-Adrián S, Blasco S, Bright SA et al (2020) Water-soluble amphiphilic ruthenium(II) polypyridyl complexes as potential light-activated therapeutic agents. Chem Commun 56:9332–9335. https://doi.org/10.1039/D0CC04397D
75. Dickerson M, Sun Y, Howerton B, Glazer EC (2014) Modifying charge and hydrophilicity of simple Ru(II) polypyridyl complexes radically alters biological activities: old complexes, surprising new tricks. Inorg Chem 53:10370–10377. https://doi.org/10.1021/ic5013796
76. Luo Z, Yu L, Yang F et al (2014) Ruthenium polypyridyl complexes as inducer of ROS-mediated apoptosis in cancer cells by targeting thioredoxin reductase. Metallomics 6:1480–1490. https://doi.org/10.1039/C4MT00044G
77. Puckett CA, Barton JK (2007) Methods to explore cellular uptake of ruthenium complexes. J Am Chem Soc 129:46–47. https://doi.org/10.1021/ja0677564
78. Boff B, Gaiddon C, Pfeffer M (2013) Cancer cell cytotoxicity of cyclometalated compounds obtained with osmium(II) complexes. Inorg Chem 52:2705–2715. https://doi.org/10.1021/ic302779q
79. Lu L, Liu L-J, Chao W et al (2015) Identification of an iridium(III) complex with anti-bacterial and anti-cancer activity. Sci Rep 5:14544. https://doi.org/10.1038/srep14544
80. Finn S, Byrne A, Gkika KS, Keyes TE (2020) Photophysics and cell uptake of self-assembled Ru(II)polypyridyl vesicles. Front Chem 8:638. https://doi.org/10.3389/fchem.2020.00638
81. Cullinane D, Gkika KS, Byrne A, Keyes TE (2020) Photostable NIR emitting ruthenium(II) conjugates; uptake and biological activity in live cells. J Inorg Biochem 207:111032. https://doi.org/10.1016/j.jinorgbio.2020.111032
82. Li SP-Y, Liu H-W, Zhang KY, Lo KK-W (2010) Modification of luminescent iridium(III) polypyridine complexes with discrete poly(ethylene glycol) (PEG) pendants: synthesis, emissive behavior, intracellular uptake, and PEGylation properties. Chem Eur J 16:8329–8339. https://doi.org/10.1002/chem.201000474
83. Li SP-Y, Lau CT-S, Louie M-W et al (2013) Mitochondria-targeting cyclometalated iridium(III)–PEG complexes with tunable photodynamic activity. Biomaterials 34:7519–7532. https://doi.org/10.1016/j.biomaterials.2013.06.028

84. Oliver C, Jamur MC (2010) Immunocytochemical methods and protocols. Humana Press, Totowa
85. Hennig S, van de Linde S, Lummer M et al (2015) Instant live-cell super-resolution imaging of cellular structures by nanoinjection of fluorescent probes. Nano Lett 15:1374–1381. https://doi.org/10.1021/nl504660t
86. Dekaliuk M, Pyrshev K, Demchenko A (2015) Visualization and detection of live and apoptotic cells with fluorescent carbon nanoparticles. J Nanobiotechnol 13:86. https://doi.org/10.1186/s12951-015-0148-7
87. Rumin J, Bonnefond H, Saint-Jean B et al (2015) The use of fluorescent Nile red and BODIPY for lipid measurement in microalgae. Biotechnol Biofuels 8:42. https://doi.org/10.1186/s13068-015-0220-4
88. Neugebauer U, Pellegrin Y, Devocelle M et al (2008) Ruthenium polypyridyl peptide conjugates: membrane permeable probes for cellular imaging. Chem Commun. https://doi.org/10.1039/B810403D
89. Tan C, Lai S, Wu S et al (2010) Nuclear permeable ruthenium(II) β-carboline complexes induce autophagy to antagonize mitochondrial-mediated apoptosis. J Med Chem 53:7613–7624. https://doi.org/10.1021/jm1009296
90. Zhao Q, Huang C, Li F (2011) Phosphorescent heavy-metal complexes for bioimaging. Chem Soc Rev 40:2508–2524. https://doi.org/10.1039/C0CS00114G
91. Baggaley E, Gill MR, Green NH et al (2014) Dinuclear ruthenium(II) complexes as two-photon, time-resolved emission microscopy probes for cellular DNA. Angew Chem Int Ed 53:3367–3371. https://doi.org/10.1002/anie.201309427
92. Gurtovenko AA, Anwar J (2007) Modulating the structure and properties of cell membranes: the molecular mechanism of action of dimethyl sulfoxide. J Phys Chem B 111:10453–10460. https://doi.org/10.1021/jp073113e
93. Dludla PV, Jack B, Viraragavan A et al (2018) A dose-dependent effect of dimethyl sulfoxide on lipid content, cell viability and oxidative stress in 3T3-L1 adipocytes. Toxicol Rep 5:1014–1020. https://doi.org/10.1016/j.toxrep.2018.10.002
94. Yuan C, Gao J, Guo J et al (2014) Dimethyl sulfoxide damages mitochondrial integrity and membrane potential in cultured astrocytes. PLoS One 9:e107447. https://doi.org/10.1371/journal.pone.0107447
95. Ellahioui Y, Patra M, Mari C et al (2019) Mesoporous silica nanoparticles functionalised with a photoactive ruthenium(II) complex: exploring the formulation of a metal-based photodynamic therapy photosensitiser. Dalton Trans 48:5940–5951. https://doi.org/10.1039/C8DT02392A
96. Caballero AB, Cardo L, Claire S et al (2019) Assisted delivery of anti-tumour platinum drugs using DNA-coiling gold nanoparticles bearing lumophores and intercalators: towards a new generation of multimodal nanocarriers with enhanced action. Chem Sci 10:9244–9256. https://doi.org/10.1039/C9SC02640A
97. Dosumu AN, Claire S, Watson LS et al (2021) Quantification by luminescence tracking of red emissive gold nanoparticles in cells. JACS Au 1:174–186. https://doi.org/10.1021/jacsau.0c00033
98. Liang J, Zheng Y, Wu X et al (2020) A tailored multifunctional anticancer nanodelivery system for ruthenium-based photosensitizers: tumor microenvironment adaption and remodeling. Adv Sci 7:1901992. https://doi.org/10.1002/advs.201901992
99. Shen J, Kim H-C, Wolfram J et al (2017) A liposome encapsulated ruthenium polypyridine complex as a theranostic platform for triple-negative breast cancer. Nano Lett 17:2913–2920. https://doi.org/10.1021/acs.nanolett.7b00132
100. Zhang KY, Liu H-W, Fong TT-H et al (2010) Luminescent dendritic cyclometalated iridium(III) polypyridine complexes: synthesis, emission behavior, and biological properties. Inorg Chem 49:5432–5443. https://doi.org/10.1021/ic902443e
101. Gottschaldt M, Schubert US, Rau S et al (2010) Sugar-selective enrichment of a D-glucose-substituted ruthenium bipyridyl complex inside HepG2 cancer cells. Chem Eur J Chem Bio 11:649–652. https://doi.org/10.1002/cbic.200900769
102. Wenzel M, de Almeida A, Bigaeva E et al (2016) New luminescent polynuclear metal complexes with anticancer properties: toward structure–activity relationships. Inorg Chem 55:2544–2557. https://doi.org/10.1021/acs.inorgchem.5b02910
103. Jakubaszek M, Rossier J, Karges J et al (2019) Evaluation of the potential of cobalamin derivatives bearing Ru(II) polypyridyl complexes as photosensitizers for photodynamic therapy. HCA. https://doi.org/10.1002/hlca.201900104

104. Zhao Z, Gao P, You Y, Chen T (2018) Cancer-targeting functionalization of selenium-containing ruthenium conjugate with tumor microenvironment-responsive property to enhance theranostic effects. Chem Eur J 24:3289–3298. https://doi.org/10.1002/chem.201705561
105. Li J, Zeng L, Xiong K et al (2019) A biotinylated ruthenium(II) photosensitizer for tumor-targeted two-photon photodynamic therapy. Chem Commun 55:10972–10975. https://doi.org/10.1039/C9CC05826E
106. Karges J, Jakubaszek M, Mari C et al (2020) Synthesis and characterization of an epidermal growth factor receptor-selective RuII polypyridyl-nanobody conjugate as a photosensitizer for photodynamic therapy. ChemBioChem 21:531–542. https://doi.org/10.1002/cbic.201900419
107. Zielonka J, Joseph J, Sikora A et al (2017) Mitochondria-targeted triphenylphosphonium-based compounds: syntheses, mechanisms of action, and therapeutic and diagnostic applications. Chem Rev 117:10043–10120. https://doi.org/10.1021/acs.chemrev.7b00042
108. Ramu V, Aute S, Taye N et al (2017) Photo-induced cytotoxicity and anti-metastatic activity of ruthenium(II)–polypyridyl complexes functionalized with tyrosine or tryptophan. Dalton Trans 46:6634–6644. https://doi.org/10.1039/C7DT00670E
109. Cosgrave L, Devocelle M, Forster RJ, Keyes TE (2010) Multimodal cell imaging by ruthenium polypyridyl labelled cell penetrating peptides. Chem Commun 46:103–105. https://doi.org/10.1039/B918611E
110. Adamson K, Dolan C, Moran N et al (2014) RGD labeled Ru(II) polypyridyl conjugates for platelet integrin alpha(IIb)beta(3) recognition and as reporters of integrin conformation. Bioconj Chem 25:928–944. https://doi.org/10.1021/bc5000737
111. Hahn EM, Estrada-Ortiz N, Han J et al (2017) Functionalization of Ruthenium(II) terpyridine complexes with cyclic RGD peptides to target integrin receptors in cancer cells: functionalization of ruthenium(II) terpyridine complexes with cyclic RGD peptides to target integrin receptors in cancer cells. Eur J Inorg Chem 2017:1667–1672. https://doi.org/10.1002/ejic.201601094
112. Martínez-Alonso M, Gasser G (2021) Ruthenium polypyridyl complex-containing bioconjugates. Coord Chem Rev 434:213736. https://doi.org/10.1016/j.ccr.2020.213736
113. Ungricht R, Kutay U (2017) Mechanisms and functions of nuclear envelope remodelling. Nat Rev Mol Cell Biol 18:229–245. https://doi.org/10.1038/nrm.2016.153
114. Freitas N, Cunha C (2009) Mechanisms and signals for the nuclear import of proteins. Curr Genom 10:550–557. https://doi.org/10.2174/138920209789503941
115. Soler M, Feliu L, Planas M et al (2016) Peptide-mediated vectorization of metal complexes: conjugation strategies and biomedical applications. Dalton Trans 45:12970–12982. https://doi.org/10.1039/C5DT04529K
116. Obitz D, Miller RG, Metzler-Nolte N (2021) Synthesis and DNA interaction studies of Ru(ii) cell penetrating peptide (CPP) bioconjugates. Dalton Trans. https://doi.org/10.1039/D1DT01776D
117. Byrne A, Burke CS, Keyes TE (2016) Precision targeted ruthenium(II) luminophores; highly effective probes for cell imaging by stimulated emission depletion (STED) microscopy. Chem Sci 7:6551–6562. https://doi.org/10.1039/C6SC02588A
118. Martin A, Byrne A, Burke CS et al (2014) Peptide-bridged dinuclear Ru(II) complex for mitochondrial targeted monitoring of dynamic changes to oxygen concentration and ROS generation in live mammalian cells. J Am Chem Soc 136:15300–15309. https://doi.org/10.1021/ja508043q
119. Burke CS, Byrne A, Keyes TE (2018) Highly selective mitochondrial targeting by a ruthenium(II) peptide conjugate: imaging and photoinduced damage of mitochondrial DNA. Angew Chem Int Ed 57:12420–12424. https://doi.org/10.1002/anie.201806002
120. Fosgerau K, Hoffmann T (2015) Peptide therapeutics: current status and future directions. Drug Discov Today 20:122–128. https://doi.org/10.1016/j.drudis.2014.10.003
121. Tesauro D, Accardo A, Diaferia C et al (2019) Peptide-based drug-delivery systems in biotechnological applications: recent advances and perspectives. Molecules 24:351. https://doi.org/10.3390/molecules24020351
122. Hussein WM, Skwarczynski M, Toth (2020) Peptide synthesis: methods and protocols. Humana Press, New York
123. Ryser HJ-P, Hancock R (1965) Histones and basic polyamino acids stimulate the uptake of albumin by tumor cells in culture. Science 150:501–503. https://doi.org/10.1126/science.150.3695.501
124. Shen WC, Ryser HJ (1978) Conjugation of poly-L-lysine to albumin and horseradish peroxidase: a novel method of enhancing the cellular uptake of proteins. Proc Natl Acad Sci 75:1872–1876. https://doi.org/10.1073/pnas.75.4.1872

125. Frankel AD, Pabo CO (1988) Cellular uptake of the tat protein from human immunodeficiency virus. Cell 55:1189–1193. https://doi.org/10.1016/0092-8674(88)90263-2
126. Green M, Loewenstein PM (1988) Autonomous functional domains of chemically synthesized human immunodeficiency virus tat trans-activator protein. Cell 55:1179–1188. https://doi.org/10.1016/0092-8674(88)90262-0
127. Mitchell DJ, Steinman L, Kim DT et al (2000) Polyarginine enters cells more efficiently than other polycationic homopolymers. J Pept Res 56:318–325. https://doi.org/10.1034/j.1399-3011.2000.00723.x
128. Puckett CA, Barton JK (2010) Targeting a ruthenium complex to the nucleus with short peptides. Bioorg Med Chem 18:3564–3569. https://doi.org/10.1016/j.bmc.2010.03.081
129. Wender PA, Mitchell DJ, Pattabiraman K et al (2000) The design, synthesis, and evaluation of molecules that enable or enhance cellular uptake: peptoid molecular transporters. Proc Natl Acad Sci 97:13003–13008. https://doi.org/10.1073/pnas.97.24.13003
130. Puckett CA, Barton JK (2009) Fluorescein redirects a ruthenium−octaarginine conjugate to the nucleus. J Am Chem Soc 131:8738–8739. https://doi.org/10.1021/ja9025165
131. Byrne A, Dolan C, Moriarty RD et al (2015) Osmium(II) polypyridyl polyarginine conjugate as a probe for live cell imaging; a comparison of uptake, localization and cytotoxicity with its ruthenium(II) analogue. Dalton Trans 44:14323–14332. https://doi.org/10.1039/C5DT01833A
132. Dmitriev RI, Kondrashina AV, Koren K et al (2014) Small molecule phosphorescent probes for O2 imaging in 3D tissue models. Biomater Sci 2:853–866. https://doi.org/10.1039/C3BM60272A
133. van Rijt SH, Kostrhunova H, Brabec V, Sadler PJ (2011) Functionalization of osmium arene anticancer complexes with (poly)arginine: effect on cellular uptake, internalization, and cytotoxicity. Bioconj Chem 22:218–226. https://doi.org/10.1021/bc100369p
134. Bouzada D, Salvadó, Barka G et al (2018) Selective G-quadruplex binding by oligoarginine-Ru(dppz) metallopeptides. Chem Commun 54:658–661. https://doi.org/10.1039/C7CC08286J
135. Herce HD, Garcia AE, Cardoso MC (2014) Fundamental molecular mechanism for the cellular uptake of guanidinium-rich molecules. J Am Chem Soc 136:17459–17467. https://doi.org/10.1021/ja507790z
136. Allolio C, Magarkar A, Jurkiewicz P et al (2018) Arginine-rich cell-penetrating peptides induce membrane multilamellarity and subsequently enter via formation of a fusion pore. Proc Natl Acad Sci USA 115:11923–11928. https://doi.org/10.1073/pnas.1811520115
137. Wadia JS, Stan RV, Dowdy SF (2004) Transducible TAT-HA fusogenic peptide enhances escape of TAT-fusion proteins after lipid raft macropinocytosis. Nat Med 10:310–315. https://doi.org/10.1038/nm996
138. Rothbard JB, Jessop TC, Lewis RS et al (2004) Role of membrane potential and hydrogen bonding in the mechanism of translocation of guanidinium-rich peptides into cells. J Am Chem Soc 126:9506–9507. https://doi.org/10.1021/ja0482536
139. Rothbard J, Jessop T, Wender P (2005) Adaptive translocation: the role of hydrogen bonding and membrane potential in the uptake of guanidinium-rich transporters into cells. Adv Drug Deliv Rev 57:495–504. https://doi.org/10.1016/j.addr.2004.10.003
140. Fuchs SM, Raines RT (2009) Polyarginine as a multifunctional fusion tag. Protein Sci 14:1538–1544. https://doi.org/10.1110/ps.051393805
141. Sakai N, Matile S (2003) Anion-mediated transfer of polyarginine across liquid and bilayer membranes. J Am Chem Soc 125:14348–14356. https://doi.org/10.1021/ja037601l
142. Nishihara M, Perret F, Takeuchi T et al (2005) Arginine magic with new counterions up the sleeve. Org Biomol Chem 3:1659. https://doi.org/10.1039/b501472g
143. Derossi D, Joliot AH, Chassaing G, Prochiantz A (1994) The third helix of the Antennapedia homeodomain translocates through biological membranes. J Biol Chem 269:10444–10450
144. Thorén PEG, Persson D, Karlsson M, Nordén B (2000) The Antennapedia peptide penetratin translocates across lipid bilayers—the first direct observation. FEBS Lett 482:265–268. https://doi.org/10.1016/S0014-5793(00)02072-X
145. Fischer PM, Zhelev NZ, Wang S et al (2000) Structure–activity relationship of truncated and substituted analogues of the intracellular delivery vector Penetratin: penetratin structure–activity relationship. J Pept Res 55:163–172. https://doi.org/10.1034/j.1399-3011.2000.00163.x
146. Owji H, Nezafat N, Negahdaripour M et al (2018) A comprehensive review of signal peptides: structure, roles, and applications. Eur J Cell Biol 97:422–441. https://doi.org/10.1016/j.ejcb.2018.06.003

147. Brandén LJ, Mohamed AJ, Smith CIE (1999) A peptide nucleic acid–nuclear localization signal fusion that mediates nuclear transport of DNA. Nat Biotechnol 17:784–787. https://doi.org/10.1038/11726
148. Ragin AD, Morgan RA, Chmielewski J (2002) Cellular import mediated by nuclear localization signal peptide sequences. Chem Biol 9:943–948. https://doi.org/10.1016/S1074-5521(02)00189-8
149. Noor F, Wüstholz A, Kinscherf R, Metzler-Nolte N (2005) A cobaltocenium-peptide bioconjugate shows enhanced cellular uptake and directed nuclear delivery. Angew Chem Int Ed 44:2429–2432. https://doi.org/10.1002/anie.200462519
150. Blackmore L, Moriarty R, Dolan C et al (2013) Peptide directed transmembrane transport and nuclear localization of Ru(ii) polypyridyl complexes in mammalian cells. Chem Commun 49:2658. https://doi.org/10.1039/c3cc40453f
151. Mari C, Pierroz V, Leonidova A et al (2015) Towards selective light-activated RuII-based prodrug candidates: towards selective light-activated RuII-based prodrug candidates. Eur J Inorg Chem 2015:3879–3891. https://doi.org/10.1002/ejic.201500602
152. Leonidova A, Pierroz V, Rubbiani R et al (2014) Photo-induced uncaging of a specific Re(I) organometallic complex in living cells. Chem Sci 5:4044. https://doi.org/10.1039/C3SC53550A
153. Day AH, Übler MH, Best HL et al (2020) Targeted cell imaging properties of a deep red luminescent iridium(III) complex conjugated with a c-Myc signal peptide. Chem Sci 11:1599–1606. https://doi.org/10.1039/C9SC05568A
154. Dang CV, Lee WM (1988) Identification of the human c-myc protein nuclear translocation signal. Mol Cell Biol 8:4048–4054. https://doi.org/10.1128/MCB.8.10.4048
155. Lee LC-C, Tsang AW-Y, Liu H-W, Lo KK-W (2020) Photofunctional cyclometalated iridium(III) polypyridine complexes bearing a perfluorobiphenyl moiety for bioconjugation, bioimaging, and phototherapeutic applications. Inorg Chem 59:14796–14806. https://doi.org/10.1021/acs.inorgchem.0c01343
156. Ypsilantis K, Plakatouras JC, Manos MJ et al (2018) Stepwise synthesis, characterization, DNA binding properties and cytotoxicity of diruthenium oligopyridine compounds conjugated with peptides. Dalton Trans 47:3549–3567. https://doi.org/10.1039/C7DT04639A
157. Horton KL, Stewart KM, Fonseca SB et al (2008) Mitochondria-penetrating peptides. Chem Biol 15:375–382. https://doi.org/10.1016/j.chembiol.2008.03.015
158. Gkika KS, Byrne A, Keyes TE (2019) Mitochondrial targeted osmium polypyridyl probe shows concentration dependent uptake, localisation and mechanism of cell death. Dalton Trans 48:17461–17471. https://doi.org/10.1039/C9DT02967B
159. Ye Y, Chen X (2011) Integrin targeting for tumor optical imaging. Theranostics 1:102–126. https://doi.org/10.7150/thno/v01p0102
160. Wang W, Wu Q, Pasuelo M et al (2005) Probing for integrin $\alpha_v \beta_3$ binding of RGD peptides using fluorescence polarization. Bioconj Chem 16:729–734. https://doi.org/10.1021/bc049763s
161. Gaertner FC, Kessler H, Wester H-J et al (2012) Radiolabelled RGD peptides for imaging and therapy. Eur J Nucl Med Mol Imaging 39:126–138. https://doi.org/10.1007/s00259-011-2028-1
162. (2021) Encyclopedia of biological chemistry. Elsevier, New York
163. Barragán F, López-Senín P, Salassa L et al (2011) Photocontrolled DNA binding of a receptor-targeted organometallic ruthenium(II) complex. J Am Chem Soc 133:14098–14108. https://doi.org/10.1021/ja205235m
164. Gandioso A, Shaili E, Massaguer A et al (2015) An integrin-targeted photoactivatable Pt(IV) complex as a selective anticancer pro-drug: synthesis and photoactivation studies. Chem Commun 51:9169–9172. https://doi.org/10.1039/C5CC03180J
165. Ma X, Jia J, Cao R et al (2014) Histidine–iridium(III) coordination-based peptide luminogenic cyclization and cyclo-RGD peptides for cancer-cell targeting. J Am Chem Soc 136:17734–17737. https://doi.org/10.1021/ja511656q
166. Xue S-S, Tan C-P, Chen M-H et al (2017) Tumor-targeted supramolecular nanoparticles self-assembled from a ruthenium-β-cyclodextrin complex and an adamantane-functionalized peptide. Chem Commun 53:842–845. https://doi.org/10.1039/C6CC08296C
167. Zhao Z, Zhang X, Li C, Chen T (2019) Designing luminescent ruthenium prodrug for precise cancer therapy and rapid clinical diagnosis. Biomaterials 192:579–589. https://doi.org/10.1016/j.biomaterials.2018.12.002
168. Ke M-R, Ng DKP, Lo P-C (2014) Synthesis and in vitro photodynamic activities of an integrin-targeting cRGD-conjugated zinc(II) phthalocyanine. Chem Asian J 9:554–561. https://doi.org/10.1002/asia.201301166

169. Cheng M, Zhou L, Ma J et al (2019) Iridium(III) and gadolinium(III) loaded and peptide-modified silica nanoparticles for photoluminescence and magnetic resonance (dual) imaging. Mater Sci Eng C 104:109972. https://doi.org/10.1016/j.msec.2019.109972
170. Lu Y, Low PS (2002) Folate-mediated delivery of macromolecular anticancer therapeutic agents. Adv Drug Deliv Rev 54:675–693. https://doi.org/10.1016/S0169-409X(02)00042-X
171. Viola-Villegas N, Rabideau AE, Cesnavicious J et al (2008) Targeting the folate receptor (FR): imaging and cytotoxicity of ReIConjugates in FR-overexpressing cancer cells. ChemMedChem 3:1387–1394. https://doi.org/10.1002/cmdc.200800125
172. van der Meel R, Vehmeijer LJC, Kok RJ et al (2013) Ligand-targeted particulate nanomedicines undergoing clinical evaluation: current status. Adv Drug Deliv Rev 65:1284–1298. https://doi.org/10.1016/j.addr.2013.08.012
173. Wang T, Zabarska N, Wu Y et al (2015) Receptor selective ruthenium-somatostatin photosensitizer for cancer targeted photodynamic applications. Chem Commun 51:12552–12555. https://doi.org/10.1039/C5CC03473F
174. McKeon AM, Noonan J, Devocelle M et al (2017) Platinum(IV) oxaliplatin–peptide conjugates targeting memHsp70+ phenotype in colorectal cancer cells. Chem Commun 53:11318–11321. https://doi.org/10.1039/C7CC04764A
175. Kitteringham E, McKeon AM, O'Dowd P et al (2020) Synthesis and characterisation of a novel mono functionalisable Pt(IV) oxaliplatin-type complex and its peptide conjugate. Inorg Chim Acta 505:119492. https://doi.org/10.1016/j.ica.2020.119492
176. Balkwill F (2004) The significance of cancer cell expression of the chemokine receptor CXCR4. Semin Cancer Biol 14:171–179. https://doi.org/10.1016/j.semcancer.2003.10.003
177. Turnbull WL, Yu L, Murrell E et al (2019) A dual modality 99mTc/Re(I)-labelled T140 analogue for imaging of CXCR4 expression. Org Biomol Chem 17:598–608. https://doi.org/10.1039/C8OB01947A
178. Tamamura H, Xu Y, Hattori T et al (1998) A low-molecular-weight inhibitor against the chemokine receptor CXCR4: a strong anti-HIV peptide T140. Biochem Biophys Res Commun 253:877–882. https://doi.org/10.1006/bbrc.1998.9871
179. Kuil J, Steunenberg P, Chin PTK et al (2011) Peptide-functionalized luminescent iridium complexes for lifetime imaging of CXCR4 expression. ChemBioChem 12:1897–1903. https://doi.org/10.1002/cbic.201100271
180. Vellaisamy K, Li G, Wang W et al (2018) A long-lived peptide-conjugated iridium(III) complex as a luminescent probe and inhibitor of the cell migration mediator, formyl peptide receptor 2. Chem Sci 9:8171–8177. https://doi.org/10.1039/C8SC02733A
181. He H-Q, Ye R (2017) The formyl peptide receptors: diversity of ligands and mechanism for recognition. Molecules 22:455. https://doi.org/10.3390/molecules22030455
182. Reubi JC, Wenger S, Schmuckli-Maurer J et al (2002) Bombesin receptor subtypes in human cancers: detection with the universal radioligand (125)I-[D-TYR(6), beta-ALA(11), PHE(13), NLE(14)] bombesin(6–14). Clin Cancer Res 8:1139–1146
183. Jaracz S, Chen J, Kuznetsova LV, Ojima (2005) Recent advances in tumor-targeting anticancer drug conjugates. Bioorg Med Chem 13:5043–5054. https://doi.org/10.1016/j.bmc.2005.04.084
184. Ranyuk E, Cauchon N, Klarskov K et al (2013) Phthalocyanine-peptide conjugates: receptor-targeting bifunctional agents for imaging and photodynamic therapy. J Med Chem 56:1520–1534. https://doi.org/10.1021/jm301311c
185. Agorastos N, Borsig L, Renard A et al (2007) Cell-specific and nuclear targeting with [M(CO)3]+ (M=99mTc, Re)-based complexes conjugated to acridine orange and bombesin. Chem Eur J 13:3842–3852. https://doi.org/10.1002/chem.200700031
186. Luengo A, Marzo, Reback M et al (2020) Luminescent bimetallic IrIII/AuIPeptide bioconjugates as potential theranostic agents. Chem Eur J 26:12158–12167. https://doi.org/10.1002/chem.202002067
187. Gasser G, Sosniak AM, Metzler-Nolte N (2011) Metal-containing peptide nucleic acid conjugates. Dalton Trans 40:7061. https://doi.org/10.1039/c0dt01706j
188. Ferri E, Donghi D, Panigati M et al (2010) Luminescent conjugates between dinuclear rhenium(i) complexes and peptide nucleic acids (PNA) for cell imaging and DNA targeting. Chem Commun 46:6255. https://doi.org/10.1039/c0cc00450b
189. Mari C, Panigati M, D'Alfonso L et al (2012) Luminescent conjugates between dinuclear rhenium complexes and peptide nucleic acids (PNA): synthesis, photophysical characterization, and cell uptake. Organometallics 31:5918–5928. https://doi.org/10.1021/om3004515

190. Gasser G, Pinto A, Neumann S et al (2012) Synthesis, characterisation and bioimaging of a fluorescent rhenium-containing PNA bioconjugate. Dalton Trans 41:2304–2313. https://doi.org/10.1039/C2DT12114J
191. Cauteruccio S, Panigati M, Veronese L et al (2019) Luminescent dinuclear rhenium(I) PNA conjugates for microRNA-21 targeting: synthesis, chemico-physical and biological characterization. J Organomet Chem 887:32–39. https://doi.org/10.1016/j.jorganchem.2019.02.020
192. Lin J-T, Chen P-C, Goud TV et al (2012) A sulfhydryl-reactive ruthenium (II) complex and its conjugation to Protein G as a universal reagent for fluorescent immunoassays. PLoS One 7:e36086. https://doi.org/10.1371/journal.pone.0036086
193. Peterson JR, Smith TA, Thordarson P (2007) Photoinduced reduction of catalytically and biologically active Ru(II)bisterpyridine–cytochrome c bioconjugates. Chem Commun. https://doi.org/10.1039/B702996A
194. Chakrabortty S, Agrawalla BK, Stumper A et al (2017) Mitochondria targeted protein-ruthenium photosensitizer for efficient photodynamic applications. J Am Chem Soc 139:2512–2519. https://doi.org/10.1021/jacs.6b13399
195. Wang X, Jia J, Huang Z et al (2011) Luminescent peptide labeling based on a histidine-binding iridium(III) complex for cell penetration and intracellular targeting studies. Chem Eur J 17:8028–8032. https://doi.org/10.1002/chem.201100568
196. Solomatina AI, Slobodina AD, Ryabova EV et al (2020) Blood-brain barrier penetrating luminescent conjugates based on cyclometalated platinum(II) complexes. Bioconjug Chem 31:2628–2637. https://doi.org/10.1021/acs.bioconjchem.0c00542
197. Leonidova A, Pierroz V, Adams LA et al (2014) Enhanced cytotoxicity through conjugation of a "Clickable" luminescent Re(I) complex to a cell-penetrating lipopeptide. ACS Med Chem Lett 5:809–814. https://doi.org/10.1021/ml500158w
198. Zeglis BM, Pierre VC, Barton JK (2007) Metallo-intercalators and metallo-insertors. Chem Commun. https://doi.org/10.1039/b710949k
199. Poynton FE, Bright SA, Blasco S et al (2017) The development of ruthenium(II) polypyridyl complexes and conjugates for in vitro cellular and in vivo applications. Chem Soc Rev 46:7706–7756. https://doi.org/10.1039/C7CS00680B
200. Saeed HK, Sreedharan S, Thomas JA (2020) Photoactive metal complexes that bind DNA and other biomolecules as cell probes, therapeutics, and theranostics. Chem Commun 56:1464–1480. https://doi.org/10.1039/C9CC09312E
201. Ghesquiere J, Le Gac S, Marcelis L et al (2012) What does the future hold for photo-oxidizing RuII complexes with polyazaaromatic ligands in medicinal chemistry? CTMC 12:185–196. https://doi.org/10.2174/156802612799079008
202. Leonidova A, Pierroz V, Rubbiani R et al (2014) Towards cancer cell-specific phototoxic organometallic rhenium(I) complexes. Dalton Trans 43:4287–4294. https://doi.org/10.1039/C3DT51817E
203. Bonam SR, Wang F, Muller S (2019) Lysosomes as a therapeutic target. Nat Rev Drug Discov 18:923–948. https://doi.org/10.1038/s41573-019-0036-1
204. Settembre C, Fraldi A, Medina DL, Ballabio A (2013) Signals from the lysosome: a control centre for cellular clearance and energy metabolism. Nat Rev Mol Cell Biol 14:283–296. https://doi.org/10.1038/nrm3565
205. Appelqvist H, Wäster P, Kågedal K, Öllinger K (2013) The lysosome: from waste bag to potential therapeutic target. J Mol Cell Biol 5:214–226. https://doi.org/10.1093/jmcb/mjt022
206. Wang F, Gómez-Sintes R, Boya P (2018) Lysosomal membrane permeabilization and cell death. Traffic 19:918–931. https://doi.org/10.1111/tra.12613
207. He L, Li Y, Tan C-P et al (2015) Cyclometalated iridium(III) complexes as lysosome-targeted photodynamic anticancer and real-time tracking agents. Chem Sci 6:5409–5418. https://doi.org/10.1039/C5SC01955A
208. Gkika KS, Kargaard A, Burke CS et al (2021) Ru(II)/BODIPY core co-encapsulated ratiometric nanotools for intracellular O_2 sensing in live cancer cells. RSC Chem Biol 2:1520–1533. https://doi.org/10.1039/D1CB00102G
209. Huang HW, Chen F-Y, Lee M-T (2004) Molecular mechanism of Peptide-induced pores in membranes. Phys Rev Lett 92:198304. https://doi.org/10.1103/PhysRevLett.92.198304
210. Subramanian A, Ma H, Dahl KN et al (2002) Adenovirus or HA-2 fusogenic peptide-assisted lipofection increases cytoplasmic levels of plasmid in nondividing endothelium with little enhancement of transgene expression. J Gene Med 4:75–83. https://doi.org/10.1002/jgm.235

211. Varkouhi AK, Scholte M, Storm G, Haisma HJ (2011) Endosomal escape pathways for delivery of biologicals. J Control Release 151:220–228. https://doi.org/10.1016/j.jconrel.2010.11.004
212. Bus T, Traeger A, Schubert US (2018) The great escape: how cationic polyplexes overcome the endosomal barrier. J Mater Chem B 6:6904–6918. https://doi.org/10.1039/C8TB00967H
213. Poynton FE, Hall JP, Keane PM et al (2016) Direct observation by time-resolved infrared spectroscopy of the bright and the dark excited states of the [Ru(phen)2(dppz)]2+ light-switch compound in solution and when bound to DNA. Chem Sci 7:3075–3084. https://doi.org/10.1039/C5SC04514B
214. Li G, Sun L, Ji L, Chao H (2016) Ruthenium(II) complexes with dppz: from molecular photoswitch to biological applications. Dalton Trans 45:13261–13276. https://doi.org/10.1039/C6DT01624C
215. Wittenberg JB (1970) Myoglobin-facilitated oxygen diffusion: role of myoglobin in oxygen entry into muscle. Physiol Rev 50:559–636. https://doi.org/10.1152/physrev.1970.50.4.559
216. Bobko AA, Dhimitruka, Eubank TD et al (2009) Trityl-based EPR probe with enhanced sensitivity to oxygen. Free Radical Biol Med 47:654–658. https://doi.org/10.1016/j.freeradbiomed.2009.06.007
217. Bussink J, Kaanders JH, Strik AM et al (2000) Optical sensor-based oxygen tension measurements correspond with hypoxia marker binding in three human tumor xenograft lines. Radiat Res 154:547–555. https://doi.org/10.1667/0033-7587(2000)154[0547:osbotm]2.0.co;2
218. Braun RD, Lanzen JL, Snyder SA, Dewhirst MW (2001) Comparison of tumor and normal tissue oxygen tension measurements using OxyLite or microelectrodes in rodents. Am J Physiol Heart Circ Physiol 280:H2533–H2544. https://doi.org/10.1152/ajpheart.2001.280.6.H2533
219. Cringle SJ, Yu PK, Su E-N, Yu D-Y (2006) Oxygen distribution and consumption in the developing rat retina. Invest Ophthalmol Vis Sci 47:4072. https://doi.org/10.1167/iovs.05-1638
220. Vaupel P, schlenger K, Knoop C, Höckel M (1991) Oxygenation of human tumors: evaluation of tissue oxygen distribution in breast cancers by computerized O2 tension measurements. Cancer Res 51:3316–3322
221. Dmitriev RI, Papkovsky DB (2012) Optical probes and techniques for O2 measurement in live cells and tissue. Cell Mol Life Sci 69:2025–2039. https://doi.org/10.1007/s00018-011-0914-0
222. Papkovsky DB (2004) Methods in optical oxygen sensing: protocols and critical analyses. Methods in enzymology. Elsevier, New York, pp 715–735
223. Dunphy I, Vinogradov SA, Wilson DF (2002) Oxyphor R2 and G2: phosphors for measuring oxygen by oxygen-dependent quenching of phosphorescence. Anal Biochem 310:191–198. https://doi.org/10.1016/S0003-2697(02)00384-6
224. Papkovsky DB, O'Riordan TC (2005) Emerging applications of phosphorescent metalloporphyrins. J Fluoresc 15:569–584. https://doi.org/10.1007/s10895-005-2830-x
225. Fercher A, Ponomarev GV, Yashunski D, Papkovsky D (2010) Evaluation of the derivates of phosphorescent Pt-coproporphyrin as intracellular oxygen-sensitive probes. Anal Bioanal Chem 396:1793–1803. https://doi.org/10.1007/s00216-009-3399-z
226. Lebedev AY, Cheprakov AV, Sakadžić S et al (2009) Dendritic phosphorescent probes for oxygen imaging in biological systems. ACS Appl Mater Interfaces 1:1292–1304. https://doi.org/10.1021/am9001698
227. Dmitriev RI, Ropiak HM, Yashunsky DV et al (2010) Bactenecin 7 peptide fragment as a tool for intracellular delivery of a phosphorescent oxygen sensor: cell-penetrating phosphorescent oxygen sensor. FEBS J 277:4651–4661. https://doi.org/10.1111/j.1742-4658.2010.07872.x
228. Dmitriev RI, Zhdanov AV, Ponomarev GV et al (2010) Intracellular oxygen-sensitive phosphorescent probes based on cell-penetrating peptides. Anal Biochem 398:24–33. https://doi.org/10.1016/j.ab.2009.10.048
229. Dmitriev RI, Ropiak HM, Ponomarev GV et al (2011) Cell-penetrating conjugates of coproporphyrins with oligoarginine peptides: rational design and application for sensing intracellular O_2. Bioconj Chem 22:2507–2518. https://doi.org/10.1021/bc200324q
230. Solomatina AI, Su S-H, Lukina MM et al (2018) Water-soluble cyclometalated platinum(ii) and iridium(iii) complexes: synthesis, tuning of the photophysical properties, and in vitro and in vivo phosphorescence lifetime imaging. RSC Adv 8:17224–17236. https://doi.org/10.1039/C8RA02742K
231. Zhang S, Hosaka M, Yoshihara T et al (2010) Phosphorescent light-emitting iridium complexes serve as a hypoxia-sensing probe for tumor imaging in living animals. Cancer Res 70:4490–4498. https://doi.org/10.1158/0008-5472.CAN-09-3948

232. Sun L, Li G, Chen X et al (2015) Azo-based iridium(III) complexes as multicolor phosphorescent probes to detect hypoxia in 3D multicellular tumor spheroids. Sci Rep 5:14837. https://doi.org/10.1038/srep14837
233. Li J, Chen H, Zeng L et al (2019) Mitochondria-targeting cyclometalated iridium(III) complexes for tumor hypoxic imaging and therapy. Inorg Chem Front 6:1003–1010. https://doi.org/10.1039/C9QI00081J
234. Yoshihara T, Murayama S, Tobita S (2015) Ratiometric molecular probes based on dual emission of a blue fluorescent coumarin and a red phosphorescent cationic iridium(III) complex for intracellular oxygen sensing. Sensors 15:13503–13521. https://doi.org/10.3390/s150613503
235. Zhang KY, Zhang T, Wei H et al (2018) Phosphorescent iridium(iii) complexes capable of imaging and distinguishing between exogenous and endogenous analytes in living cells. Chem Sci 9:7236–7240. https://doi.org/10.1039/C8SC02984A
236. Yoshihara T, Yamaguchi Y, Hosaka M et al (2012) Ratiometric molecular sensor for monitoring oxygen levels in living cells. Angew Chem Int Ed 51:4148–4151. https://doi.org/10.1002/anie.201107557
237. Yasukagawa M, Yamada K, Tobita S, Yoshihara T (2019) Ratiometric oxygen probes with a cell-penetrating peptide for imaging oxygen levels in living cells. J Photochem Photobiol A 383:111983. https://doi.org/10.1016/j.jphotochem.2019.111983
238. Hara D, Umehara Y, Son A et al (2018) Tracking the oxygen status in the cell nucleus with a hoechst-tagged phosphorescent ruthenium complex. ChemBioChem 19:956–962. https://doi.org/10.1002/cbic.201700685
239. Zhong W, Urayama P, Mycek M-A (2003) Imaging fluorescence lifetime modulation of a ruthenium-based dye in living cells: the potential for oxygen sensing. J Phys D Appl Phys 36:1689–1695. https://doi.org/10.1088/0022-3727/36/14/306
240. Martin A, Byrne A, Dolan C et al (2015) Solvent switchable dual emission from a bichromophoric ruthenium–BODIPY complex. Chem Commun 51:15839–15841. https://doi.org/10.1039/C5CC07135F
241. Byrne A, Jacobs J, Burke CS et al (2017) Rational design of polymeric core shell ratiometric oxygen-sensing nanostructures. Analyst 142:3400–3406. https://doi.org/10.1039/C7AN00753A
242. Koo Lee Y-E, Smith R, Kopelman R (2009) Nanoparticle PEBBLE sensors in live cells and in vivo. Annu Rev Anal Chem 2:57–76. https://doi.org/10.1146/annurev.anchem.1.031207.112823
243. Koo Lee Y-E, Ulbrich EE, Kim G et al (2010) Near infrared luminescent oxygen nanosensors with nanoparticle matrix tailored sensitivity. Anal Chem 82:8446–8455. https://doi.org/10.1021/ac1015358
244. Coogan MP, Court JB, Gray VL et al (2010) Probing intracellularoxygen by quenched phosphorescence lifetimes of nanoparticles containing polyacrylamide-embedded $[Ru(dpp(SO_3Na)_2)_3]Cl_2$. Photochem Photobiol Sci 9:103–109. https://doi.org/10.1039/B9PP00071B
245. Wu C, Bull B, Christensen K, McNeill J (2009) Ratiometric single-nanoparticle oxygen sensors for biological imaging. Angew Chem Int Ed 48:2741–2745. https://doi.org/10.1002/anie.200805894
246. Wang X, Gorris HH, Stolwijk JA et al (2011) Self-referenced RGB colour imaging of intracellular oxygen. Chem Sci 2:901. https://doi.org/10.1039/c0sc00610f
247. Kondrashina AV, Dmitriev RI, Borisov SM et al (2012) A phosphorescent nanoparticle-based probe for sensing and imaging of (intra)cellular oxygen in multiple detection modalities. Adv Funct Mater 22:4931–4939. https://doi.org/10.1002/adfm.201201387
248. Dmitriev RI, Borisov SM, Düssmann H et al (2015) Versatile conjugated polymer nanoparticles for high-resolution O_2 imaging in cells and 3D tissue models. ACS Nano 9:5275–5288. https://doi.org/10.1021/acsnano.5b00771
249. Shi H, Ma X, Zhao Q et al (2014) Ultrasmall phosphorescent polymer dots for ratiometric oxygen sensing and photodynamic cancer therapy. Adv Funct Mater 24:4823–4830. https://doi.org/10.1002/adfm.201400647
250. Koo Y-EL, Cao Y, Kopelman R et al (2004) Real-time measurements of dissolved oxygen inside live cells by organically modified silicate fluorescent nanosensors. Anal Chem 76:2498–2505. https://doi.org/10.1021/ac035493f
251. Dobroschke M, Geldmacher Y, Ott et al (2009) Cytotoxic Rhodium(III) and Iridium(III) polypyridyl complexes: structure–activity relationships, antileukemic activity, and apoptosis induction. ChemMedChem 4:177–187. https://doi.org/10.1002/cmdc.200800311

252. Zhang C, Qiu K, Liu C et al (2018) Tracking mitochondrial dynamics during apoptosis with phosphorescent fluorinated iridium(III) complexes. Dalton Trans 47:12907–12913. https://doi.org/10.1039/C8DT02918K
253. Leung PK-K, Lee LC-C, Ip TK-Y et al (2021) Luminescent rhenium(I) perfluorobiphenyl complexes as site-specific labels for peptides to afford photofunctional bioconjugates. Chem Commun. https://doi.org/10.1039/D1CC04740J
254. Stephenson KA, Banerjee SR, Besanger T et al (2004) Bridging the gap between in vitro and in vivo imaging: isostructural Re and ^{99m} Tc complexes for correlating fluorescence and radioimaging studies. J Am Chem Soc 126:8598–8599. https://doi.org/10.1021/ja047751b
255. Morais M, Zamora-Carreras H, Raposinho PD et al (2017) NMR insights into the structure-function relationships in the binding of melanocortin analogues to the MC1R receptor. Molecules 22:1189. https://doi.org/10.3390/molecules22071189
256. Rosita D, Dewit MA, Luyt LG (2009) Fluorine and rhenium substituted ghrelin analogues as potential imaging probes for the growth hormone secretagogue receptor. J Med Chem 52:2196–2203. https://doi.org/10.1021/jm8014519
257. Alves S, Paulo A, Correia JDG et al (2005) Pyrazolyl derivatives as bifunctional chelators for labeling tumor-seeking peptides with the fac-[M(CO)3]+ moiety (M = 99mTc, Re): synthesis, characterization, and biological behavior. Bioconjug Chem 16:438–449. https://doi.org/10.1021/bc0497968
258. Jiang H, Kasten BB, Liu H et al (2012) Novel, cysteine-modified chelation strategy for the incorporation of [M(I)(CO)(3)](+) (M = Re, (99m)Tc) in an α-MSH peptide. Bioconjug Chem 23:2300–2312. https://doi.org/10.1021/bc300509k
259. Li F, Liu Q, Liang Z et al (2016) Synthesis and biological evaluation of peptide-conjugated phthalocyanine photosensitizers with highly hydrophilic modifications. Org Biomol Chem 14:3409–3422. https://doi.org/10.1039/C6OB00122J
260. Ashki N, Hayes KC, Bao F (2008) The peroxynitrite donor 3-morpholinosydnonimine induces reversible changes in electrophysiological properties of neurons of the guinea-pig spinal cord. Neuroscience 156:107–117. https://doi.org/10.1016/j.neuroscience.2008.06.050
261. Bist I, Song B, Mosa IM et al (2016) Electrochemiluminescent array to detect oxidative damage in ds-DNA using $[Os(bpy)_2(phen\text{-}benz\text{-}COOH)]^{2+}$ /nafion/graphene films. ACS Sens 1:272–278. https://doi.org/10.1021/acssensors.5b00189
262. Kankanamage RNT, Ghosh AB, Jiang D et al (2020) Metabolites of tobacco- and E-cigarette-related nitrosamines can drive Cu^{2+}-mediated DNA oxidation. Chem Res Toxicol 33:2072–2086. https://doi.org/10.1021/acs.chemrestox.0c00027
263. Zhang W, Liu Y, Gao Q et al (2018) A ruthenium(II) complex–cyanine energy transfer scaffold based luminescence probe for ratiometric detection and imaging of mitochondrial peroxynitrite. Chem Commun 54:13698–13701. https://doi.org/10.1039/C8CC08061E
264. Zhang W, Zhao D, Zhang R et al (2011) A ruthenium(ii) complex based turn-on electrochemiluminescence probe for the detection of nitric oxide. Analyst 136:1867. https://doi.org/10.1039/c0an01003k
265. Vidanapathirana AK, Pullen BJ, Zhang R et al (2019) A novel ruthenium-based molecular sensor to detect endothelial nitric oxide. Sci Rep 9:1720. https://doi.org/10.1038/s41598-019-39123-3
266. Zhao N, Wu Y-H, Wang R-M et al (2011) An iridium(iii) complex of oximated 2,2′-bipyridine as a sensitive phosphorescent sensor for hypochlorite. Analyst 136:2277. https://doi.org/10.1039/c1an15030h

Topics in Current Chemistry (2022) 380:46
https://doi.org/10.1007/s41061-022-00400-x

REVIEW

Luminescent Metal Complexes as Emerging Tools for Lipid Imaging

Bradley J. Schwehr[1] · **David Hartnell**[1,2] · **Massimiliano Massi**[1] · **Mark J. Hackett**[1,2]

Received: 5 November 2021 / Accepted: 20 June 2022

Abstract

Fluorescence microscopy is a key tool in the biological sciences, which finds use as a routine laboratory technique (e.g., epifluorescence microscope) or more advanced confocal, two-photon, and super-resolution applications. Through continued developments in microscopy, and other analytical methods, the importance of lipids as constituents of subcellular organelles, signalling or regulating molecules continues to emerge. The increasing recognition of the importance of lipids to fundamental cell biology (in health and disease) has prompted the development of protocols and techniques to image the distribution of lipids in cells and tissues. A diverse suite of spectroscopic and microscopy tools are continuously being developed and explored to add to the "toolbox" to study lipid biology. A relatively recent breakthrough in this field has been the development and subsequent application of metal-based luminescent complexes for imaging lipids in biological systems. These metal-based compounds appear to offer advantages with respect to their tunability of the photophysical properties, in addition to capabilities centred around selectively targeting specific lipid structures or classes of lipids. The presence of the metal centre also opens the path to alternative imaging modalities that might not be applicable to traditional organic fluorophores. This review examines the current progress and developments in metal-based luminescent complexes to study lipids, in addition to exploring potential new avenues and challenges for the field to take.

Keywords Metal complexes · Lipid droplets · Phospholipids · Fluorescence · Microscopy · Cell imaging

Chapter 3 was originally published as Schwehr, B. J., Hartnell, D., Massi, M. & Hackett, M. J. Topics in Current Chemistry (2022) 380: 46. https://doi.org/10.1007/s41061-022-00400-x.

✉ Massimiliano Massi
m.massi@curtin.edu.au

✉ Mark J. Hackett
mark.j.hackett@curtin.edu.au

Extended author information available on the last page of the article

Published online: 17 August 2022

Abbreviations

cryo-TEM	Cryogenic transmission electron microscopy
DAPI	4′,6-Diamidino-2-phenylindole
DSPC	1,2-Distearoyl-sn-glycero-3-phosphocholine
ER	Endoplasmic reticulum
FAD	Flavin adenine dinucleotide
FLIM	Fluorescence lifetime imaging microscopy
FTIR	Fourier transform infrared
ICP-MS	Inductively coupled plasma mass spectrometry
LC	Ligand-centred
LD	Lipid droplet
LF	Ligand field
MLCT	Metal-to-ligand charge transfer
MRI	Magnetic resonance imaging
NAD	Nicotinamide adenine dinucleotide
nanoSIMS	Nanoscale secondary ion mass spectrometry
PLIM	Phosphorescence lifetime imaging microscopy
POSS	Polyhedral oligomeric silsesquioxane
SAXS	Small-angle X-ray scattering
STED	Stimulated emission depletion

1 Introduction

Lipids are ubiquitous across all cell systems, serving a wide range of biological functions; for example, they modulate cell and organelle membrane structure and fluidity, are a vital metabolic energy source, and hold important inflammatory signalling roles [1, 2]. Yet, despite the established importance of lipids, many specific pathways of lipid synthesis, storage and metabolism remain only partly understood. Apart from key roles for healthy tissue and cell function, lipids may be central to disease pathways, and indeed lipid biology is implicated in inflammatory disorders [3–5], neurodegenerative diseases [1, 6, 7] and conditions that result from lysosomal storage disorders [8–10]. With increasing recognition of the role that lipids hold in the fundamental function and health of cells and tissues, it is critical to develop robust and accessible imaging tools to study lipid biology at the cellular and subcellular level.

Traditional histochemical methods typically incorporate lipophilic dyes (stains) such as Oil Red O [11, 12], Sudan Black B [13, 14] and Luxol Fast Blue [15] (Fig. 1), and have long been used to study lipid distribution in cells and tissues. Such methods have provided and continue to provide substantial and valuable insight into lipid biology. Unfortunately, the majority of these stains are not compatible for imaging live cells and are only applicable to chemically fixed cells and tissues. Further, histochemical methods offer relatively limited options for visualising lipid co-localisation with other biochemical-, organelle- or cell-specific markers. Detection limits associated with light microscopy and histochemical stains are also typically poorer than fluorescence microscopy methods, which can largely be attributed to the

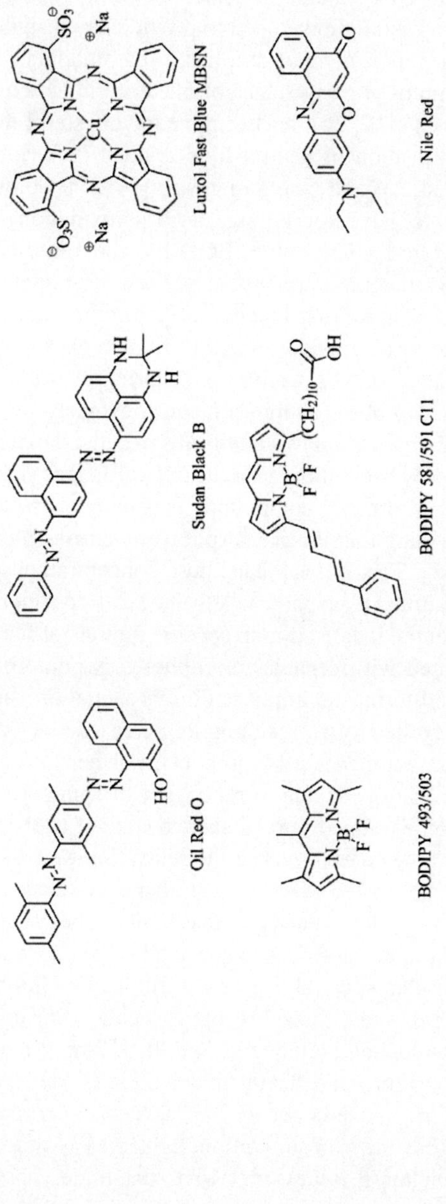

Fig. 1 Chemical structures of traditional histochemical dye examples, commercially available BODIPY dyes, and Nile Red

compatibility of the latter with the use of lasers. Lastly, while histochemical methods are well suited to detect neutral lipids, they are not well suited to detect polar lipids or to differentiate between lipid classes.

More recently, fluorescent probes have enabled scientists to greatly expand the fundamental knowledge of lipid biology. Boron-dipyrromethene-based molecular probes, more widely known as BODIPY (Fig. 1), have paved the way for countless explorations of lipid biochemistry in cells. BODIPY-based probes are relatively easy to solubilise in cell culture media and are generally compatible with live cell imaging. They have been successfully developed to provide key insight into lipid biosynthesis [16] and the connection of lipid droplets to the endoplasmic reticulum (ER) [17], lipid droplet dynamics at early stages of bacterial infection [18], lipid droplet function in skeletal muscle [19, 20], analysing endocytic sterol trafficking [21], and visualisation and quantification of neutral lipid content in various microalgae [22, 23], mammalian cells [24, 25] and other organisms [26, 27]. Commercially available BODIPY 581/591 C11 (Fig. 1) is used as a lipid peroxidation sensor in cells and membranes [28, 29]. More recently, other BODIPY analogues have been shown to be compatible with lipid imaging in ex vivo tissue sections, opening new opportunities to study lipid biology in animal models [30]. Another cause for recent excitement in the field of lipid-tissue imaging was the re-examination of Nile Red (Fig. 1), finding use as a fluorescent stain of myelin lipids during multiple sclerosis [31].

While BODIPY-based probes, along with the previously mentioned stains, are currently of widespread application to study lipids in cells, these compounds are still associated with some drawbacks that are intrinsic to their organic chemical nature [32]. The photophysics of organic compounds is usually characterised by typically small Stokes shifts, meaning that their absorption and emission profiles are overlapping to some extent [33]. This overlap can cause concentration quenching by reabsorption of emitted photons, especially when the local concentration of the probe increases due to preferential localisation in specific regions of the cell. Another typical issue often associated with organic fluorophores is photo-bleaching [34]. The amount of energy added to the fluorophore during excitation causes the species to become electronically excited, often resulting in increased reactivity, which can lead to the formation of non-luminescent by-products. Further to this, a spin-forbidden intersystem crossing might occur due to relaxation of spin-selection rules, causing the fluorophore to become trapped into an excited state of triplet multiplicity. These states are usually very long-lived, because the relaxation to the singlet ground state is also spin forbidden. Unwanted reactions can therefore occur from the long-lived triplet excited state. Lastly, the capacity to discriminate between the probe fluorescence signal and cellular autofluorescence is often problematic, especially if the two emissions occur in similar spectral regions. Cellular autofluorescence originates from endogenous fluorophores; these are biomolecules that contain π-conjugated systems such as the tryptophan amino acid ($\approx$400–450 nm), nicotinamide dinucleotide (NAD^+; $\approx$420–550 nm) and flavin dinucleotide (FAD; $\approx$475–575 nm) [35]. These drawbacks are often overcome by appropriate experimental design and/or instrumental features, but they are limitations nonetheless that at times cannot be avoided. To circumvent these issues, metal-based luminescent probes are emerging tools as alternatives to the use of organic fluorophores [36–39]. Designing and leveraging coordination chemistry to prepare luminescent metal-based compounds is an exciting research proposition, currently explored by many. Within the life sciences, the ability to design organic ligands to selectively coordinate transition metal ions has most frequently been harnessed to detect labile metal ions in biological systems. The literature contains an abundance of information on this topic, which is beyond the scope of this review. However, some prominent examples include "switch-on probes" that become luminescent upon coordination to Fe [40–42], Cu [43–45] or Zn [46, 47], or "emission-shift probes" which exhibit a shift in emission upon coordination to Fe [48], Cu [49] or Zn [50–53]. There are several excellent reviews on luminescent metal probes for the interested reader [54–57].

In addition to providing the capability to detect labile metal ions in biological systems, coordination chemistry can be leveraged to design or "build in" desired photophysical properties. In this respect, the typical photophysical features of luminescent metal-based complexes can be utilised to overcome the drawbacks associated with organic probes. However, it is important to point out that luminescent metal complexes are not necessarily advantageous or superior to organic fluorophores; instead, they can offer an alternative to current imaging tools. They are indeed less explored, and research in this area is expected to result in important development for expanding the optical imaging toolbox beyond the exclusive capability offered by organic fluorophores.

A summary of the main features that characterise the photophysical properties of metal complexes will be presented here. However, the reader is encouraged to consult some of the detailed reviews that have been published in the past years, if in need of a greater level of detail [58–60]. Luminescent metal complexes are characterised by more diverse photophysical properties, which are mainly dependent on the identity of the metal centre and the chemical structure of the ligands. Transition metal complexes bound to π-conjugated ligands can emit from excited states localised within the ligand (π–π*) or involving both the ligand(s) and the metal centre. The latter are known as charge transfer excited states, usually characterised by the promotion of an electron from the *d* orbitals of the metal centre to the empty π* orbitals of the ligands. These states are commonly referred to as metal-to-ligand charge transfer (MLCT) states. Depending on the metal complex, these charge transfer states can also be ligand-to-metal (LMCT), albeit these appear to be less common in the field of luminescent transition metal complexes. Depending on the nature of the conjugated ligand, ligand-to-ligand charge transfer (LLCT) states are also possible. An important feature of heavy metals is their relatively strong spin–orbit coupling, which tends to increase from the first to the third transition periods. A strong spin–orbit coupling favours intersystem crossing, which facilitates the transition from excited states of singlet multiplicity to excited states of triplet multiplicity. Intersystem crossing competes with fluorescent decay, meaning that a strong spin–orbit coupling favours phosphorescent emission, although in some cases a certain degree of fluorescence can still be observed. The emission from phosphorescent metal complexes is therefore characterised by a relatively large Stokes shift. This shift originates from the fact that the intersystem crossing between the singlet and triplet state of the same excited electronic configuration is accompanied by a stabilisation (Hund's rule) [61]. This stabilisation means that the energy gap between the excited state and ground state is reduced, and hence the emission profile is red-shifted. Another feature of phosphorescent decay is the relatively longer excited state lifetime when compared to fluorescence. This elongation originates from the fact that the transition between the triplet excited state and the singlet ground state is spin-forbidden, and hence it takes longer to occur with respect to the spin-allowed fluorescent transition between a singlet excited state and a singlet ground state. It should be noted though that the spin–orbit coupling of the metal centre also facilitates phosphorescent decay [61].

Along transition metals, lanthanoids have also been widely investigated as luminescent complexes [62]. The photophysical properties of lanthanoid complexes are however distinctly different. The electronic transitions involved in lanthanoid photophysics are intraconfigurational *f–f* transitions. While other transitions are possible, such as *f–d* or charge transfer, their discussion goes beyond the scope of this review. *f–f* Transitions are symmetry- and often spin-forbidden, the latter depending on the specific electronic configuration of the lanthanoid element, meaning that trivalent lanthanoids do not absorb photons efficiently. To overcome this issue, the adopted strategy is to coordinate chromophoric π-conjugated ligands to the lanthanoid centre to activate sensitisation via the so-called antenna effect. The conjugated ligand absorbs photons through π–π* electronic transitions, populating singlet excited states. The strong spin–orbit coupling of the lanthanoid centre promotes efficient intersystem crossing, populating the ligand triplet excited state, which then transfers its energy to the lanthanoid centre. Emission from lanthanoid elements spans from the visible (e.g. Eu, Tb) to the near infrared (e.g. Yb, Nd, Er). The emission profile, usually appearing as line-like peaks rather than broad bands, is dependent on the identity of the lanthanoid centre and relatively unaffected by the coordinated ligands (e.g. Eu is always red-emitting, while Tb is always green-emitting), due to the inner core nature of the valence 4*f* electrons. This is very advantageous because structural ligand variations to tune biological properties of the complex can be optimised independently from the photophysical properties, contrary to the general case of transition metal complexes. In common with the luminescence of transition metals, lanthanoid complexes are characterised by large Stokes shifts and relatively long excited lifetime decays. While lanthanoids offer considerable advantages, their labile nature requires the design of ligands with high denticity to confer stability in cellular environments.

The above-listed photophysical characteristics that distinguish metal complexes from organic species can be exploited to alleviate the drawbacks of traditional fluorophores. A large Stokes shift means that concentration quenching phenomena can be minimised and background autofluorescence can be conveniently excluded with the use of appropriate spectral filters. The capability for custom-designing probes with a large Stokes shift can enable multi-modal imaging with direct spectroscopic analyses, such as Raman microscopy, which would otherwise be inhibited by fluorescence overlap with the relatively weak Raman scattering bands [63]. Furthermore, the relatively long excited state lifetime decay is amenable to time-gated microscopy techniques, whereby the emission from the sample is collected after a time-delay from excitation. Since endogenous autofluorescence occurs within few tens of nanoseconds, time gating the signal acquisition ensures the detected emission mostly, if not exclusively, belongs to the probe [64]. The unique photophysical properties of luminescent metal complexes make them suitable for alternative imaging techniques such as super-resolution microscopy. Some notable techniques include stimulated emission depletion (STED) microscopy, stochastic optical reconstruction microscopy (STORM) and structured illumination microscopy (SIM). These techniques might demand more specialised probes but offer superior sub-diffraction resolution. Notable examples which demonstrate applicability with these techniques include Ru(II) complexes and, specifically for lipid imaging, Mn(II) complexes. An alternative technique is two-photon microscopy, where increased imaging depth and reduced phototoxicity are at the expense of decreased resolution in comparison to confocal microscopy. Two-photon excitation requires the simultaneous absorption of two photons with approximately twice the wavelength required for single-photon excitation. The use of longer-wavelength photons reduces photodamage to the sample, and provides greater penetration depth. However, due to the comparatively weak emission process, suitable probes must have high quantum yield, which might reduce the number of suitable metal probes.

The metal centre also offers a multi-modal analytical advantage by providing the ability to detect the metal through techniques such as MRI [65], nanoSIMS [66] and ICP-MS [67, 68]. Although MRI does not typically offer cellular spatial resolution, it is compatible with whole-body or whole-organ imaging in live animals, which can facilitate correlation between biochemical studies at the tissue level in vivo and subsequent subcellular studies ex vivo. Probes containing metal centres have increased electron density associated with the metal centre; therefore, detection with electron microscopies [69, 70] or ion microscopies (nanoSIMS) [71] is possible, which allows for imaging with spatial resolution below 100 nm. Quantification using these microscopies is difficult; however, opportunities for quantitative measurements such as cellular uptake are enabled using ICP-MS or other ICP analysis [67, 68].

Finally, and of most importance, through careful selection of ligands, one can enable differential targeting of the probe to a variety of regions within the cell. Multiple examples already exist in the literature for designing metal-based probes to target specific cellular organelles [72]. More recently, several groups have reported metal-based probes that are selective to lipids or to different lipid species (e.g. neutral vs polar lipids). Metal-based luminescent lipid probes now appear poised to contribute substantially to our understanding of lipid biology and the mechanisms through which lipids promote health or disease. In this review, we examine recent work on the serendipitous discovery or rational development of luminescent metal complexes as lipid probes. These probes have facilitated knowledge advancement of lipid biology and have provided future directions and opportunities for the field of lipid imaging. The review is structured with respect to biological application, specifically describing the use of metal complexes for the study of lipid droplets, phospholipid membranes and lastly generic lipid staining (not specific to any subcellular organelle or compartment). A summary of all the data collected from the literature regarding metal complexes for the imaging of lipids is reported in Table 1. The table is structured to capture a broad overview of the photophysical properties of the complexes and their use in imaging. From a certain point of view, the data highlight the infancy of this research, for which standardised protocols of testing are

Table 1 Summary of photophysical properties, imaging/staining conditions and biological properties of the complexes

Probe	Metal	Photophysical properties in solution					Imaging and staining conditions					Biological properties			Xlog P^a	References
		λabs (nm)	λem (nm)	Φaer	Taer	Solvent	Biological staining	Target	λex for imaging (nm)	Staining concentration	Staining media	Uptake mechanism	Cytotoxicity	log *P*		
1	Re	266	569	0.103	2373 (86%) ns, 575 (14%) ns	1% DMSO/water	Live cells, fixed cells, fixed tissue	Lipid droplets	830 (TPE)	10 µM	PBS	–	No signs of cytotoxicity (MTS)	–	1.601	[102]
2	Ir	470	616	0.078	–	MeCN	Living cells, living organism	Lipid droplets	580–670	10 µM	Serum-free media	Energy-dependent pathway	> 40 µM (IC50, 48 h, A549, HeLa, HepG2 and LO2 cells)	2.15	5.249	[67]
3	Ir	470	618	0.056	–	MeCN	Living cells, living organism	Lipid droplets	580–670	10 µM	Serum-free media	–	> 40 µM (IC50, 48 h, A549, HeLa, HepG2 and LO2 cells)	2.19	5.799	[67]
4	Ir	275	612, 664	0.09	0.83 µs	DCM	Live cells	Lipid droplets	405	10 µM	Pluronic F127/water (1/100)	Energy-dependent, mixed caveolae- and clathrin-mediated endocytosis	52.9 ± 3.7 µM (dark) 1.1 ± 0.1 µM (light) (IC50, 24 h, HeLa cells)	1.08 ± 0.05	–	[104]

Table 1 (continued)

Probe	Metal	Photophysical properties in solution					Imaging and staining conditions					Biological properties			Xlog P^a	References
		λabs (nm)	λem (nm)	Φaer	Taer	Solvent	Biological staining	Target	λex for imaging (nm)	Staining concentration	Staining media	Uptake mechanism	Cytotoxicity	log *P*		
5	Cu	309	510	0.005	250 (76%) ps, 2.79 (24%) ns	MeCN	Live cells	Lipid droplets	488	25 μM	–	–	–	–	5.915	[105]
6	Zn	390	630	0.44	0.49 (80%), 1.55 (20%) ns	DMSO	Live cells, fixed tissue	Lipid droplets	543, 790 (TPE)	2 μM	DMSO/culture medium (1/250)	Energy-dependent, clathrin-mediated endocytosis	8 μM (CCK-8, 90% cell viability)	1.8 ± 0.1	8.336	[108]
7	Au	256	580	0.045 ± 0.005	0.45 ± 0.02 μs	Hexane	Fixed cells, fixed tissue	Lipid droplets	405, 710 (TPE)	1 mg/mL	Isopropanol/water (1/1)	–	–	–	–	[109]
8	Ru	–	655	0.0044	117	Liposome	–	Phospholipids	440	–	Acetonitrile	–	–	–	5.398	[113]
9	Ru	–	–	–	–	–	Live cells, fixed cells	Phospholipids	488	2 μM	Buffer containing preformed LUVs or calf thymus D	–	–	–	6.830	[125]

Table 1 (continued)

Probe	Metal	Photophysical properties in solution					Imaging and staining conditions					Biological properties			Xlog P^a	References
		λabs (nm)	λem (nm)	Φaer	Taer	Solvent	Biological staining	Target	λex for imaging (nm)	Staining concentration	Staining media	Uptake mechanism	Cytotoxicity	log *P*		
10	Ru	–	655	0.0186	155	Liposome	–	Phospholipids	440	–	Acetonitrile	–	–	–	8.262	[113]
11	Ru	–	655	0.011	158	Liposome	–	Phospholipids	440	–	Acetonitrile	–	–	–	11.126	[113]
12	Ru	–	660	0.0031	86	Liposome	–	Phospholipids	–	–	–	–	–	–	6.003	[113]
13	Ru	–	660	0.0073	97	Liposome	–	Phospholipids	440	–	Acetonitrile	–	–	–	4.571	[113]
14	Ru	–	660	0.0104	105	Liposome	–	Phospholipids	440	–	Acetonitrile	–	–	–	7.435	[113]
15	Ru	–	–	–	–	–	–	Phospholipids	–	–	–	–	–	–	8.422	[114]
16	Ru	285	655	0.027	368 ns	Liposome in PBS	–	Phospholipids	–	–	–	–	–	–	7.258	[115]
17	Ir	250	600	0.084	610, 161, 70 ns	Liposome in PBS	–	Phospholipids	–	–	–	–	–	–	10.654	[115]

Table 1 (continued)

Probe	Metal	Photophysical properties in solution					Imaging and staining conditions					Biological properties			Xlog P^a	References
		λabs (nm)	λem (nm)	Φaer	Taer	Solvent	Biological staining	Target	λex for imaging (nm)	Staining concentration	Staining media	Uptake mechanism	Cytotoxicity	log *P*		
18	Pt	ca. 375	528	0.30	1.95 µs	HEPES buffer	Live cells	Phospholipids	400, 720 (TPE)	1 µg/mL	Culture media	–	10 µg/mL HeLa cells, ca. 90% cell viability, 25 h 10 µg/mL epithelial cells, ca. 70% cell viability 25 h	–	14.303	[116]
19	Ru	258	613	–	0.68 ± 0.017 µs	9/1 MeCN/DMSO	Live cells	Phospholipids	458	35 µM	Culture media	–	35 µM SP2 myeloma cells, ca. 70% cell viability	–	5.129	[63]
20	Ru	258	614	–	0.78 ± 0.022 µs	9/1 MeCN/DMSO	Live cells	Phospholipids	458	35 µM	Culture media	–	35 µM SP2 myeloma cells, ca. 70% cell viability	–	6.674	[63]
21	Re	380	550	–	–	–	Live cells	Phospholipids	405	100 µg/mL	–	–	–	–	1.343	[118]
22	Re	380	550	–	–	–	Live cells	Phospholipids	405	100 µg/mL	–	–	100 µg/mL MFC-7 cells, ca. 25% cell viability	–	3.491	[118]

Table 1 (continued)

Probe	Metal	Photophysical properties in solution					Imaging and staining conditions					Biological properties			Xlog P^a	References
		λabs (nm)	λem (nm)	Φaer	Taer	Solvent	Biological staining	Target	λex for imaging (nm)	Staining concentration	Staining media	Uptake mechanism	Cytotoxicity	log *P*		
23	Re	380	550	–	–	–	Live cells	Phospholipids	405	100 μg/mL	–	–	–	–	4.923	[118]
24	Ir	ca. 295	587	0.18	0.68 μs	DCM	Live cells	Phospholipids	488	5 μM	–	Energy-dependent, endocytosis	–	9.89	13.300	[68]
25	Al	394	643	0.42	4.3 ns	DMSO	Live cells	Phospholipids	561	1 μM	DMSO/culture medium (1/2000)	Membrane potential-dependent passive diffusion	2 μM HeLa cells, ca. 90% cell viability, 48 h	–	7.174	[119]
26	Tb	–	–	–	–	–	Live cells	Phospholipids	–	–	–	–	–	–	–	[120]
27	Ir	263	618	0.055	131 ns	DCM	Live cells	General lipophilic staining	403	20 μM	DMSO/culture medium (1/500)	Energy dependent pathway	Low (40 μM, 24 h) (MTS)	2.09 ± 0.06	3.904	[68]

Table 1 (continued)

Probe	Metal	Photophysical properties in solution					Imaging and staining conditions					Biological properties			Xlog P^a	References
		λabs (nm)	λem (nm)	Φaer	Taer	Solvent	Biological staining	Target	λex for imaging (nm)	Staining concentration	Staining media	Uptake mechanism	Cytotoxicity	log *P*		
28	Ir	256	542	0.042	140 ns	DCM	Live cells	General lipophilic staining	403	20 μM	DMSO/culture medium (1/500)	Energy dependent pathway	Low (40 μM, 24 h) (MTS)	2.01 ± 0.05	4.548	[68]
29	Ir	268	560	0.023	114 ns	DCM	Live cells	General lipophilic staining	403	20 μM	DMSO/culture medium (1/500)	Energy dependent pathway	Low (40 μM, 24 h) (MTS)	2.68 ± 0.08	5.8444	[68]
30	Ir	263	580	0.021	153 ns	DCM	Live cells	General lipophilic staining	403	20 μM	DMSO/culture medium (1/500)	Energy dependent pathway	Low (40 μM, 24 h) (MTS)	2.23 ± 0.04	5.484	[68]
31	Ir	260	575	0.036	497 ns	DCM	Live cells	General lipophilic staining	403	20 μM	DMSO/culture medium (1/500)	Energy dependent pathway	Low (40 μM, 24 h) (MTS)	2.57 ± 0.05	5.484	[68]
32	Ir	257	635	0.028	218 ns	DCM	Live cells	General lipophilic staining	403	20 μM	DMSO/culture medium (1/500)	Energy dependent pathway	High (40 μM, 24 h) (MTS)	1.68 ± 0.05	3.985	[68]

Table 1 (continued)

Probe	Metal	Photophysical properties in solution					Imaging and staining conditions					Biological properties			Xlog P^a	References
		λabs (nm)	λem (nm)	Φaer	Taer	Solvent	Biological staining	Target	λex for imaging (nm)	Staining concentration	Staining media	Uptake mechanism	Cytotoxicity	log *P*		
33	Ir	–	520	0.062	332 (88%) ns, 38 (12%)	0.1% DMSO/water	Live bacteria	General lipophilic staining	405	20 μM	DMSO/culture media (1/1000)	–	Low	–	4.907	[66]
34	Ir	–	552	0.057	628 (59%) ns, 189 (41%) ns	0.1% DMSO/water	Live bacteria	General lipophilic staining	405	20 μM	DMSO/culture media (1/1000)	–	Low	–	4.548	[66]
35	Ir	–	600	0.039	928 (75%) ns, 305 (25%) ns	0.1% DMSO/water	Live bacteria	General lipophilic staining	405	20 μM	DMSO/culture media (1/1000)	–	Low	–	6.128	[66]
36	Re	276	517	0.25	5 (57%) ns, 10 (43%) ns	DCM	Live bacteria, fixed bacteria	General lipophilic staining	–	–	–	–	–	–	4.963	[71]
37	Pt	282	515	0.28	11 ns	DCM	Live bacteria, fixed bacteria	General lipophilic staining	488	10 μM	DMSO/nutrient broth (1/1000)	–	–	–	10.580	[71]

Table 1 (continued)

Probe	Metal	Photophysical properties in solution					Imaging and staining conditions					Biological properties			Xlog P^a	References
		λabs (nm)	λem (nm)	Φaer	Taer	Solvent	Biological staining	Target	λex for imaging (nm)	Staining concentration	Staining media	Uptake mechanism	Cytotoxicity	log *P*		
38	Pd	256	585, 638	0.12	0.88 ns	Cyclohexane	–	General lipophilic staining	–	–	–	–	–	–	4.978	[121]
39	Pd	258	595, 650	0.50	4.46 ns	Cyclohexane	–	General lipophilic staining	–	–	–	–	–	–	6.376	[121]
40	Pd	253	587, 636	0.14	1.05 ns	Cyclohexane	–	General lipophilic staining	–	–	–	–	–	–	8.406	[121]
41	Pd	760	660	–	–	5% DMSO/water	Live cells, living organism	General lipophilic staining	561, 760 (TPE)	5 μM	DMSO/culture medium (1/200)	–	Low (< 20 mM) (MTT)	–	8.839	[123]
42	Cr	337	395, 415	–	0.2 (78.5%) ns, 6.0 (21.5%) ns	DCM	Live cells	General lipophilic staining	405, 810 (TPE)	50 μM	Ethanol/culture medium (1/50)	–	100 μM ca. 10% viability 4 h (Alamar Blue) and 15 min (LDH)	–	8.153	[124]

Table 1 (continued)

Probe	Metal	Photophysical properties in solution					Imaging and staining conditions					Biological properties			Xlog P^a	References
		λabs (nm)	λem (nm)	Φaer	Taer	Solvent	Biological staining	Target	λex for imaging (nm)	Staining concentration	Staining media	Uptake mechanism	Cytotoxicity	log *P*		
43	Mn	455	480	0.29	4.51 ns	DCM	Live cells, fixed tissue	General lipophilic staining	405, 760 (TPE)	5–10 μM	–	–	Low (80 μM, 24 h) (MTT)	–	12.560	[65]

[a]Values calculated using the Xlog*P* v2.0 method

not yet available. Therefore, probes are often tested on different cell lines, at different concentrations and with various incubation times. For the future development of this research field, some consistency will be required, especially in view of trying to establish structure–activity relationships that will guide an optimisation of metal-based probes based on rational design rather than serendipity.

2 Discussion

Phosphorescent metal complexes can be broadly divided into four classes. Hexa-coordinated second- and third-row transition metal complexes of nd^6 electronic valence configuration are amongst the most popular. Classical examples for this type of complexes are Re(I) tricarbonyl diimine [73–76], Ru(II)/Os(II) polypyridine [77–80] and cyclometalated Ir(III) complexes [81–83]. These complexes have received a great deal of focus in the past decades for their application both as diagnostic and therapeutic agents. The latter application can exploit photoactivated ligand exchange reactions as well as photo-generation of toxic reactive oxygen species such as $^1O_2$ within cells [84]. A second class of metal-based luminophores is represented by complexes of Pt(II), Au(III) and Au(I) [85–89]. For this class, most of the effort has been centred around Pt(II) complexes; however, complexes of Au are starting to emerge as promising candidates for imaging. The photophysical properties of these complexes are well known, and these species usually emit from metal-to-ligand charge transfer states of triplet multiplicity (^{3}MLCT) or metal-perturbed π–π^* ligand-centred emission (^{3}LC). This knowledge allows for fine-tuning of their photophysical and photochemical properties to optimise their use in life sciences. Complexes of first-row transition metals are usually non-luminescent, aside from tetrahedral complexes of Cu(I). The lack of luminescence is attributed to low-lying ligand field excited states (^{3}LF) that are non-emissive. While paradigm shifts in the use of first-row transition metal luminophores have occurred very recently, these developments are still too young to have found tangible applications in the life sciences [90, 91]. However, examples of lipid-based probes that contain metals such as Cr, Mn, Cu and Zn have been reported and will be discussed in this review. The final class of metal-based probes is represented by lanthanoid complexes [92–96].

Metal-based complexes can be conjugated to biological vectors to impart cellular specificity [77, 97]. Alternatively, their chemical nature can be fine-tuned in order to define biological properties such as membrane permeability, organelle localisation and/or interaction with specific biomolecules. There are currently rules of thumb available for the design of probes to target various regions of the cell, and these usually rely on charge and lipophilicity. The charge is important for solubility in aqueous medium, which is essential during the incubation stage, and to favour internalisation by exploiting membrane potentials. Many investigated probes are cationic in nature, albeit more recently neutral probes have started to gain attention due to the fact that their biological properties and cytotoxicity might be fundamentally different. The lipophilicity also can assist during internalisation, for example by favouring passive diffusion through the cell membrane, and preferential interaction with specific organelles.

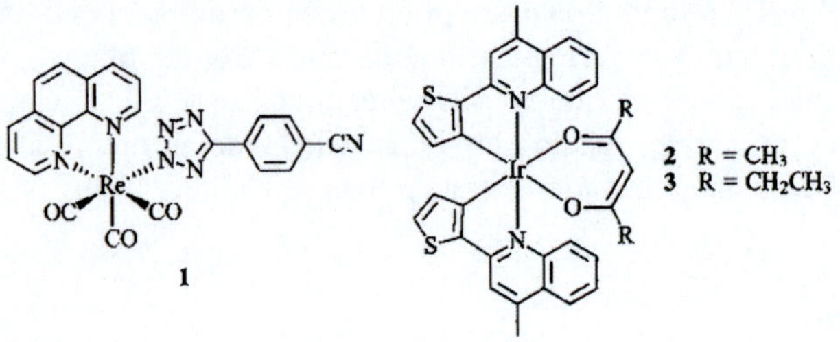

Fig. 2 Chemical structures of **1–3**

2.1 Lipid Droplets

Lipid droplets (LDs) are cytoplasmic organelles that store excess lipids such as sterol esters and triacylglycerols [98]. Their exterior consists of a phospholipid monolayer, and depending on droplet depletion or growth, metabolic intermediates such as fatty acids and unesterified sterols [99]. At their core, LDs contain neutral lipids such as triglycerides and cholesterol esters [100]. LDs and their associated processes are vital for lipid homeostasis, and as such, LDs have been implicated in conditions such as type 2 diabetes, cardiovascular disease, Alzheimer's disease and a variety of cancers, which all exhibit lipid imbalances and alterations during their progression [101]. Therefore, it is critical to have accessible methods to study LD composition and their roles in lipid homeostasis.

Targeting LDs can be achieved with lipophilic probes that are able to penetrate cells, where localisation within the neutral core of LDs occurs via hydrophobic interactions. Probes which target LDs have log*P* values spanning a large range, between 1.5 and 8.5, with a general trend of high-lipophilicity probes displaying specificity for LDs due to localisation within the core of LDs. As lipophilicity decreases, specificity to LDs is reduced, and more diffuse localisation with targeting of other organelles is observed. On the other hand, highly lipophilic compounds are characteristically difficult to solubilise in culture media and often require the use of cytotoxic solvents (such as dimethyl sulfoxide [DMSO], acetonitrile or ethanol), which can damage the cell membrane and produce misleading results.

2.1.1 Imaging Lipid Droplets with Metal-Based Probes

2.1.1.1 Probes Incorporating Third-Row Transition Metals (Ir, Re) In 2014, the luminescent metal complex **1** was reported by Bader et al. to stain LDs (Fig. 2) [102]. The neutral rhenium(I) tricarbonyl 1,10-phenanthroline complex bound to 4-cyanophenyltetrazolate was shown to localise in regions with high concentrations of polar lipids such as phosphatidylethanolamine, sphingomyelin, sphingosine and lysophosphatidic acid in the LDs of mammalian adipocytes [103]. The low toxicity combined with the high photostability of **1** enabled its use for live imaging of *Drosophila* larval adipose tissues. The moderately lipophilic probe revealed lipid distribution when imaged by confocal microscopy, displaying higher affinity for polar lipids at

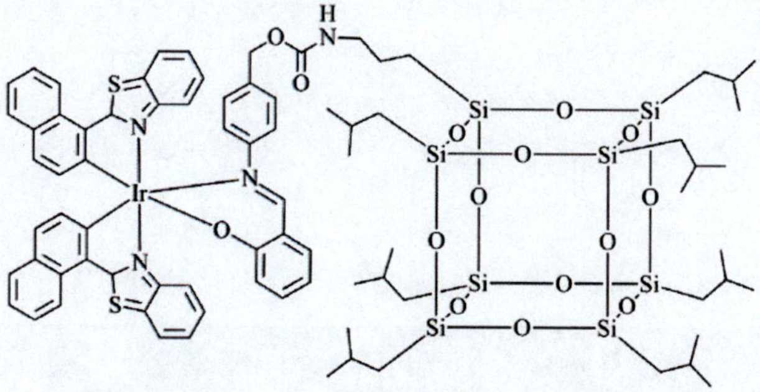

4

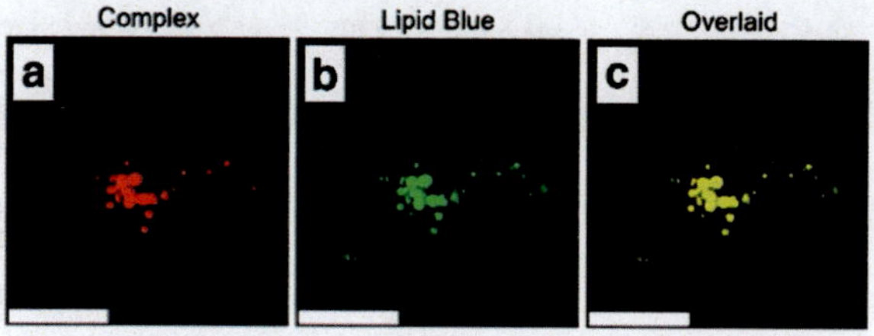

Fig. 3 Chemical structure of **4**. Bottom panel shows **4** (red) (**a**) incubated in 3T3-L1 differentiated adipocytes and co-stained with Lipid Blue (green) (**b**) and merged image (**c**). Scale bar: 25 μm. Reprinted (adapted) with permission from Zhu et al. [104]. Copyright 2021 American Chemical Society

the edges of the LDs compared to neutral lipids of the LD core. Correlative imaging using Raman and FTIR microscopy confirmed this finding, where a strong correlation between the probe and polar lipids of the LDs was shown.

Two lipophilic neutral cyclometalated Ir(III) complexes, **2** and **3**, have shown to be suitable probes for LD visualisation in live cells and organisms [67]. When applied to A549 cells and co-stained with BODIPY 493/503, significant co-localisation was observed, confirming LD accumulation. Featuring large two-photon absorption cross sections and long excited state lifetimes, the suitability of these probes for two-photon phosphorescence lifetime imaging microscopy (PLIM) was demonstrated, allowing for the elimination of background autofluorescence. Optical imaging followed by ICP-MS measurements reveals higher uptake efficiency of **2** in comparison to **3**. Complex **2** was shown to be compatible with living organisms such as zebrafish for the visualisation of lipid metabolism in vivo using optical imaging.

Probe **4** is another example of a neutral cyclometalated Ir(III) complex that is specific to LDs (Fig. 3) [104]. The probe features a bulky polyhedral oligomeric silsesquioxane (POSS) cage, which, when compared to the POSS-free counterpart, has a higher lipophilicity (experimentally determined $\log P = 1.08$ and 0.63, respectively). When applied to HeLa cells, the POSS unit showed unaltered specificity to lipid droplets, which is demonstrated by co-localisation studies using Lipid Blue. However, **4** has low water solubility, requiring the use of a triblock copolymer to make a solid dispersion in aqueous solution for cell incubation. Despite the addition of the bulky POSS unit, **4** retains high cell uptake via energy-dependent pathways.

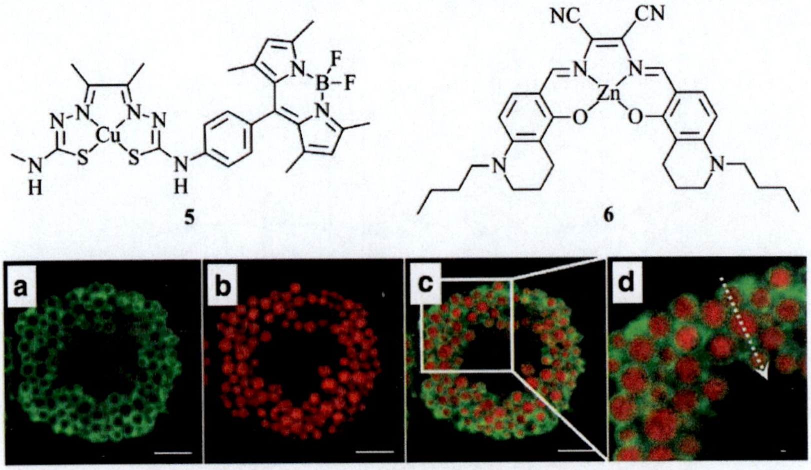

Fig. 4 Chemical structures of **5–6**. Analysis of adipocytes with immunofluorescence (green, **a**) and uptake of compound **5** (red, **b**), merged images of (**a**, **b**) (**c**, **d**). Reproduced with permission from Tang et al. [108]

The attachment of the POSS unit resulted in a reduced cytotoxicity of the complex in the dark, but its phototoxicity upon radiation provides potential use in photodynamic therapy. Complex **4** was applied to cancer cells such as Neuro-2a, a fast-growing neuroblastoma cell line, revealing significantly increased emission from LDs in comparison to normal cells under the same experimental conditions.

2.1.1.2 Probes Incorporating First-Row Transition Metals (Cu, Zn) Hickey et al. report the synthesis of probe **5**, a BODIPY-labelled copper bis(thiosemicarbonato) complex (Fig. 4) [105]. These complexes are generally weakly fluorescent, but have gained interest due to their potential neuroprotective capability [106]. Enabling the detection of this probe often requires the inclusion of a fluorescent tag [107] such as BODIPY, providing **5** with high quantum yield and high stability under physiological conditions. Fluorescence lifetime imaging microscopy (FLIM) is enabled due to the difference in emission lifetime of **5** compared to the copper-free ligand. It is shown that the probe localises within lysosomes of neuronal cells and both LDs and lysosomes in secondary M17 neuroblastoma cells. Once accumulated within LDs, FLIM reveals that some Cu(II) ions are released and the liberated ligand coordinates to bioavailable Zn(II).

Probe **6** (Fig. 4) features Zn(II) bound by a salen (*N*,*N*′-bis(salicylidene)ethylenediamine) ligand, a non-fluorescent complex in an aqueous environment with significant increase of fluorescence in lipophilic environments [108]. Low cytotoxicity and high photostability is reported, enabling its imaging in living adipose cells. Immunofluorescence using perilipin-1 antibodies co-stained with probe **6** in HeLa cells shows that **6** specifically localises within the hydrophobic core of LDs, and co-staining with BODIPY 493/503 shows comparatively higher specificity of **6** for LDs. This behaviour is surprising when considering the reported log*P* value is 1.8;

the general trend would anticipate non-specific staining of LDs and polar lipids. Interestingly, when the log*P* is calculated, a value of 8.3 is obtained, which is consistent with the trend, but there is a large discrepancy with the experimental value. The suitability to image LDs stained with **5** in tissues is demonstrated in rat subcutaneous adipose tissue incubated with the probe and subsequent imaging by two-photon microscopy to reveal LDs. Further, the potential for tracking LD biogenesis during adipogenesis of rat preadipocytes is demonstrated, indicating potential for further understanding of LD homeostasis.

2.1.1.3 Metal Clusters Mononuclear metal complexes used for biological imaging have received criticism due to being prone to quenching by oxygen and/or water, both which are unavoidable in biological systems [109]. Koshel et al. have synthesised polynuclear homo- and heterometallic gold–alkynyl clusters, such as **7**, which are shown to be insensitive to oxygen and water quenching, while maintaining the favourable spectral properties of mononuclear metal complexes (Fig. 5) [110]. The clusters were evaluated in a broad range of biological applications including various adipose tissues from a range of animals (mouse, chicken, pigeon) as well as Hep G2 cells. The samples were counterstained with DAPI and Oil Red O before being imaged by two-photon microscopy. Despite having a relatively high molecular size, **7** is able to penetrate cells to reveal high affinity to LDs in several tissue types. The suitability of the probe for phosphorescence lifetime imaging is demonstrated in chicken subcutaneous adipose tissues, allowing for the elimination of background signal.

2.2 Phospholipids

Phospholipids are the most abundant lipids comprising the bilayer structure responsible for cell membranes [111]. Additionally, phospholipids also facilitate signalling and energy storage. Phospholipids have two hydrophobic chains that are joined at a polar head. Variations in chain length, saturation and incorporation of cholesterol create alterations in molecular packing and consequently alter the fluidity of the membrane. Mimicking these structural features in the design of probes has been successful in furthering the understanding of this complex relationship. Typically, probes targeting phospholipids incorporate long hydrophobic alkyl chains appended to charged metal complexes, favouring the incorporation of the probe within the lipids of the bilayer. Probes with high affinity for membranes are unlikely to diffuse across the plasma membrane; however, if the probe is internalised via energy-dependent pathways, other lipids and organelles within the cell can be targeted, such as the intracellular membranes associated with the Golgi apparatus, ER and lysosomes.

Bolstered by advancements in phospholipidomics, new evidence indicates that changes in phospholipid composition, metabolism and distribution within tissues, cells and fluids are linked to diabetes, cancer and cardiovascular diseases [112]. To further explore these links, visualisation methods are vital to study phospholipids and uncover their connection to these conditions.

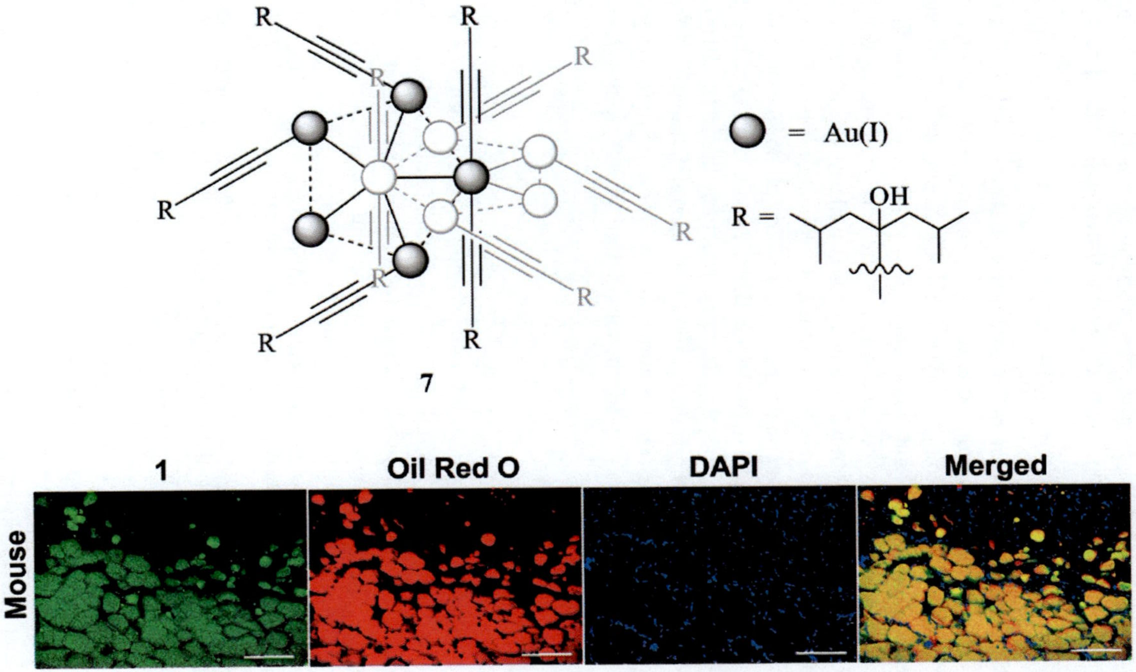

Fig. 5 Chemical structure of **7**. The decanuclear molecule contains two interlocked, perpendicular 5-membered rings. The ring which is perpendicular to the page has been greyed out for clarity. Bottom panel shows co-localisation of Oil Red O (red) with **7** (green), counterstained with DAPI (blue) n subcutaneous adipose tissues of mouse. Scale bars: 100 µm [109]. Reprinted (adapted) with permission from Koshevoy et al. [110]. Copyright 2021 American Chemical Society

2.2.1 Imaging Phospholipids with Metal-Based Probes

2.2.1.1 Probes Incorporating Second- and Third-Row Transition Metals (Ir, Re, Ru, Pt) The targeting of lipid bilayers such as the plasma membrane employs the use of ligands with long alkyl chains bound to cationic metal complexes. Such probes based on cationic Ru(II) complexes have been investigated in liposomes for their interactions with phospholipid bilayers in order to further understand the structure and dynamics of lipid membranes (Fig. 6). Ru(II) bound to dipyridophenazine-type ligands **8–14** feature varying alkyl chain lengths ($n = 2, 4, 6, 10$), and the number of alkyl chains (one or two) was investigated using flow linear dichroism with phospholipid vesicles as membrane models [113]. Probe **13**, with alkyl chain length of six C atoms, was found to be optimal where the alkyl chain inserts parallel to the lipid chain of the liposome bilayer, and the charged Ru(II) acts as a polar head group.

Bonnet et al. [114] report the synthesis of **15**, a thioether–cholestanol hybrid ligand coordinated to Ru(II) via its sulfur atom (Fig. 6). Complex **15** was applied to liposomes containing negatively charged lipids. Photophysical measurements and cryo-TEM reveal that electrostatic interactions of the probe as well as the insertion of the nonpolar tail are involved in the probe–membrane interaction. Cleavage of the Ru–S bond can be triggered by irradiation, causing the Ru(II) complex to diffuse away. The authors conclude that this may be a useful method for drug delivery. Mechler et al. [115] synthesise Ru(II) and Ir(III) lipid-mimetic surfactants **16** and **17** (Fig. 6), which are designed to more closely mimic the phospholipid membrane, avoiding the alteration of its biophysical properties. Complexes **16** and **17** were applied to dimyristoyl-phosphatidylcholine liposomes and imaged by confocal microscopy, highlighting membrane localisation. Small-angle X-ray scattering (SAXS) measurements and atomic force microscopy (AFM) imaging are used to demonstrate that the bilayer is unaltered by the addition of the probes. The mimetic approach is demonstrated as an effective tool for biophysical characterisation of lipid membranes.

A water-soluble and amphiphilic Pt(II) complex, **18**, has been developed by Koo et al. [116] for in vitro plasma membrane staining (Fig. 7). The complex features a trisulfonated triphenylphosphine bound to the Pt(II) for enhancement of water solubility, and a C18 alkyl chain on the cyclometalated ligand to increase lipophilicity. Two-photon confocal microscopy of the complex incubated within live HeLa cells reveals exclusive plasma membrane staining, observing no subcellular migration after 3 h. The poor cellular uptake and plasma membrane specificity may be rationalised by the negatively charged sulfonate groups, which inhibit internalisation. On the other hand, the long hydrophobic alkyl chain favourably interacts with the membrane.

Cosgrave et al. [63] report probes **19** and **20**, which are Ru(II) complexes featuring dipyridophenazine ligands and an ancillary 2-(4-carboxyphenyl)-1H-imidazo[4,5-*f*][1,10]phenanthroline ligand (Fig. 8). When applied to live SP2 myeloma cells, **19** does not enter the cell, and exclusive staining of the outer cell membrane is observed. However, **20** includes a cell-penetrating polyarginine peptide which is shown to facilitate passage into the cell via endocytosis, resulting in internal membrane staining of subcellular structures. Detection of the probe using

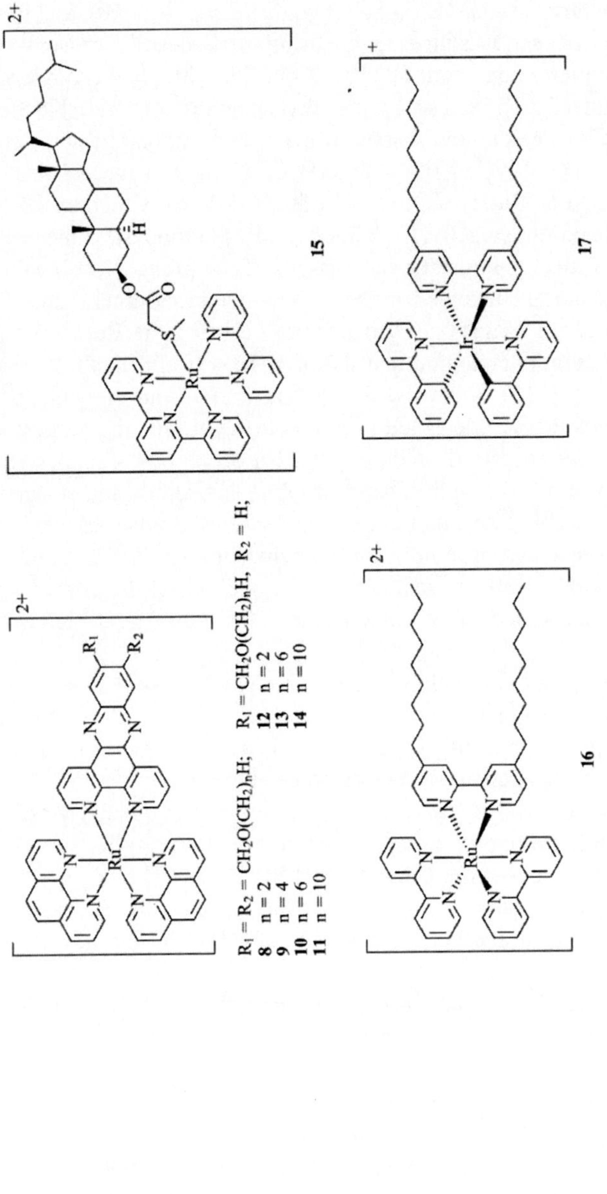

Fig. 6 Chemical structures of **8–17**

Fig. 7 Chemical structure of **18**

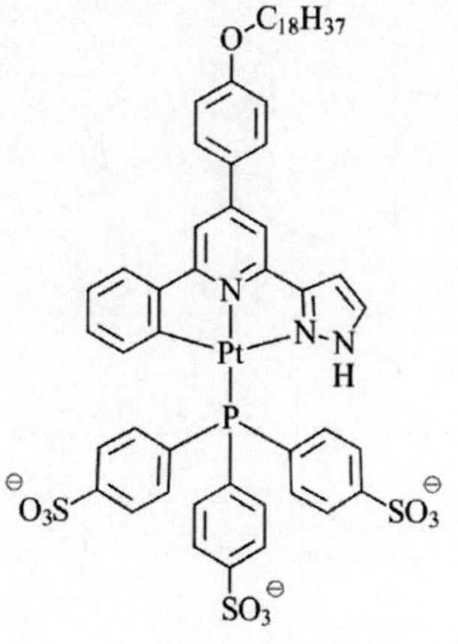

18

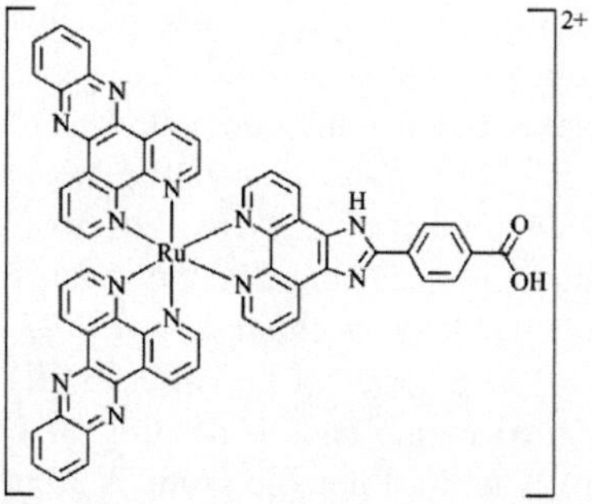

19

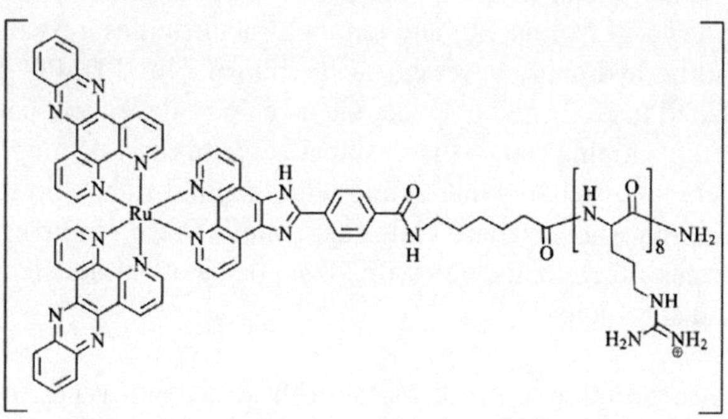

20

Fig. 8 Chemical structures of **19–20**

resonance Raman spectroscopy is enabled due to its large Stokes shift, whereby an interference-free resonance Raman spectra can be collected. Applying this technique to live myeloma cells validates the localisation of **19** and **20** within the cell membranes. In a more recent study, **20** was utilised as the first example of a metal

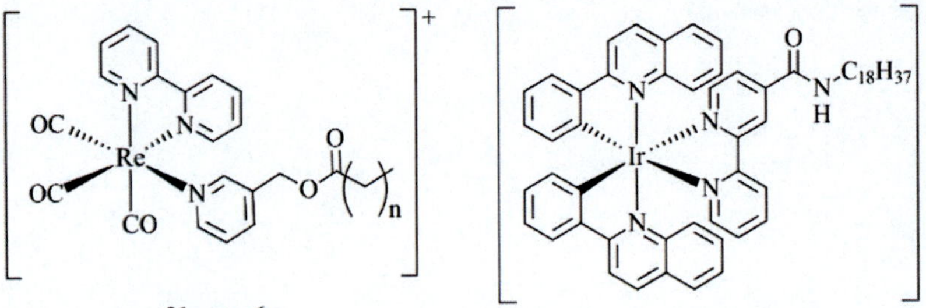

21 n = 6
22 n = 12
23 n = 16

24

Fig. 9 Chemical structures of **21–24**

complex for STED microscopy, owing to the high photostability, red-shifted emission, and low cytotoxicity [117]. By performing STED imaging of HeLa cells, **20** highlights the tubular structure of the ER with high resolution and with improved performance in comparison to currently used organic fluorophores. This example demonstrates the use of cell-penetrating peptides as a means to enhance internalisation for those metal complexes that display poor cellular uptake.

Fernández-Moreira et al. [118] report a series of cationic Re(I) probes, **21–23**, with varying alkyl chain lengths ($n=6$, 12, 16) (Fig. 9). When applied to MCF-7 cells, the shorter chain analogues, **21** and **22**, show general cytoplasmic staining, whereas the complex with the longest chain ($n=16$), **23**, shows high membrane specificity. Lo et al. [69] report a series of lipophilic Ir(III) complexes, which were applied to artificial biological membranes consisting of DSPC (1,2-distearoyl-*sn*-glycero-3-phosphocholine). The most notable probe is **24**, a cationic Ir(III) complex containing a C18 alkyl chain (Fig. 9). The probe was shown to incorporate within the liposomes by cryo-TEM imaging and photophysical studies, revealing that **24** is embedded within the hydrophobic region of the phospholipid DSPC. Interestingly, when **24** is incubated in HeLa cells, uptake via an energy-dependent pathway results in general lipophilic staining rather than exclusive membrane staining. The authors propose that the complex is likely interacting with the lipid-rich regions of the endoplasmic reticulum, mitochondria and Golgi apparatus. These examples nicely illustrate the link between functionalisation with alkyl chains of variable lengths, cellular internalisation and specificity.

2.2.1.2 Probes Incorporating p-Block Metals (Al) A salen ligand coordinated to Al(III), **25**, has been developed as a probe with selectivity towards phospholipids (Fig. 10) [119]. The selective binding of **25** to various phospholipids was investigated, revealing the Al(III) complex covalently binds to mono-negatively charged and oxo-containing lipids, likely due to the hard Lewis acidity and high oxophilicity of Al(III). This specificity is further demonstrated by the addition of the probe to an artificial fluorescent liposome, whereby the green fluorescence of the liposome showed strong correlation with the red fluorescence of **25**. When applied to living HeLa cells, it was found that the probe binds to the phospholipids of the Golgi apparatus before

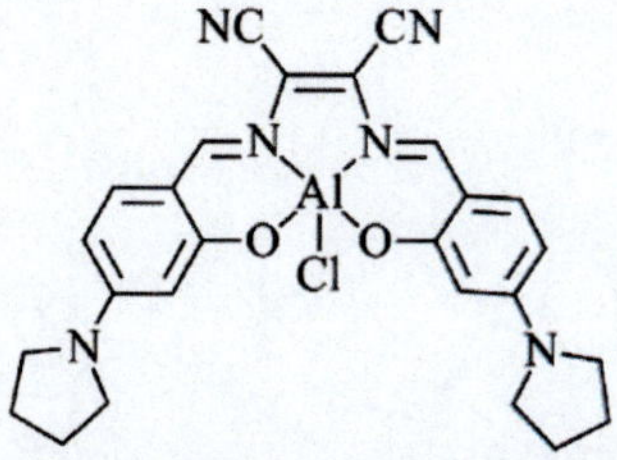

25

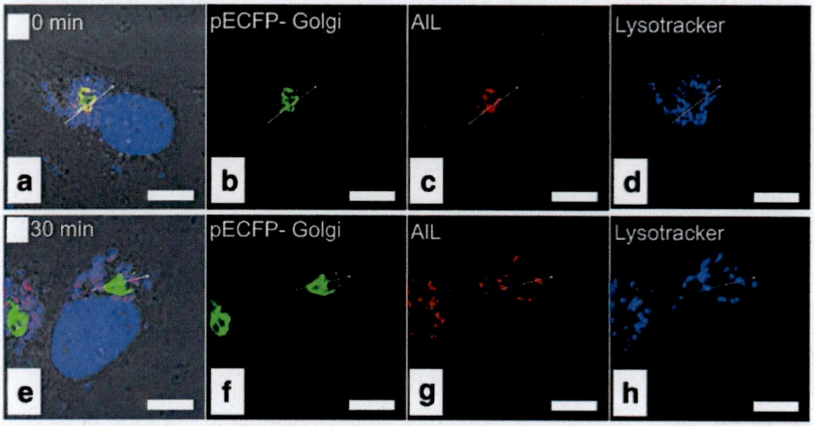

Fig. 10 Chemical structure of **25**. Bottom panel showing merged brightfield images of HeLa cells, incubated with (**a**, **e**) of pECFP-Golgi (green) (**b**, **f**), **25** (red) (**c**, **g**) and LysoTracker® Deep Red (blue) (**d**, **h**) at 0 min (**a–d**) and 30 min (**e–h**) [119]. Scale bar = 10 μm. Figure reproduced from Tang et al. [119]. Under the Creative Commons Attribution 3.0 Unported Licence

it is transported to the lysosomes via membrane vesicle trafficking. The potential of **25** as an alternative to the use of fluorescent proteins is highlighted, specifically as a tool for the investigation of the vesicle trafficking from the Golgi apparatus to the lysosomes.

2.2.1.3 Probes Incorporating Lanthanoids (Tb) An emissive Tb(III) complex, **26**, featuring an azaxanthone chromophore as the light-harvesting group featuring a dodecylamine side chain has been investigated for applications in live cell imaging (Fig. 11) [120]. It was found that the complex is cytotoxic when applied to NIH-3T3 cells, which was ascribed to a disruption of membrane integrity. This feature hinders the application of the probe in live NIH-3T3 cell studies. However, lower cytotoxicity in CHO cells was observed, indicating the potential use as a general membrane stain. This example, along with previous ones, highlights the importance of assessing the properties of molecular probes in multiple cell lines, rather than drawing conclusions on structure–activity relationships after incubation within a single cell type.

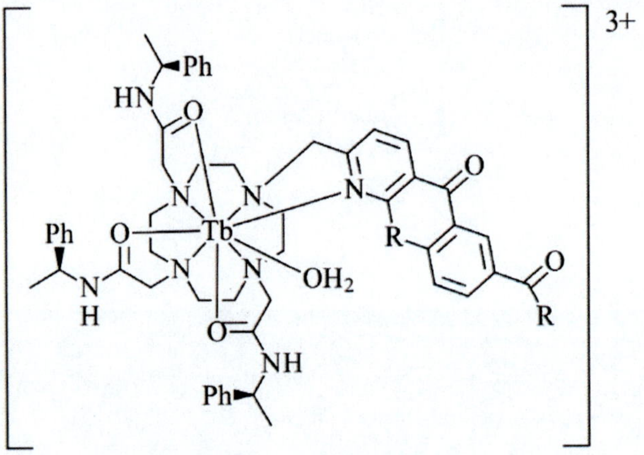

26 R = $NHC_{12}H_{25}$

Fig. 11 Chemical structure of **26**

2.3 General Lipophilic Compounds

The design of lipid-targeting metal complexes often results in non-specific lipid binding. While lack of specificity might be regarded as a negative feature, non-specific lipid staining is desirable for studying biochemical processes in injury and disease. Non-specific lipid staining is useful to contrast lipidic alterations with cellular processes. Organic probes such as BODIPY [30] and Nile Red [31] have been used to measure bulk lipid alterations in brain white matter in traumatic brain injuries and multiple sclerosis. However, the emission of these compounds occupies a populated area of the spectrum and are not always suited for specific applications due to emission overlap with autofluorescence. Therefore, a platform of molecular probes displaying non-specific lipid binding with tunable emission across the entire visible range would be highly desirable.

2.3.1 Staining of General Lipophilic Compounds

2.3.1.1 Probes Incorporating Second- and Third-Row Transition Metals (Ir, Re, Pt, Pd) While trends in localisation of probes are observed based on physical properties such as lipophilicity and charge, these properties alone are not enough to rationalise the behaviour when applied to complex systems such as cells and tissues. Caporale et al. report the synthesis and biological investigation of probes **27–32** (Fig. 12), a family of neutral cyclometalated Ir(III) tetrazolato complexes, as well as their methylated cationic analogues [68]. It was found that variations to lipophilicity and charge by altering the cyclometalated phenylpyridine and tetrazolato ligands do not follow expected trends in localisation or uptake behaviour in live cells. For example, the measurement of cellular uptake by ICP-MS shows comparable uptake of the positively charged probe, **32**, in comparison to the neutral probe, **30**, whereas **27** shows a significantly lower uptake. When applied to live H9c2 cells, two-photon confo-

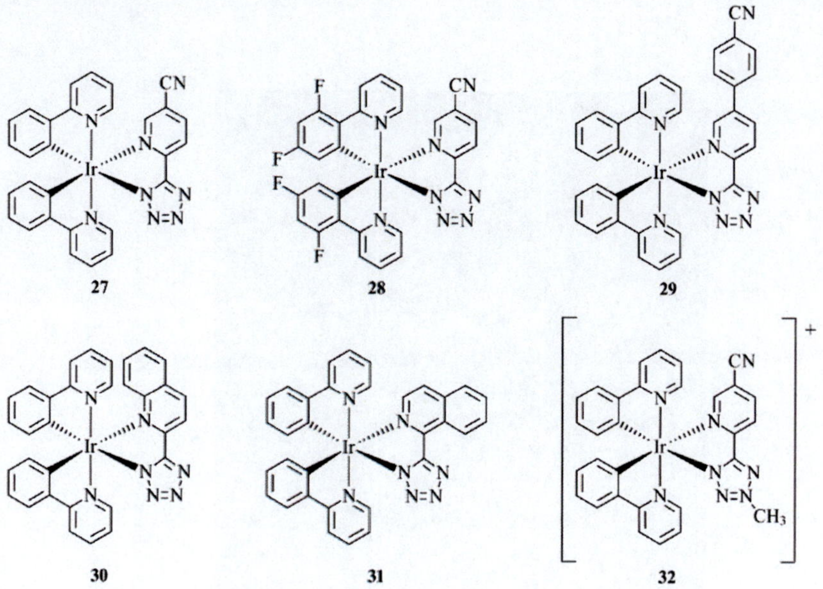

Fig. 12 Chemical structures of **27–32**

cal microscopy reveals that the complexes **27–32** co-stained with BODIPY 500/510 or ER-Tracker display both LD and ER localisation. Comparing the experimentally determined log*P* of **27** and **30**, 2.09 and 2.23, respectively, it would be expected that the higher log*P* would result in higher affinity to LDs; however, the opposite was found, where **30** displayed a lower affinity for LDs and a higher affinity for the ER. In addition, the charged probes exhibited mitochondrial localisation and higher cytotoxicity compared to the neutral counterparts, indicating a trend that utilising charged species for enhanced membrane permeability may be at the expense of cytotoxicity and alteration of specificity.

In addition to targeting lipids in mammalian cells, organometallic probes for targeting lipids in bacteria based on Re(I), Ir(III) and Pt(II) have been explored (Fig. 13). Neutral Ir(III) tetrazolato complexes (**33–35**) with tunable emission spanning from 520 to 600 nm with the same excitation wavelength have been shown to efficiently label lipid vacuoles in live *B. cereus* [66]. Subcellular localisation of the nontoxic probes was confirmed with confocal Raman imaging in addition to co-staining studies using BODIPY 493/503. Probes based on Re(I) (**36**) and Pt(II) (**37**) containing a 4-amino-1,8-naphthalimide moiety have been designed and synthesised as potential multi-functional probes for bacteria imaging [71]. However, the low stability of **36** rendered it unsuitable for imaging. The suitability of **37** for structured illumination microscopy (SIM) on live *Bacillus cereus* was demonstrated, revealing LD localisation. The presence of Pt(II) in the probe allowed for subsequent correlative analysis of the sample by nano-scale secondary ion mass spectrometry (nanoSIMS).

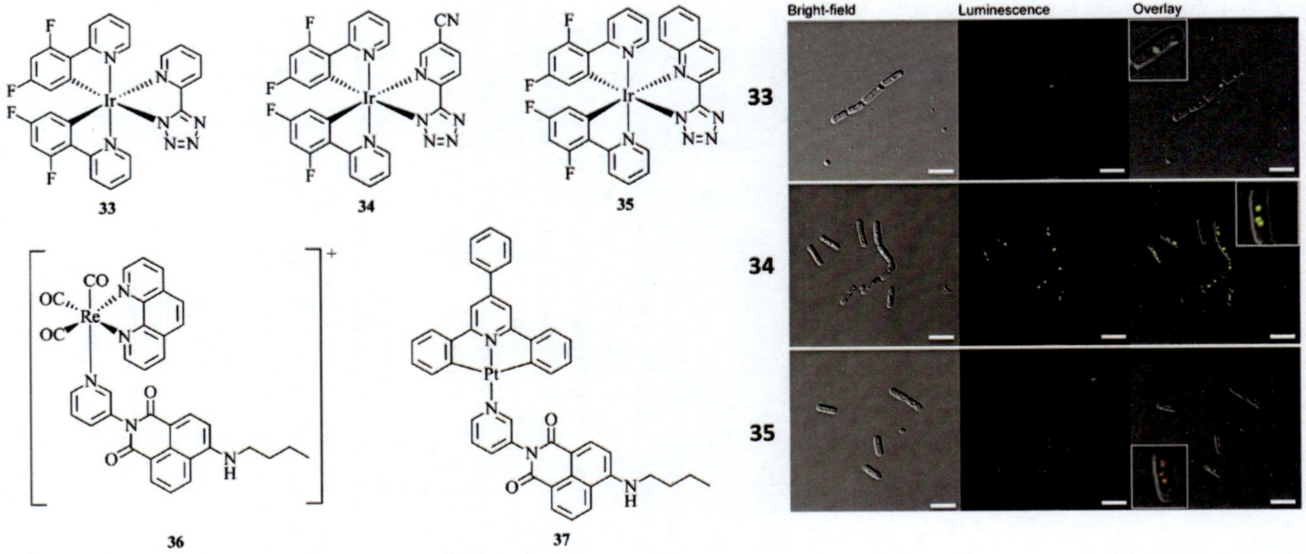

Fig. 13 Chemical structures of **33**–**37**. Application of compounds **33**–**35** (yellow) to image lipophilic structures in living bacteria (*B. cereus*) using confocal fluorescence microscopy [66]. Scale bar = 5 μm. Reproduced with permission from Ranieri et al. [66]

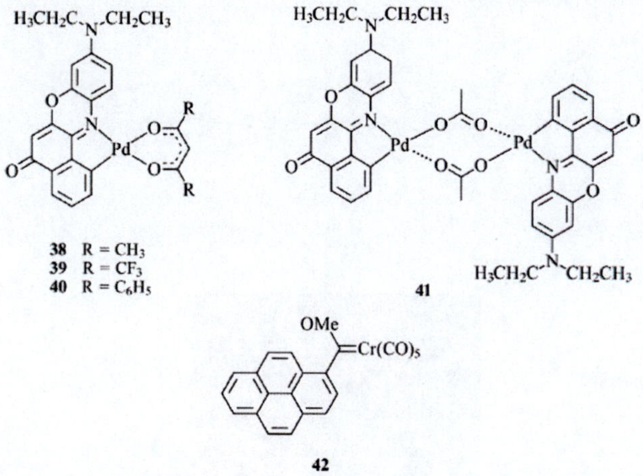

Fig. 14 Chemical structures of **38–42**

Probes which display solvatochromism change their emission maxima in response to the polarity of their environment. This feature enables monitoring of the lipid polarity surrounding the probe. Nile Red cyclometalated with disubstituted acetylacetonato square planar Pd(II) complexes, such as **38–40**, displays efficient red emission and similar solvatochromic behaviour to Nile Red, indicating a potential ability to distinguish between polar and non-polar lipids (Fig. 14) [121]. The analogous dinuclear complex **41** showed two-photon excitation and was used as CO sensor in vitro and in vivo; however, no investigation on the interaction of this probe with lipids has been reported [122, 123]. A carbon monoxide-releasing Fischer carbene chromium complex (**42**) has been applied to HeLa cells, and co-localisation was observed with both Mitotracker Red and Nile Red [124]. This data suggests the probe localises non-specifically in both the mitochondria and LDs.

2.3.1.2 Probes Incorporating First-Row Transition Metals (Mn) Tian et al. report a manganese(II) terpyridine metal complex, **43**, which is suitable for both two-photon fluorescence microscopy and MRI (Fig. 15) [65]. Furthermore, it is demonstrated that **43** is suited for stimulated emission depletion (STED). This presents a powerful and unique combination of imaging modalities where super-resolution microscopy reveals structural details at the cellular level, and MRI offers deeper tissue penetration on a larger scale. Complex **43** demonstrates high membrane permeability and low cytotoxicity. Live HepG2 cells were incubated with **43** and co-stained with ER-Tracker Green, showing high co-localisation. When applied to brain tissue, **43** displays strong affinity for parenchymal cells and the myelin sheath within the hippocampus. Myelin specificity was confirmed by co-staining with antibody microtubule-associated protein (MAP2) which label neuronal axons and imaging using STED.

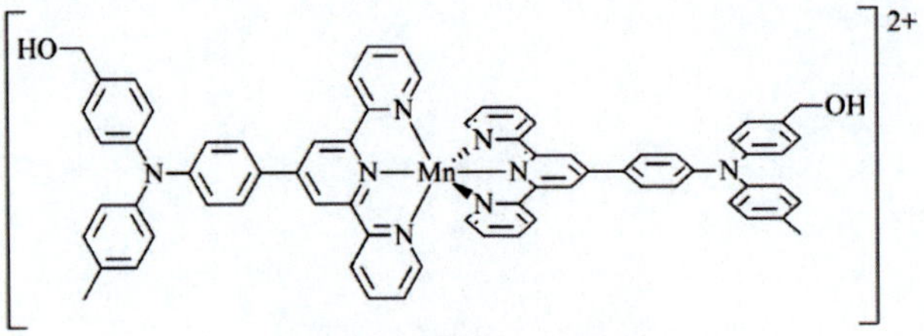

43

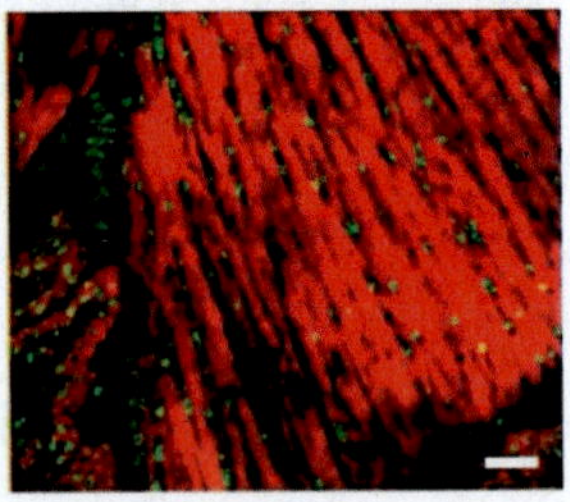

Fig. 15 Chemical structure of **43**. Mouse brain incubated with 43 (red) and sectioned into 20 μm slices, co-stained with NucRed (green) and imaged using confocal microscopy. Scale bar = 20 μm. Reproduced with permission from Tian et al. [65]

3 Conclusion

The microscopy tools available to study lipids in biological systems are rapidly expanding, allowing lipid biology to be explored in unprecedented detail. A critical component of the toolbox to study lipids is the diverse range of lipophilic stains and luminescent probes to reveal cellular and subcellular distributions of lipids. Careful exploitation of the diverse chemistry associated with metal ions, selective tuning of photophysical properties, and analyte specificity has now been achieved to study lipid distribution in cells by numerous research groups. In particular, the use of metal ions provides opportunities to utilise phosphorescence, with associated large Stokes shifts, minimising interference of endogenous autofluorescence encountered in biological samples with time-gated imaging techniques. The high photostability of metal probes which can be exploited for super resolution techniques is beginning to be recognised. While the choice of metal ion is critical to the ability to tune photophysical properties, target specificity is often controlled through selection of ligands. Generic rules for targeting lipids using metal complexes are established, but the detailed relationships between the physicochemical properties and structure of the complex (e.g. charge, lipophilicity and functionalisation) with lipid specificity still requires extensive investigation. What we identify lacking in this area is a set of standard protocols to assess metal-based complexes as lipid probes, or more generally as molecular probes. Experimentations are carried out in different cell lines and conditions, making it very difficult to make comparisons and draw nearer to detailed structure–activity relationships. Furthermore, most newly synthesised compounds and their associated protocols for imaging are optimised for cells, and they do not necessarily translate to the staining of tissue. Given the importance and usefulness of tissue staining, it would be ideal if the set of established standard protocols for the assessment of new molecular probes includes both cells and tissue substrates.

Acknowledgements BS and DH both acknowledge support from the Australian Government through their Research Training Program Scholarships. MH acknowledges support from the Australian Research Council through a Future Fellowship (FT190100017).

Author contributions BS was the major student author contributing to the majority of reviewing, writing, editing and figure making. DH was the minor student author, contributing to the minority of reviewing, writing and editing. MH and MM are primary supervisors of DH and BS, respectively. MH and MM were primary editors of writing.

Funding Open Access funding enabled and organized by CAUL and its Member Institutions.

Declarations

Conflict of interest The authors do not have any conflicts of interest to declare.

References

1. Adibhatla RM, Hatcher JF (2010) Lipid oxidation and peroxidation in CNS health and disease: from molecular mechanisms to therapeutic opportunities. Antioxid Redox Signal 12(1):125–169
2. Dietschy JM, Turley SD (2004) Thematic review series: brain lipids. Cholesterol metabolism in the central nervous system during early development and in the mature animal. J Lipid Res 45(8):1375–1397
3. Zhang C, Wang K, Yang L, Liu R, Chu Y, Qin X et al (2018) Lipid metabolism in inflammation-related diseases. Analyst 143(19):4526–4536. https://doi.org/10.1039/C8AN01046C
4. Feingold KR, Grunfeld C (2000) The effect of inflammation and infection on lipids and lipoproteins [Updated 2022 Mar 7]. In: Feingold KR, Anawalt B, Boyce A et al (eds) Endotext [Internet]. MDText.com Inc., South Dartmouth. https://www.ncbi.nlm.nih.gov/books/NBK326741/
5. Robertson J, Peters MJ, McInnes IB, Sattar N (2013) Changes in lipid levels with inflammation and therapy in RA: a maturing paradigm. Nat Rev Rheumatol 9(9):513–523. https://doi.org/10.1038/nrrheum.2013.91
6. Liu L, Zhang K, Sandoval H, Yamamoto S, Jaiswal M, Sanz E et al (2015) Glial lipid droplets and ROS induced by mitochondrial defects promote neurodegeneration. Cell 160(1–2):177–190
7. Reed TT (2011) Lipid peroxidation and neurodegenerative disease. Free Radic Biol Med 51(7):1302–1319. https://doi.org/10.1016/j.freeradbiomed.2011.06.027
8. Norton W, Poduslo SE (1973) Myelination in rat brain: changes in myelin composition during brain maturation 1. J Neurochem 21(4):759–773
9. Kutzelnigg A, Lucchinetti CF, Stadelmann C, Brück W, Rauschka H, Bergmann M et al (2005) Cortical demyelination and diffuse white matter injury in multiple sclerosis. Brain 128(11):2705–2712

10. Futerman AH, Van Meer G (2004) The cell biology of lysosomal storage disorders. Nat Rev Mol Cell Biol 5(7):554–565
11. Mehlem A, Hagberg CE, Muhl L, Eriksson U, Falkevall A (2013) Imaging of neutral lipids by oil red O for analyzing the metabolic status in health and disease. Nat Protoc 8(6):1149–1154. https://doi.org/10.1038/nprot.2013.055
12. Frösen J, Tulamo R, Heikura T, Sammalkorpi S, Niemelä M, Hernesniemi J et al (2013) Lipid accumulation, lipid oxidation, and low plasma levels of acquired antibodies against oxidized lipids associate with degeneration and rupture of the intracranial aneurysm wall. Acta Neuropathol Commun 1(1):71. https://doi.org/10.1186/2051-5960-1-71
13. Bayliss OB, Adams CWM (1972) Bromine-Sudan Black: a general stain for lipids including free cholesterol. Histochem J 4(6):505–515. https://doi.org/10.1007/BF01011130
14. Olson KR, Traub RK (1990) Visual enhancement of myelinated tissues in the central nervous system of the rat using Sudan Black B. Stain Technol 65(3):151–153. https://doi.org/10.3109/10520299009139928
15. Maganti RJ, Hronowski XL, Dunstan RW, Wipke BT, Zhang X, Jandreski L et al (2019) Defining changes in the spatial distribution and composition of brain lipids in the shiverer and cuprizone mouse models of myelin disease. J Histochem Cytochem 67(3):203–219. https://doi.org/10.1369/0022155418815860
16. Kuerschner L, Moessinger C, Thiele C (2008) Imaging of lipid biosynthesis: how a neutral lipid enters lipid droplets. Traffic 9(3):338–352. https://doi.org/10.1111/j.1600-0854.2007.00689.x
17. Jacquier N, Choudhary V, Mari M, Toulmay A, Reggiori F, Schneiter R (2011) Lipid droplets are functionally connected to the endoplasmic reticulum in *Saccharomyces cerevisiae*. J Cell Sci 124(14):2424–2437. https://doi.org/10.1242/jcs.076836
18. Barisch C, Paschke P, Hagedorn M, Maniak M, Soldati T (2015) Lipid droplet dynamics at early stages of *Mycobacterium marinum* infection in *Dictyostelium*. Cell Microbiol 17(9):1332–1349. https://doi.org/10.1111/cmi.12437
19. Spangenburg EE, Pratt SJP, Wohlers LM, Lovering RM (2011) Use of BODIPY (493/503) to visualize intramuscular lipid droplets in skeletal muscle. J Biomed Biotechnol 2011:598358. https://doi.org/10.1155/2011/598358
20. Strauss JA, Shepherd DA, Macey M, Jevons EFP, Shepherd SO (2020) Divergence exists in the subcellular distribution of intramuscular triglyceride in human skeletal muscle dependent on the choice of lipid dye. Histochem Cell Biol 154(4):369–382. https://doi.org/10.1007/s00418-020-01898-2
21. Maxfield FR, Wüstner D (2012) Analysis of cholesterol trafficking with fluorescent probes. Methods Cell Biol 108:367–393. https://doi.org/10.1016/B978-0-12-386487-1.00017-1
22. Mutanda T, Ramesh D, Karthikeyan S, Kumari S, Anandraj A, Bux F (2011) Bioprospecting for hyper-lipid producing microalgal strains for sustainable biofuel production. Bioresour Technol 102(1):57–70. https://doi.org/10.1016/j.biortech.2010.06.077
23. Rumin J, Bonnefond H, Saint-Jean B, Rouxel C, Sciandra A, Bernard O et al (2015) The use of fluorescent Nile red and BODIPY for lipid measurement in microalgae. Biotechnol Biofuels 8(1):42. https://doi.org/10.1186/s13068-015-0220-4
24. Qiu B, Simon MC (2016) BODIPY 493/503 staining of neutral lipid droplets for microscopy and quantification by flow cytometry. Bio-Protoc 6(17):e1912. https://doi.org/10.21769/BioProtoc.1912
25. Brasaemle DL, Rubin B, Harten IA, Gruia-Gray J, Kimmel AR, Londos C (2000) Perilipin A increases triacylglycerol storage by decreasing the rate of triacylglycerol hydrolysis. J Biol Chem 275(49):38486–38493. https://doi.org/10.1074/jbc.M007322200
26. Zhang SO, Trimble R, Guo F, Mak HY (2010) Lipid droplets as ubiquitous fat storage organelles in *C. elegans*. BMC Cell Biol 11(1):96. https://doi.org/10.1186/1471-2121-11-96
27. Anderson JL, Carten JD, Farber SA (2016) Using fluorescent lipids in live zebrafish larvae: from imaging whole animal physiology to subcellular lipid trafficking. Methods Cell Biol 133:165–178. https://doi.org/10.1016/bs.mcb.2016.04.011
28. Awda BJ, Mackenzie-Bell M, Buhr MM (2009) Reactive oxygen species and boar sperm function. Biol Reprod 81(3):553–561. https://doi.org/10.1095/biolreprod.109.076471
29. Invitrogen™: BODIPY™ 581/591 C11 (Lipid Peroxidation Sensor). https://www.thermofisher.com/order/catalog/product/D3861?SID=srch-srp-D3861#/D3861?SID=srch-srp-D3861

30. Hartnell D, Schwehr BJ, Gillespie-Jones K, Alwis D, Rajan R, Hou H et al (2020) Imaging lipophilic regions in rodent brain tissue with halogenated BODIPY probes. Analyst 145(11):3809–3813. https://doi.org/10.1039/D0AN00099J
31. Teo W, Caprariello AV, Morgan ML, Luchicchi A, Schenk GJ, Joseph JT et al (2021) Nile Red fluorescence spectroscopy reports early physicochemical changes in myelin with high sensitivity. Proc Natl Acad Sci USA 118(8):e2016897118. https://doi.org/10.1073/pnas.2016897118
32. Lo KK-W (2010) Exploitation of luminescent organometallic rhenium(I) and iridium(III) complexes in biological studies. In: Lees AJ (ed) Photophysics of organometallics. Springer, Berlin, pp 73–114
33. Gao Z, Hao Y, Zheng M, Chen Y (2017) A fluorescent dye with large Stokes shift and high stability: synthesis and application to live cell imaging. RSC Adv 7(13):7604–7609. https://doi.org/10.1039/C6RA27547H
34. Eggeling C, Widengren J, Rigler R, Seidel CAM (1998) Photobleaching of fluorescent dyes under conditions used for single-molecule detection: evidence of two-step photolysis. Anal Chem 70(13):2651–2659. https://doi.org/10.1021/ac980027p
35. Cordina NM, Sayyadi N, Parker LM, Everest-Dass A, Brown LJ, Packer NH (2018) Reduced background autofluorescence for cell imaging using nanodiamonds and lanthanide chelates. Sci Rep 8(1):4521. https://doi.org/10.1038/s41598-018-22702-1
36. Fernández-Moreira V, Thorp-Greenwood FL, Coogan MP (2010) Application of d6 transition metal complexes in fluorescence cell imaging. Chem Commun 46(2):186–202. https://doi.org/10.1039/B917757D
37. Zhao Q, Huang C, Li F (2011) Phosphorescent heavy-metal complexes for bioimaging. Chem Soc Rev 40(5):2508–2524. https://doi.org/10.1039/C0CS00114G
38. Smith DG, Pal R, Parker D (2012) Measuring equilibrium bicarbonate concentrations directly in cellular mitochondria and in human serum using europium/terbium emission intensity ratios. Chem Eur J 18(37):11604–11613. https://doi.org/10.1002/chem.201201738
39. Lo KK-W (2020) Molecular design of bioorthogonal probes and imaging reagents derived from photofunctional transition metal complexes. Acc Chem Res 53(1):32–44. https://doi.org/10.1021/acs.accounts.9b00416
40. Hirayama T, Okuda K, Nagasawa H (2013) A highly selective turn-on fluorescent probe for iron(II) to visualize labile iron in living cells. Chem Sci 4(3):1250–1256. https://doi.org/10.1039/C2SC21649C
41. Au-Yeung HY, Chan J, Chantarojsiri T, Chang CJ (2013) Molecular imaging of labile iron(II) pools in living cells with a turn-on fluorescent probe. J Am Chem Soc 135(40):15165–15173. https://doi.org/10.1021/ja4072964
42. Khatun S, Biswas S, Binoy A, Podder A, Mishra N, Bhuniya S (2020) Highly chemoselective turn-on fluorescent probe for ferrous (Fe^{2+}) ion detection in cosmetics and live cells. J Photochem Photobiol B 209:111943. https://doi.org/10.1016/j.jphotobiol.2020.111943
43. Shi Z, Tang X, Zhou X, Cheng J, Han Q, Zhou J-A et al (2013) A highly selective fluorescence "turn-on" probe for Cu(II) based on reaction and its imaging in living cells. Inorg Chem 52(21):12668–12673. https://doi.org/10.1021/ic401865e
44. Yuan Y, Sun S, Liu S, Song X, Peng X (2015) Highly sensitive and selective turn-on fluorescent probes for Cu^{2+} based on rhodamine B. J Mater Chem B 3(26):5261–5265. https://doi.org/10.1039/C5TB00423C
45. Zeng L, Miller EW, Pralle A, Isacoff EY, Chang CJ (2006) A selective turn-on fluorescent sensor for imaging copper in living cells. J Am Chem Soc 128(1):10–11. https://doi.org/10.1021/ja055064u
46. Yang Y-S, Ma C-M, Zhang Y-P, Xue Q-H, Ru J-X, Liu X-Y et al (2018) A highly selective "turn-on" fluorescent sensor for zinc ion based on a cinnamyl pyrazoline derivative and its imaging in live cells. Anal Methods 10(16):1833–1841. https://doi.org/10.1039/C8AY00037A
47. Zastrow ML, Radford RJ, Chyan W, Anderson CT, Zhang DY, Loas A et al (2016) Reaction-based probes for imaging mobile zinc in live cells and tissues. ACS Sens 1(1):32–39. https://doi.org/10.1021/acssensors.5b00022
48. Feng X, Li Y, He X, Liu H, Zhao Z, Kwok RTK et al (2018) A substitution-dependent light-up fluorescence probe for selectively detecting Fe^{3+} ions and its cell imaging application. Adv Funct Mater 28(35):1802833. https://doi.org/10.1002/adfm.201802833
49. Chung CY-S, Posimo JM, Lee S, Tsang T, Davis JM, Brady DC et al (2019) Activity-based ratiometric FRET probe reveals oncogene-driven changes in labile copper pools induced by altered

glutathione metabolism. Proc Natl Acad Sci USA 116(37):18285–18294. https://doi.org/10.1073/pnas.1904610116
50. Ryu SY, Huh M, You Y, Nam W (2015) Phosphorescent zinc probe for reversible turn-on detection with bathochromically shifted emission. Inorg Chem 54(20):9704–9714. https://doi.org/10.1021/acs.inorgchem.5b00967
51. Garwin SA, Kelley MS, Sue AC, Que EL, Schatz GC, Woodruff TK et al (2019) Interrogating intracellular zinc chemistry with a long stokes shift zinc probe ZincBY-4. J Am Chem Soc 141(42):16696–16705. https://doi.org/10.1021/jacs.9b06442
52. Lu Z, Fan W, Lu Y, Fan C, Zhao H, Guo K et al (2018) A highly sensitive fluorescent probe for bioimaging zinc ion in living cells and zebrafish models. New J Chem 42(14):12198–12204. https://doi.org/10.1039/C8NJ02197J
53. Liu Z, Zhang C, Chen Y, Qian F, Bai Y, He W et al (2014) In vivo ratiometric Zn^{2+} imaging in zebrafish larvae using a new visible light excitable fluorescent sensor. Chem Commun 50(10):1253–1255. https://doi.org/10.1039/C3CC46262E
54. Monteiro JHSK (2020) Recent advances in luminescence imaging of biological systems using lanthanide(III) luminescent complexes. Molecules 25(9):2089. https://doi.org/10.3390/molecules25092089
55. Lo KK-W (2007) Luminescent transition metal complexes as biological labels and probes. In: Yam VWW (ed) Photofunctional transition metal complexes. Springer, Berlin, pp 205–245
56. Caporale C, Massi M (2018) Cyclometalated iridium(III) complexes for life science. Coord Chem Rev 363:71–91. https://doi.org/10.1016/j.ccr.2018.02.006
57. Ma D-L, He H-Z, Leung K-H, Chan DS-H, Leung C-H (2013) Bioactive luminescent transition-metal complexes for biomedical applications. Angew Chem Int Ed 52(30):7666–7682. https://doi.org/10.1002/anie.201208414
58. Lakowicz JR (2006) Principles of fluorescence spectroscopy, 3rd edn. Springer, Berlin. https://doi.org/10.1007/978-0-387-46312-4
59. (2007) Photochemistry and photophysics of coordination compounds I. 1 ed. Top. Curr. Chem. Springer, Berlin
60. (2007) Photochemistry and photophysics of coordination compounds II. 1 ed. Top. Curr. Chem. Springer, Berlin
61. Chou PT, Chi Y, Chung M-W, Lin C-C (2011) Harvesting luminescence via harnessing the photophysical properties of transition metal complexes. Coord Chem Rev 255:2653–2665
62. Eliseeva SV, Bünzli J-CG (2010) Lanthanide luminescence for functional materials and biosciences. Chem Soc Rev 39(1):189–227. https://doi.org/10.1039/B905604C
63. Cosgrave L, Devocelle M, Forster RJ, Keyes TE (2010) Multimodal cell imaging by ruthenium polypyridyl labelled cell penetrating peptides. Chem Commun 46(1):103–105. https://doi.org/10.1039/B918611E
64. Murphy L, Congreve A, Pålsson L-O, Williams JAG (2010) The time domain in co-stained cell imaging: time-resolved emission imaging microscopy using a protonatable luminescent iridium complex. Chem Commun 46(46):8743–8745. https://doi.org/10.1039/C0CC03705B
65. Tian X, Xiao L, Shen Y, Luo L, Zhang G, Zhang Q et al (2019) A combination of super-resolution fluorescence and magnetic resonance imaging using a Mn(II) compound. Inorg Chem Front 6(10):2914–2920. https://doi.org/10.1039/C9QI00895K
66. Ranieri AM, Caporale C, Fiorini V, Hubbard A, Rigby P, Stagni S et al (2019) Complementary approaches to imaging subcellular lipid architectures in live bacteria using phosphorescent iridium complexes and raman spectroscopy. Chem Eur J 25(45):10566–10570. https://doi.org/10.1002/chem.201902023
67. He L, Cao J-J, Zhang D-Y, Hao L, Zhang M-F, Tan C-P et al (2018) Lipophilic phosphorescent iridium(III) complexes as one- and two-photon selective bioprobes for lipid droplets imaging in living cells. Sens Actuators B 262:313–325. https://doi.org/10.1016/j.snb.2018.02.022
68. Caporale C, Bader CA, Sorvina A, MaGee KDM, Skelton BW, Gillam TA et al (2017) Investigating intracellular localisation and cytotoxicity trends for neutral and cationic iridium tetrazolato complexes in live cells. Chem Eur J 23(62):15666–15679. https://doi.org/10.1002/chem.201701352
69. Lo KK-W, Lee P-K, Lau JS-Y (2008) Synthesis, characterization, and properties of luminescent organoiridium(III) polypyridine complexes appended with an alkyl chain and their interactions with lipid bilayers, surfactants, and living cells. Organometallics 27(13):2998–3006. https://doi.org/10.1021/om800212t

70. Suwalsky M, Ungerer B, Villena F, Norris B, Cárdenas H, Zatta P (1999) Interactions of Al(acac)3 with cell membranes and model phospholipid bilayers. J Inorg Biochem 75(4):263–268. https://doi.org/10.1016/s0162-0134(99)00097-5
71. Maria Ranieri A, Vezzelli M, Leslie KG, Huang S, Stagni S, Jacquemin D et al (2021) Structure illumination microscopy imaging of lipid vesicles in live bacteria with naphthalimide-appended organometallic complexes. Analyst 146(12):3818–3822. https://doi.org/10.1039/D1AN00363A
72. Qiu K, Chen Y, Rees TW, Ji L, Chao H (2019) Organelle-targeting metal complexes: from molecular design to bio-applications. Coord Chem Rev 378:66–86. https://doi.org/10.1016/j.ccr.2017.10.022
73. Langdon-Jones EE, Jones AB, Williams CF, Hayes AJ, Lloyd D, Mottram HJ et al (2017) Anticancer, azonafide-inspired fluorescent ligands and their rhenium(I) complexes for cellular imaging. Eur J Inorg Chem 2017(3):759–766. https://doi.org/10.1002/ejic.201601271
74. Lo KK-W (2015) Luminescent rhenium(I) and iridium(III) polypyridine complexes as biological probes, imaging reagents, and photocytotoxic agents. Acc Chem Res 48(12):2985–2995. https://doi.org/10.1021/acs.accounts.5b00211
75. Lee LC-C, Leung K-K, Lo KK-W (2017) Recent development of luminescent rhenium(i) tricarbonyl polypyridine complexes as cellular imaging reagents, anticancer drugs, and antibacterial agents. Dalton Trans 46(47):16357–16380. https://doi.org/10.1039/C7DT03465B
76. Otero C, Carreño A, Polanco R, Llancalahuen FM, Arratia-Pérez R, Gacitúa M et al (2019) Rhenium (I) complexes as probes for prokaryotic and fungal cells by fluorescence microscopy: do ligands matter? Front Chem 7:454. https://doi.org/10.3389/fchem.2019.00454
77. Burke CS, Byrne A, Keyes TE (2018) Highly selective mitochondrial targeting by a ruthenium(II) peptide conjugate: imaging and photoinduced damage of mitochondrial DNA. Angew Chem Int Ed 57(38):12420–12424. https://doi.org/10.1002/anie.201806002
78. Newton MD, Fairbanks SD, Thomas JA, Rueda DS (2021) A minimal load-and-lock RuII luminescent DNA probe. Angew Chem Int Ed 60(38):20952–20959. https://doi.org/10.1002/anie.202108077
79. Gkika KS, Byrne A, Keyes TE (2019) Mitochondrial targeted osmium polypyridyl probe shows concentration dependent uptake, localisation and mechanism of cell death. Dalton Trans 48(47):17461–17471. https://doi.org/10.1039/C9DT02967B
80. Munteanu A-C, Notaro A, Jakubaszek M, Cowell J, Tharaud M, Goud B et al (2020) Synthesis, characterization, cytotoxic activity, and metabolic studies of ruthenium(II) polypyridyl complexes containing flavonoid ligands. Inorg Chem 59(7):4424–4434. https://doi.org/10.1021/acs.inorgchem.9b03562
81. Yang C, Mehmood F, Lam TL, Chan SL-F, Wu Y, Yeung C-S et al (2016) Stable luminescent iridium(III) complexes with bis(*N*-heterocyclic carbene) ligands: photo-stability, excited state properties, visible-light-driven radical cyclization and CO_2 reduction, and cellular imaging. Chem Sci 7(5):3123–3136. https://doi.org/10.1039/C5SC04458H
82. Tso KK-S, Lo KK-W (2017) Strategic applications of luminescent iridium(III) complexes as biomolecular probes, cellular imaging reagents, and photodynamic therapeutics. Iridium(III) in optoelectronic and photonics applications, pp 415–477
83. Jiang W, Gao Y, Sun Y, Ding F, Xu Y, Bian Z et al (2010) Zwitterionic iridium complexes: synthesis, luminescent properties, and their application in cell imaging. Inorg Chem 49(7):3252–3260. https://doi.org/10.1021/ic9021239
84. Liang P, Kolodieznyi D, Creeger Y, Ballou B, Bruchez MP (2020) Subcellular singlet oxygen and cell death: location matters. Front Chem 8:1045. https://doi.org/10.3389/fchem.2020.592941
85. Wu P, Wong EL-M, Ma D-L, Tong GS-M, Ng K-M, Che C-M (2009) Cyclometalated platinum(II) complexes as highly sensitive luminescent switch-on probes for practical application in protein staining and cell imaging. Chem Eur J 15(15):3652–3656. https://doi.org/10.1002/chem.200802707
86. Barnard PJ, Wedlock LE, Baker MV, Berners-Price SJ, Joyce DA, Skelton BW et al (2006) Luminescence studies of the intracellular distribution of a dinuclear gold(I) *N*-heterocyclic carbene complex. Angew Chem Int Ed 45(36):5966–5970. https://doi.org/10.1002/anie.200601526
87. Barnard PJ, Berners-Price SJ (2007) Targeting the mitochondrial cell death pathway with gold compounds. Coord Chem Rev 251(13):1889–1902. https://doi.org/10.1016/j.ccr.2007.04.006
88. Botchway SW, Charnley M, Haycock JW, Parker AW, Rochester DL, Weinstein JA et al (2008) Time-resolved and two-photon emission imaging microscopy of live cells with inert platinum complexes. Proc Natl Acad Sci USA 105(42):16071–16076. https://doi.org/10.1073/pnas.0804071105

89. Mauro M, Aliprandi A, Septiadi D, Kehr NS, De Cola L (2014) When self-assembly meets biology: luminescent platinum complexes for imaging applications. Chem Soc Rev 43(12):4144–4166. https://doi.org/10.1039/C3CS60453E
90. Herr P, Kerzig C, Larsen CB, Häussinger D, Wenger OS (2021) Manganese(I) complexes with metal-to-ligand charge transfer luminescence and photoreactivity. Nat Chem 13(10):956–962. https://doi.org/10.1038/s41557-021-00744-9
91. Büldt LA, Guo X, Vogel R, Prescimone A, Wenger OS (2017) A tris(diisocyanide)chromium(0) complex is a luminescent analog of Fe(2,2′-bipyridine). J Am Chem Soc 139(2):985–992. https://doi.org/10.1021/jacs.6b11803
92. Cotruvo JA (2019) The chemistry of lanthanides in biology: recent discoveries, emerging principles, and technological applications. ACS Cent Sci 5(9):1496–1506. https://doi.org/10.1021/acscentsci.9b00642
93. New EJ, Congreve A, Parker D (2010) Definition of the uptake mechanism and sub-cellular localisation profile of emissive lanthanide complexes as cellular optical probes. Chem Sci 1(1):111–118. https://doi.org/10.1039/C0SC00105H
94. Bünzli J-CG, Eliseeva SV (2013) Intriguing aspects of lanthanide luminescence. Chem Sci 4(5):1939–1949. https://doi.org/10.1039/C3SC22126A
95. Thibon A, Pierre VC (2009) Principles of responsive lanthanide-based luminescent probes for cellular imaging. Anal Bioanal Chem 394(1):107–120. https://doi.org/10.1007/s00216-009-2683-2
96. Yang X, Wang S, Zhang Y, Liang G, Zhu T, Zhang L et al (2018) A self-assembling luminescent lanthanide molecular nanoparticle with potential for live cell imaging. Chem Sci 9(20):4630–4637. https://doi.org/10.1039/C8SC00650D
97. Connell TU, Donnelly PS (2018) Labelling proteins and peptides with phosphorescent d6 transition metal complexes. Coord Chem Rev 375:267–284. https://doi.org/10.1016/j.ccr.2017.12.001
98. Fujimoto T, Ohsaki Y, Cheng J, Suzuki M, Shinohara Y (2008) Lipid droplets: a classic organelle with new outfits. Histochem Cell Biol 130(2):263–279. https://doi.org/10.1007/s00418-008-0449-0
99. Thiam AR, Farese RV Jr, Walther TC (2013) The biophysics and cell biology of lipid droplets. Nat Rev Mol Cell Biol 14(12):775–786. https://doi.org/10.1038/nrm3699
100. Murphy S, Martin S, Parton RG (2010) Quantitative analysis of lipid droplet fusion: inefficient steady state fusion but rapid stimulation by chemical fusogens. PLoS One 5(12):e15030. https://doi.org/10.1371/journal.pone.0015030
101. Dalhaimer P (2019) Lipid droplets in disease. Cells 8(9):974. https://doi.org/10.3390/cells8090974
102. Bader CA, Brooks RD, Ng YS, Sorvina A, Werrett MV, Wright PJ et al (2014) Modulation of the organelle specificity in Re(I) tetrazolato complexes leads to labeling of lipid droplets. RSC Adv 4(31):16345–16351. https://doi.org/10.1039/C4RA00050A
103. Bader CA, Carter EA, Safitri A, Simpson PV, Wright P, Stagni S et al (2016) Unprecedented staining of polar lipids by a luminescent rhenium complex revealed by FTIR microspectroscopy in adipocytes. Mol BioSyst 12(7):2064–2068. https://doi.org/10.1039/C6MB00242K
104. Zhu J-H, Yiu S-M, Tang BZ, Lo KK-W (2021) Luminescent neutral cyclometalated iridium(III) complexes featuring a cubic polyhedral oligomeric silsesquioxane for lipid droplet imaging and photocytotoxic applications. Inorg Chem 60(15):11672–11683. https://doi.org/10.1021/acs.inorgchem.1c01728
105. Hickey JL, James JL, Henderson CA, Price KA, Mot AI, Buncic G et al (2015) Intracellular distribution of fluorescent copper and zinc bis(thiosemicarbazonato) complexes measured with fluorescence lifetime spectroscopy. Inorg Chem 54(19):9556–9567. https://doi.org/10.1021/acs.inorgchem.5b01599
106. Hung LW, Villemagne VL, Cheng L, Sherratt NA, Ayton S, White AR et al (2012) The hypoxia imaging agent CuII(atsm) is neuroprotective and improves motor and cognitive functions in multiple animal models of Parkinson's disease. J Exp Med 209(4):837–854. https://doi.org/10.1084/jem.20112285
107. Lim S, Price KA, Chong S-F, Paterson BM, Caragounis A, Barnham KJ et al (2010) Copper and zinc bis(thiosemicarbazonato) complexes with a fluorescent tag: synthesis, radiolabelling with copper-64, cell uptake and fluorescence studies. JBIC J Biol Inorg Chem 15(2):225–235. https://doi.org/10.1007/s00775-009-0587-4
108. Tang J, Zhang Y, Yin H-Y, Xu G, Zhang J-L (2017) Precise labeling and tracking of lipid droplets in adipocytes using a luminescent ZnSalen complex. Chem Asian J 12(19):2533–2538. https://doi.org/10.1002/asia.201701010

109. Koshel EI, Chelushkin PS, Melnikov AS, Serdobintsev PY, Stolbovaia AY, Saifitdinova AF et al (2017) Lipophilic phosphorescent gold(I) clusters as selective probes for visualization of lipid droplets by two-photon microscopy. J Photochem Photobiol A Chem 332:122–130. https://doi.org/10.1016/j.jphotochem.2016.08.017
110. Koshevoy IO, Chang Y-C, Karttunen AJ, Selivanov SI, Jänis J, Haukka M et al (2012) Intensely luminescent homoleptic alkynyl decanuclear gold(I) clusters and their cationic octanuclear phosphine derivatives. Inorg Chem 51(13):7392–7403. https://doi.org/10.1021/ic300856h
111. Alberts B, Johnson A, Lewis J, Raff M, Roberts K, Walter P (2002) Molecular biology of the cell, 4th edn. Garland Science, New York
112. Cai T, Yang F (2018) Phospholipid and phospholipidomics in health and diseases. In: Wang X, Wu D, Shen H (eds) Lipidomics in health and disease: methods and application. Springer, Singapore, pp 177–202
113. Svensson FR, Li M, Nordén B, Lincoln P (2008) Luminescent dipyridophenazine-ruthenium probes for liposome membranes. J Phys Chem B 112(35):10969–10975. https://doi.org/10.1021/jp803964x
114. Bonnet S, Limburg B, Meeldijk JD, Klein Gebbink RJM, Killian JA (2011) Ruthenium-decorated lipid vesicles: light-induced release of $[Ru(terpy)(bpy)(OH_2)]^{2+}$ and thermal back coordination. J Am Chem Soc 133(2):252–261. https://doi.org/10.1021/ja105025m
115. Mechler A, Stringer BD, Mubin MSH, Doeven EH, Phillips NW, Rudd-Schmidt J et al (2014) Labeling phospholipid membranes with lipid mimetic luminescent metal complexes. Biochim Biophys Acta 1838(11):2939–2946. https://doi.org/10.1016/j.bbamem.2014.08.005
116. Koo C-K, Wong K-L, Man CW-Y, Tam H-L, Tsao S-W, Cheah K-W et al (2009) Two-photon plasma membrane imaging in live cells by an amphiphilic, water-soluble cyclometalated platinum(II) complex. Inorg Chem 48(16):7501–7503. https://doi.org/10.1021/ic9007679
117. Byrne A, Burke CS, Keyes TE (2016) Precision targeted ruthenium(II) luminophores; highly effective probes for cell imaging by stimulated emission depletion (STED) microscopy. Chem Sci 7(10):6551–6562. https://doi.org/10.1039/C6SC02588A
118. Fernández-Moreira V, Thorp-Greenwood FL, Amoroso AJ, Cable J, Court JB, Gray V et al (2010) Uptake and localisation of rhenium fac-tricarbonyl polypyridyls in fluorescent cell imaging experiments. Org Biomol Chem 8(17):3888–3901. https://doi.org/10.1039/C004610H
119. Tang J, Yin H-Y, Zhang J-L (2018) A luminescent aluminium salen complex allows for monitoring dynamic vesicle trafficking from the Golgi apparatus to lysosomes in living cells. Chem Sci 9(7):1931–1939. https://doi.org/10.1039/c7sc04498d
120. Kielar F, Congreve A, Law G-L, New EJ, Parker D, Wong K-L et al (2008) Two-photon microscopy study of the intracellular compartmentalisation of emissive terbium complexes and their oligo-arginine and oligo-guanidinium conjugates. Chem Commun 21:2435–2437. https://doi.org/10.1039/B803864C
121. Deda ML, Ghedini M, Aiello I, Pugliese T, Barigelletti F, Accorsi G (2005) Organometallic emitting dyes: palladium(II) nile red complexes. J Organomet Chem 690(4):857–861. https://doi.org/10.1016/j.jorganchem.2004.10.028
122. Liu K, Kong X, Ma Y, Lin W (2018) Preparation of a Nile Red-Pd-based fluorescent CO probe and its imaging applications in vitro and in vivo. Nat Protoc 13(5):1020–1033. https://doi.org/10.1038/nprot.2018.013
123. Liu K, Kong X, Ma Y, Lin W (2017) Rational design of a robust fluorescent probe for the detection of endogenous carbon monoxide in living zebrafish embryos and mouse tissue. Angew Chem Int Ed 56(43):13489–13492. https://doi.org/10.1002/anie.201707518
124. Arrowsmith RL, Atkin AJ, Botchway SW, Fairlamb IJS, Lynam JM, Moir JWB et al (2015) Confocal and fluorescence lifetime imaging sheds light on the fate of a pyrene-tagged carbon monoxide-releasing Fischer carbene chromium complex. Dalton Trans 44(11):4957–4962. https://doi.org/10.1039/C4DT03312D
125. Svensson FR, Matson M, Li M, Lincoln P (2010) Lipophilic ruthenium complexes with tuned cell membrane affinity and photoactivated uptake. Biophys Chem 149(3):102–106. https://doi.org/10.1016/j.bpc.2010.04.006

Authors and Affiliations

Bradley J. Schwehr[1] · **David Hartnell**[1,2] · **Massimiliano Massi**[1] · **Mark J. Hackett**[1,2]

[1] School of Molecular and Life Sciences, Curtin University, Perth, WA 6845, Australia

[2] Curtin Health Innovation Research Institute, Curtin University, Perth, WA 6845, Australia

Topics in Current Chemistry (2022) 380:35
https://doi.org/10.1007/s41061-022-00389-3

REVIEW

Phosphorescent Ir(III) Complexes for Biolabeling and Biosensing

Byung Hak Jhun[1,2] · Dayoon Song[1,3] · Soo Young Park[2] · Youngmin You[1,3]

Received: 29 November 2021 / Accepted: 27 May 2022 / Published online: 10 August 2022

Abstract

Cyclometalated Ir(III) complexes exhibit strong phosphorescence emission with lifetime of submicroseconds to several microseconds at room temperature. Their synthetic versatility enables broad control of physical properties, such as charge and lipophilicity, as well as emission colors. These favorable properties have motivated the use of Ir(III) complexes in luminescent bioimaging applications. This review examines the recent progress in the development of phosphorescent biolabels and sensors based on Ir(III) complexes. It begins with a brief introduction about the basic principles of the syntheses and photophysical processes of cyclometalated Ir(III) complexes. Focus is placed on illustrating the broad imaging utility of Ir(III) complexes. Phosphorescent labels illuminating intracellular organelles, including mitochondria, lysosomes, and cell membranes, are summarized. Ir(III) complexes capable of visualization of tumor spheroids and parasites are also introduced. Facile chemical modification of the cyclometalating ligands endows the Ir(III) complexes with strong sensing ability. Sensors of temperature, pH, CO_2, metal ions, anions, biosulfur species, reactive oxygen species, peptides, and viscosity have recently been added to the molecular imaging tools. This diverse utility demonstrates the potential of phosphorescent Ir(III) complexes toward bioimaging applications.

Chapter 4 was originally published as Jhun, B. H., Song, D., Park, S. Y. & You, Y. Topics in Current Chemistry (2022) 380: 35. https://doi.org/10.1007/s41061-022-00389-3.

✉ Soo Young Park
parksy@snu.ac.kr

✉ Youngmin You
odds2@ewha.ac.kr

1 Division of Chemical Engineering and Materials Science, Ewha Womans University, Seoul 03760, Republic of Korea

2 Center for Supramolecular Optoelectronic Materials (CSOM) and Department of Materials Science and Engineering, Seoul National University, Seoul 08826, Republic of Korea

3 Graduate Program for System Health Science and Engineering, Ewha Womans University, Seoul 03760, Republic of Korea

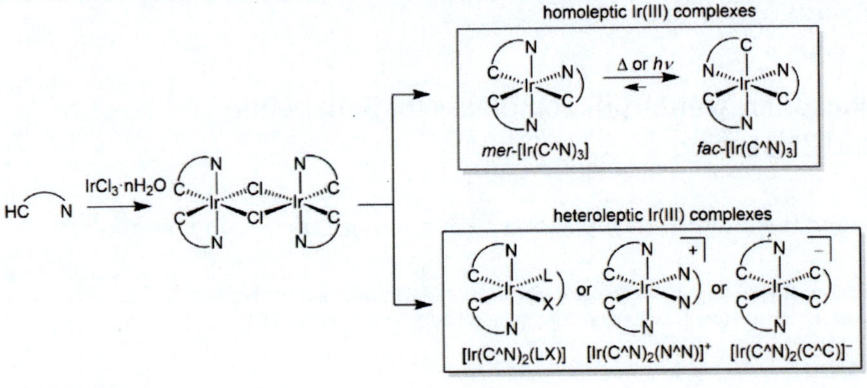

Scheme 1 Chemical structures and syntheses of cyclometalated Ir(III) complexes having bidentate ligands

Keywords Phosphorescence · Ir(III) complex · Biolabels · Biosensors

1 Introduction

A cyclometalated Ir(III) complex contains multiple iridacycles with bi- or tridentate ligands. The preparation of a cyclometalated complex having bidentate ligands typically employs the standard two-step procedure which includes the formation of a bis(μ-Cl) dinuclear Ir(III) complex having cyclometalating ligands, followed by dissociative substitution of the chloro ligands with the third ligands (Scheme 1). A variety of ligands, including a monoanionic C^N, a monoanionic ancillary ligand (LX), a charge-neutral diimine (N^N) ligand, and a dianionic cyclometalating ligand (C^C), can be used as the third ligand. In addition to the classical method, recent research has established alternative protocols for the preparation of cyclometalated Ir(III) complexes. Interested readers are strongly recommended to refer to the excellent reviews of You, Chi, and López [1–3]. The synthetic versatility enables facile control of the physical properties of Ir(III) complexes. Charge, lipophilicity, and environment-sensitivity of phosphorescence emission are highly dependent on the structure of the ligands.

Photophysical processes of the phosphorescence emission of an Ir(III) complex include multiple steps:1 as shown in Fig. 1, photoexcitation promotes electronic transition to the singlet excited states with ligand-centered π–π* (^{1}LC) or metal-to-ligand charge-transfer (^{1}MLCT) characteristics. The initially populated ^{1}LC and ^{1}MLCT transition states are converted to their triplet manifolds (i.e., ^{3}LC and ^{3}MLCT). This intersystem crossing is ultrafast due to large spin–orbit coupling exerted by the central Ir(III) atom. The spin–orbit coupling also facilitates formerly forbidden phosphorescence emission from the ^{3}LC or ^{3}MLCT state. The phosphorescence transition of cyclometalated Ir(III) complexes typically proceeds at the time-scale of sub- to several microseconds [4–6]. Thus, spin–orbit coupling

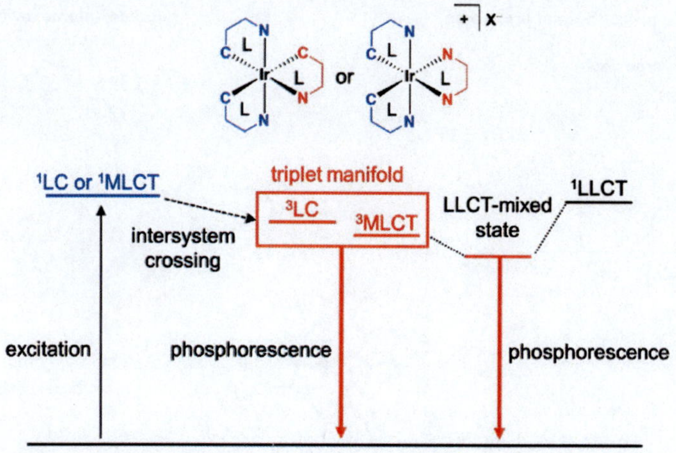

Fig. 1 Photophysical processes of cyclometalated Ir(III) complexes. Refer to the main text for the abbreviations

drives the two spin-flipping steps (i.e., intersystem crossing and phosphorescence transition).

The presence of the third ligand different from the initially coordinated C^N ligands provides rich photophysics. If an N^N ligand is used as the third ligand, ligand-to-ligand charge-transfer (LLCT) transition occurs from the monoanionic C^N ligand to the charge-neutral N^N ligand. The LLCT transition state is usually mixed with the LC and MLCT states, which substantially increases the phosphorescence responses to environmental changes and analyte binding.

The synthetic versatility and the rich photophysics make phosphorescent cyclometalated Ir(III) complexes useful platforms for creating biolabels and sensors as shown in Fig. 2. The emission wavelength is tunable over a broad range by changing ligand structures. In addition, emission performance, such as photoluminescence quantum yield and Stokes shift, can also be controlled by ligand modifications. Cyclometalated Ir(III) complexes usually remain stable against prolonged exposure to photoexcitation beams. This photostability is particularly advantageous for live bioimaging experiments where high-intensity pulsed laser excitation beams are focused for a certain period of time. Ir(III) complexes are compatible with mammalian cells, and their intracellular uptake behaviors can be readily tailored by changing charges and ligand modifications. Several cyclometalated Ir(III) complexes have been demonstrated to produce two-photon-induced phosphorescence [7–11].

This ability draws particular attention toward diagnostic and therapeutic purposes, as it enables deep penetration of an excitation beam with high spatial precision. The greatest advantage of using phosphorescent Ir(III) complexes is their ability in enabling time-gated luminescence imaging which benefits from long-lived phosphorescence emission. The long-lifetime phosphorescence is easily discriminated from short-lived background noises of autofluorescence and scattered lights [12]. Time-gated acquisition of the phosphorescence signals thus enables complete elimination of the noise in the images, producing high spatial resolution.

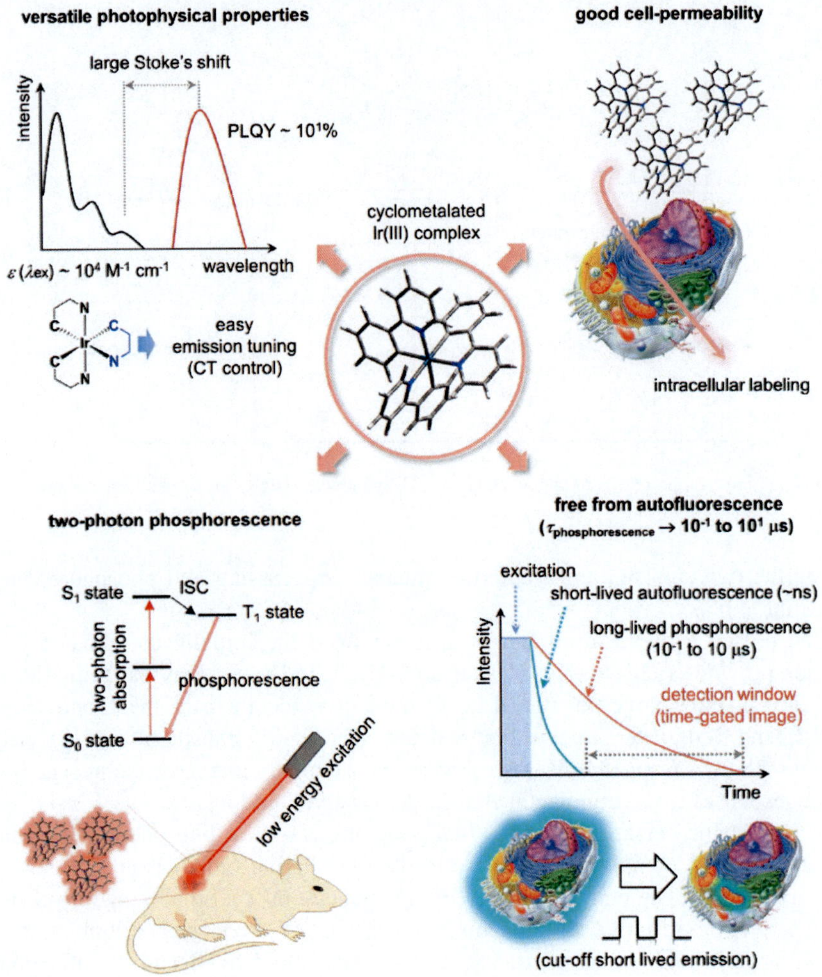

Fig. 2 Benefits of using phosphorescent Ir(III) complexes for theranostic applications

The following parts summarize the recent advances in the bioimaging utility of phosphorescent cyclometalated Ir(III) complexes. The first part examines the recently developed phosphorescence labels of intracellular organelles. Phosphorescence labels of tumor spheroids and animals are also introduced. The second part illustrates phosphorescence sensing utility of Ir(III) complexes. This part is presented following target analytes which include temperature, CO_2, pH, metal ions, anions, biosulfur species, oxygen and reactive oxygen species, peptides, and viscosity. Focus is on the structures and phosphorescence behaviors of the labels and sensors.

2 Phosphorescent Biolabels Based on Cyclometalated Ir(III) Complexes

2.1 Intracellular Organelle Labels

Phosphorescent cyclometalated Ir(III) complexes exhibit unique advantages over fluorescent molecules for biolabeling applications. The phosphorescence emission exhibits large Stokes shifts, avoiding self-absorption. In addition, the long-lived phosphorescence can be temporally discriminated from autofluorescence, which enables visualization of target organelles without suffering from diffuse background fluorescence. The ability of subcellular organelle localization further provides therapeutic utility because the phosphorescent state can generate cytotoxic reactive oxygen species. Table 1 lists the photophysical properties and labeling targets of phosphorescent Ir(III) complexes.

Selective labeling of intracellular organelles requires judicious control over the chemical structures of Ir(III) complexes. The size, charge, hydrophobicity, hydrophilicity, and substituents govern the entry, trafficking and localization of Ir(III) complexes. Previous reviews discussed the structure-localization relationship of phosphorescent Ir(III) complexes [5, 13]. The following parts introduce the recently added strategies of cyclometalated Ir(III) complexes toward selective targeting intracellular organelles.

2.1.1 Mitochondrion Labels

The mitochondrion, the core subcellular organelle in eukaryotic cells, plays a key role in energy production. Mitochondria also store a variety of proteins, such as apoptosis-signaling proteins [14, 15]. Visualization of mitochondria is, therefore, of fundamental importance for understanding normal cellular function and pathogenesis. Since the mitochondrial membrane retains a negative potential, the molecular design for mitochondrion labeling should include chemical structures that generate cationic charge. Certain amphiphilicity is also crucial for crossing the hydrophobic mitochondrial membrane and for accumulation within mitochondria [16–18]. Mao's group recently developed several mitochondrial labels based on phosphorescent cyclometalated Ir(III) complexes (**1** and **2** in Fig. 3a) [19, 20]. *N*-heterocyclic carbene (NHC) ligands were used to localize their complexes within mitochondria. In phosphate-buffered saline (PBS) solution, complex **1** exhibited blue emission with moderate photoluminescence quantum yield of 0.07. This complex could photosensitize the singlet oxygen generation with high quantum yields (Φ_Δ, 0.58), which demonstrated its potential for photodynamic activity. Complex **2** exhibited bathochromically shifted emission relative to that of **1** (peak emission wavelength (λ_{em}) = 590 nm), because they incorporated more π-extended C^N ligands. The ligand modifications did not exert notable changes in Φ_Δ of 0.62 [20]. Dual-emissive phosphorescence mitochondrial labels were developed by Lo's group (**3**) [21].

Table 1 Phosphorescent labels based on cyclometalated Ir(III) complexes

	Labeling targets	λ_{ex} (nm)[a]	λ_{em} (nm)[b]	PLQY[c]	References
1	Mitochondria	380	475	0.073	[19]
2	Mitochondria	405	590	0.14	[20]
3	Mitochondria	–	567	0.50	[21]
4	Mitochondria	405	609	0.0020	[22]
5	Mitochondria	405	642	0.011	[23]
6	Mitochondria	405	600	0.024	[24]
7	Mitochondria	405	604	0.012	[24]
8	Mitochondria	365	540	0.16	[25]
9	Mitochondria	405	630	0.017	[26]
10	Mitochondria	405	630	0.010	[26]
11	Mitochondria	411	527	0.039	[27]
12	Mitochondria	366	506	0.57	[28]
13	Mitochondria	405	490	0.022	[31]
14	Mitochondria	–	610	0.039	[32]
15	Mitochondria	405	619	0.0060	[33]
16	Mitochondria	350	627	0.0013	[34]
17	Mitochondria	405	671	0.039	[35]
18	Mitochondria	382	597	0.122	[36]
19	Mitochondria	405	623	0.0090	[37]
20	Mitochondria	405	692	0.064	[38]
21	Mitochondria	–	700	–	[39]
22	Mitochondria	405	526	0.15	[40]
23	Mitochondria	365	589	0.045	[41]
24	Lysosome	375	520	–	[43]
25	Lysosome	375	580	–	[43]
26	Lysosome	398	581	0.021	[44]
27	Lysosome	398	612	0.0056	[44]
28	Lysosome	400	598	0.14	[9, 45]
29	Lysosome	366	497, 544	–	[49]
30	Lysosome	–	432, 479	–	[50]
31	Lysosome	405	600	0.021	[51]
32	Lysosome	405	553	–	[52]
33	Lysosome	405	534	0.00040	[53]
34	Lysosome	405	551	0.00060	[53]
35	Lysosome	405	600	0.00040	[53]
36	Cytoplasm	365	598	–	[55]
37	Cytoplasm	355	632	0.13	[10]
38	Cytoplasm	359	595	0.023	[10]
39	Cytoplasm	365	612	0.13	[56]
40	Cytoplasm	405	514	0.051	[57]
41	Cytoplasm	405	511	0.050	[57]
42	Cytoplasm	405	514	0.065	[57]

Table 1 (continued)

	Labeling targets	λ_{ex} (nm)[a]	λ_{em} (nm)[b]	PLQY[c]	References
43	Cytoplasm	420	683	0.010	[58]
44	Cytoplasm	405	498	0.044	[59]
45	Cytoplasm	405	530	0.094	[59]
46	Cytoplasm	405	534	0.059	[59]
47	Cytoplasm	405	523	0.029	[59]
48	Cytoplasm	405	662	0.0070	[59]
49	Cytoplasm	405	650	0.098	[59]
50	Cytoplasm	350	526	0.019	[60]
51	Cytoplasm	350	594	0.004	[60]
52	Cytoplasm	350	582	0.015	[60]
53	Cytoplasm	350	649	0.0070	[60]
54	Cytoplasm	350	556	0.0049	[61]
55	Cytoplasm	350	601	0.0008	[61]
56	Cytoplasm	350	561	0.0062	[61]
57	Cytoplasm	350	647	0.0013	[61]
58	Cell membrane	315	475	0.15	[65]
59	Cell membrane	365	506	0.33	[66]
60	Cell membrane	366	506	0.35	[67]
61	Cell membrane	–	–	–	[68]
62	Cell membrane	–	–	–	[68]
63	Cell membrane	–	–	–	[68]
64	Cell membrane	350	514	0.0099	[69]
65	Cell membrane	–	509	0.082	[70]
66	Cell membrane	405	593	0.0040	[71]
67	Tumor spheroid	–	524	0.15	[11]
68	Tumor spheroid	405	536	0.30	[74]
69	Tumor spheroid	405	551	0.35	[74]
70	Tumor spheroid	405	560	0.21	[74]
71	Tumor spheroid	405	578	0.10	[74]
72	Tumor spheroid	405	626	0.08	[74]
73	Tumor spheroid	405	544	0.068	[75]
74	Tumor spheroid	405	556	0.15	[75]
75	Tumor spheroid	405	594	0.30	[76]
76	Tumor spheroid	405	592	0.42	[76]
77	*Schistosoma japonicum*	405	519	0.15	[77]
78	*Schistosoma japonicum*	405	561	0.096	[77]

[a]Excitation wavelength. [b]Emission wavelength. [c]Photoluminescence quantum yield

An Ir(III) complex bearing coumarin exhibited phosphorescence emission in the red regions ($\lambda_{em} = 609$ nm) in PBS solution (**4**) [22]. This complex could stain mitochondria of HeLa cells and exhibited antiproliferative activity. Time-dependent phosphorescence tracking of the mitochondrial morphology enabled

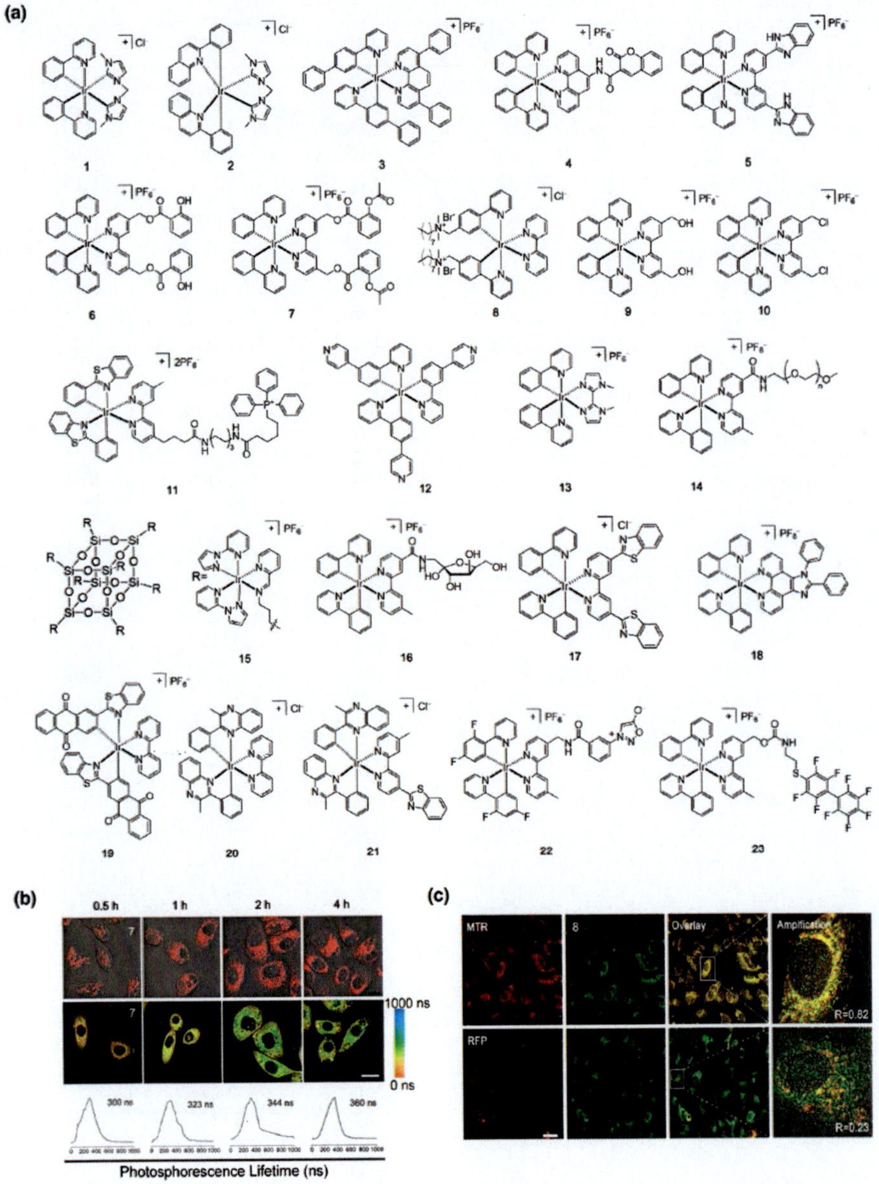

Fig. 3 **a** Phosphorescence mitochondrial labels based on cyclometalated Ir(III) complexes. **b** Phosphorescence mitochondrial tracking of prostate carcinoma (PC3) cells stained with **7**. Reproduced from Ref. [24]. Copyright 2019, WILEY-VCH Verlag GmbH & Co. KGaA, Weinheim. **c** Phosphorescence mitochondrial staining of HepG2 cells treated with **8**. Reproduced from Ref. [25]. Copyright 2019, American Chemical Society

investigations of anticancer effects of the complex. Red-emitting complex **5** having benzimidazole ancillary ligands was also found to enter mitochondria of A549 cells [23]. The complexes elicited apoptosis of the cells, demonstrating the potential for anticancer effects. Ir(III) complexes **6** and **7** bearing salicylate antiphlogistic moieties were demonstrated for their capacity for use in mitochondrial imaging and anti-inflammatory and antitumor activity [24]. These complexes exhibited long-lived red phosphorescence emission with λ_{em} of 600 and 604 nm, enabling visualization of mitochondrial changes with minimal background noise (Fig. 3b). The combination of the antiphlogistic structures elicited strong cytotoxicity and anti-inflammatory action. The amphiphilic Ir(III) complex developed by Hou's group (**8**) could be localized within mitochondria [25]. Notably, **8** showed high singlet oxygen photosensitization ability with Φ_{Δ} of 0.0054. These properties enabled the mitochondria-selective generation of singlet oxygen in HepG2 cells, hindering cellular growth (Fig. 3c).

To further increase the mitochondrial affinity, complexes **9** and **10** were designed to incorporate a chloromethyl moiety [26]. This modification enabled mitochondrial fixation due to the bond-forming interaction with cysteine. Complexes **9** and **10** exhibited red emission with λ_{em} of 630 nm and photoluminescence quantum yields of 0.010–0.017. As expected, the complexes remained within the mitochondria upon repeated washing and mitochondrial membrane potential decreases. Complex **11** bearing lipophilic cationic triphenylphosphonium (TPP) easily penetrated the mitochondrial membrane and remained within the mitochondria [27]. Complex **11** exhibited phosphorescence emission at λ_{em} of 527 nm. Notably, it exhibited the higher photostability than a commercial fluorescent mitochondrial stain, MitoTracker.

Aoki's group developed an Ir(III) complex bearing basic pyridines (**12**). Under acidic conditions, **12** was protonated and exhibited very weak red emission ($\lambda_{em}=593$ nm; photoluminescence quantum yield of 0.0081) in aqueous DMSO. Increasing pH elicited phosphorescence turn-on responses due to deprotonation of **12** (photoluminescence quantum yield of 0.57). This pH-dependent emission of **12** served to visualize mitochondrial pH [28].

Mitophagy is an autophagic process which removes damaged mitochondria from a cell and thus maintains normal mitochondrial activity [29, 30]. Since mitophagy is suggested to be involved in the genesis of various pathological conditions, such as Parkinson's disease, cancer, and cardiovascular diseases, it is important to monitor mitophagy. Complex **13** were found to accumulate within mitochondria and cause mitophagy [31]. The strong phosphorescence emission of **13** enabled monitoring of the morphological changes in the mitochondria during the progression of mitophagy.

Lo's group developed several mitochondria targeting Ir(III) complexes tethering various units. An Ir(III) complex having phenylpyridine ligands with poly(ethylene glycol) (PEG) units (**14**) exhibited phosphorescence emission in the red region ($\lambda_{em}=610$ nm) in acetonitrile. The balanced amphiphilicity exerted by the PEG unit and the phenylpyridine units enabled mitochondrial uptake by live HeLa cells and zebrafish [32]. A similar approach based on polyhedral oligomeric silsesquioxanes has also been examined. The Ir(III) complex (**15**) could be successfully internalized within the mitochondria of the live HeLa cell and their localization was validated with MitoTracker Deep Red [33]. An Ir(III) complex with fructose (**16**) was also found to stain mitochondria [34]. A phosphorescent Ir(III) complex with benzothiazole (**17**) could visualize mitochondria, presumably due to its amphiphilic nature. This complex exhibited red phosphorescence ($\lambda_{em} = 671$ nm) with a photoluminescence quantum yield of 0.039 in PBS solution [35].

Chao's group developed a phosphorescent label based on a cyclometalated Ir(III) complex (**18**). This complex bearing imidazophenanthroline could be localized within mitochondria in living HeLa cells [36]. A complex having anthraquinone moieties (**19**) could label mitochondria under hypoxic conditions through phosphorescence signaling upon two-photon absorption (TPA) at a wavelength of 720 nm [37]. An Ir(III) complex (**20**) based on 3-methyl-2-phenylquinoline produced phosphorescence emission in the near-infrared (NIR) region within mitochondria of HeLa cells [38]. Complex **21** was also employed to visualize mitochondria. In particular, the super-resolution microscopic technique permitted a spatial resolution as narrow as 80 nm [39].

The biorthogonal approach is advantageous for reliable targeting of certain biospecies. An Ir(III) complex having sydnone units (**22**) could react selectively with bovine serum albumin modified with strained alkyne, through the strained-promoted sydnone-alkyne cycloaddition reaction. This reaction-based targeting afforded high-fidelity phosphorescence labeling of mitochondria ($\lambda_{em} = 526$ nm) [40]. The conjugate formation by the reaction between the cysteine residue of peptides and an Ir(III) complex having perfluorobiphenyl (PFBP) units (**23**) was also found effective toward mitochondrial labeling. The benefit of this approach is the versatility of targeting organelles [41].

2.1.2 Lysosome Labels

Lysosomes are a subcellular organelle regulated at low pH (4.5–5.5). Lysosomes play pivotal roles in a variety of physiological processes, including intracellular transportation and degradation and recycling of proteins. Lysosome labels are typically charge-neutral and retain high lipophilicity. The design of lysosome labels takes advantage of the degradation mechanisms of the lysosome. It was shown that labels conjugated with certain biomacromolecules could easily enter lysosomes [42]. Keilar's group developed cell-permeable Ir(III) complexes bearing amphiphilic alkyl chains (Fig. 4) [43]. Complexes **24** and **25** formed aggregates in PBS solutions and exhibited phosphorescence emission at λ_{em} of 520 nm and 580 nm, respectively. **24** and **25** readily entered into NIH-3T3 cells and were localized within the lysosome. Ward's group also developed phosphorescence lysosome labels based on Ir(III) complexes of the $[Ir(N^\wedge N^\wedge N)(C^\wedge N)^+X^-]$ structure (N^N^N is the triimine ligand) (**26** and **27**) [44].

The labels showing pH-responsive luminescence changes are valuable for the selective visualization of acidic lysosomes. Mao's group prepared an Ir(III) complex bearing β-carboline (**28**) [9]. Complex **28** exhibited pH-dependent luminescence changes due to the presence of nitrogen atoms in β-carboline: decreasing pH from 7.4 to 5.4 elicited a gradual hypsochromic shift in λ_{em} from 645 to 598 nm. Interestingly, the chromic shift accompanied a 30-fold increase in the phosphorescence intensity. These dual responses were useful for lysosome

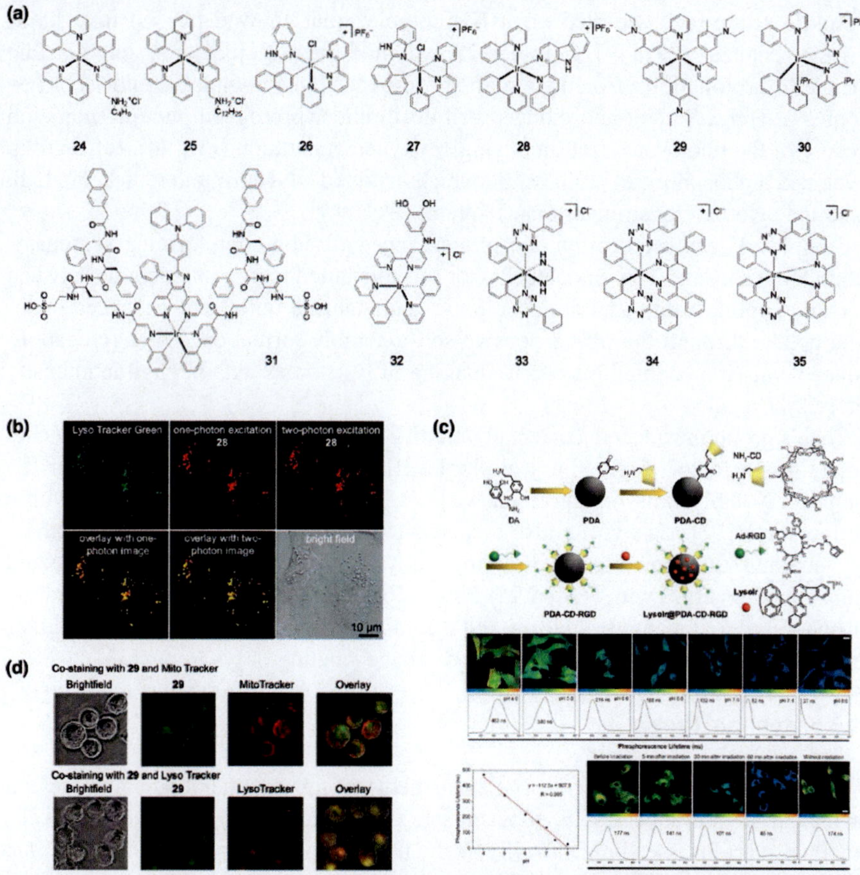

Fig. 4 **a** Phosphorescence lysosome labels based on cyclometalated Ir(III) complexes. **b** Single- and two-photon-excitation phosphorescence labeling of lysosomes of A549 cells treated with **28**. Reproduced from Ref. [9]. Copyright 2014, WILEY–VCH Verlag GmbH & Co. KGaA, Weinheim. **c** Two-photon-excitation phosphorescence lifetime imaging of lysosomes of U-87 cells treated with nanoparticles of **28**@PDA-CD-RGD (see main text for the details). Reproduced from Ref. [45]. Copyright 2018, WILEY–VCH Verlag GmbH & Co. KGaA, Weinheim. **d** Phosphorescence labeling of lysosomes of HeLa-S3 cells incubated with **29**. Reproduced from Ref. [49]. Copyright 2012, American Chemical Society

labeling. Complex **28** exhibited a TPA cross-section of 972 GM at 830 nm. This property enabled Mao's group to visualize lysosomes of A549 cells that experienced autophagy (Fig. 4b). Complex **28** could be further encapsulated within nanoparticles consisting of polydopamine (PDA), β-cyclodextrin, and arginine-glycine-aspartic acid (RGD) [45]. PDA and RGD are known to exhibit a high photothermal property and cancer-targeting ability, respectively [46–48]. The multi-doped nanoparticles could label lysosomes and exhibited photothermal activity under photoirradiation at a wavelength of 810 nm (Fig. 4c).

Aoki's group also reported an Ir(III) complex that showed pH-sensitive luminescence changes (**29**) [49]. Complex **29** exhibited intense green emission at acidic pH due to protonation of diethylamino substituents. This complex could label lysosomes of HeLa-S3 cell and induced cell death under prolonged photoillumination because of the photosensitization of singlet oxygen generation (Fig. 4d). Liu's group developed a phosphorescence label **30** which involved a NHC-imino moiety [50]. **30** exhibited cytotoxicity stronger than cisplatin.

Complex **31** showed turn-on phosphorescence emission upon forming aggregates within the lysosome. The unique behavior was attributed to π–π interactions between aromatic peptides units and aminosulfonic acids-taurine units, and enhanced endocytic uptake through the pH-responsive self-assembly formation. The aggregation-induced emission enabled long-term imaging of lysosomes with negligible intensity decreases [51].

Chao's group introduced a catechol unit into an Ir(III) complex (**32**). The catechol moiety was oxidized by free intracellular Fe(III) ions to be liberated from the Ir(III) complex. Notably, this oxidative cleavage is accelerated under acidic pH within the lysosome [52]. Lo's group also prepared phosphorescent Ir(III) complexes having tetrazine that could react biorthogonally (**33–35**). Inverse electron-demand Diels–Alder reaction with (1*R*,8*S*,9*S*)-bicyclo[6.1.0]non-4-yne (BCN) enhanced the phosphorescence quantum yield by 3900-folds. Biorthogonal labeling was further demonstrated to be effective for phosphorescence labeling of lysosome [53].

2.1.3 Cytoplasm Labels

Cytoplasm is the fluidic species that exists between the cell membrane and the cell nucleus of a eukaryotic cell. Cytoplasm serves as a storage for biogenic chemicals, and thus, various metabolic activity is found within cytoplasm. Charge-cationic Ir(III) complexes with certain substituents are found to remain within the cytoplasm [54]. Several phosphorescence Ir(III) complexes capable of staining the cytoplasm have been reported. Leung's group synthesized an Ir(III) solvent complex with good cell permeability (**36** in Fig. 5) [55]. This complex showed phosphorescence emission at λ_{em} of 598 nm, and the emission intensity increased upon interacting with bovine serum albumin and histidine. In contrast, the phosphorescence emission remained unaltered in the presence of calf thymus DNA. Zhao's group also developed cell-permeable complexes **37** and **38** [10]. Complex **37** formed nanoparticles with an average diameter of 56 nm in aqueous solutions. The nanoparticles could penetrate into the cell membranes. Two-photon excitation at wavelengths of 690 and 800 nm produced phosphorescence emission from the cytoplasm.

In order to produce strong red phosphorescence, Zhao's group designed complex **39** containing *o*-carbolane [56]. **39** exhibited red emission with a λ_{em} of 612 nm and a photoluminescence quantum yield of 0.13. The cytoplasm of HeLa cells could be readily stained with **39**, which enabled the quantification of cytoplasmic O_2 levels by phosphorescent lifetime imaging experiments.

Mao's group also prepared complexes **40–42** which could visualize cytoplasm [57]. The complexes also exhibited cytotoxic activity which was ascribed to the singlet oxygen generation and the photodegradation of labile ligands. The complexes

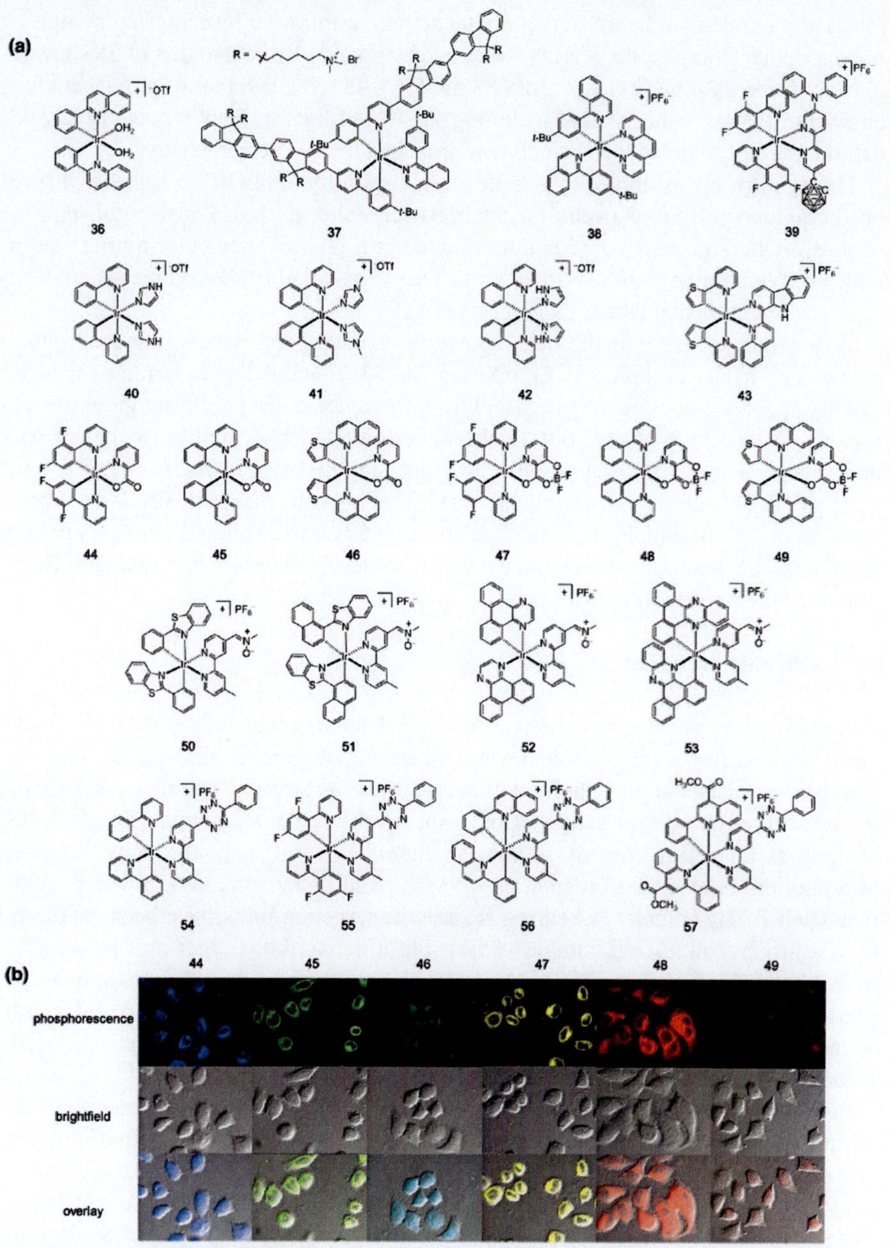

Fig. 5 **a** Phosphorescence cytoplasm labels based on cyclometalated Ir(III) complexes. **b** Multi-color cytoplasm staining of HeLa cells using **44–49**. Reproduced from Ref. [59]. Copyright 2013, Royal Society of Chemistry

released monodentate ligands upon photoexcitation, followed by capturing cytoplasmic histidine. Complex **43** bearing the ancillary ligand identical to that of **28** showed phosphorescence emission at λ_{em} of 683 nm [58]. **43** remained in the cytoplasm after entry into human cancer cells, including A549, A549cisR, HepG2, and HeLa. **43** exhibited a cytotoxic activity which was proposed to involve autophagy.

Huang's group examined organelle-staining ability of Ir(III) complexes based on the archetype blue-phosphorescent Ir(III) complex [59]. The structural modifications, including boryl coordination, elicited phosphorescence color-tuning over a wide emission range from 498 to 662 nm. The color-tuned Ir(III) complexes enabled multi-color cytoplasm labeling of HeLa cells (Fig. 5b).

Lo's group focused on the development of biorthogonal labels based on phosphorescent Ir(III) complexes. Complexes **50–53** contained polypyridine nitrone. The complexes exhibited very weak phosphorescence emission in the green to red regions (λ_{em} = 526–649 nm) due to the emission-quenching ability of the nitrone unit. The biorthogonal reaction with a cytoplasmic benzylguanine substrate modified with BCN turned on the phosphorescence intensity dramatically [60]. Phosphorescent Ir(III) complexes with tetrazine (**54–57**) also exhibited similar labeling behaviors. The reaction with cytoplasmic bovine serum albumin modified with BCN improved the phosphorescence intensity over a 50-fold [61].

2.1.4 Cell Membrane Labels

Cell membranes are the outermost structures of mammalian cells and exhibit various types of activity, such as binding and trafficking of extracellular species. The cell membrane consists of phospholipid bilayers, where a variety of peptides and ligands are present at the interior and exterior, respectively, of the membrane. The molecular designs for cell membrane staining include the use of lipophilic units targeting phospholipid bilayers, cell membrane proteins, and ligands [62–64]. Several cyclometalated Ir(III) complexes bearing membrane targeting units have been reported. Gray's group synthesized complex **58** having a deoxyribose sugar unit in order to track nucleoside transporters (Fig. 6) [65]. **58** exhibited a bathochromic shift in its phosphorescence emission upon incubation with KB3-1 cells. Pretreatment with adenosine retarded the intracellular uptake of **58**, which indicated the transport of **58** mediated by nucleoside transporters.

Aoki's group synthesized complexes **59–63** which tethered peptides at their peripheries [66–68]. They found that **59**, having a cyclic peptide capable of binding to a death receptor (DR), could anchor onto the membrane of Jurkat cells. Cell death could be visualized by phosphorescence imaging experiments. The tumor necrosis factor-related apoptosis-inducing ligand (TRAIL) structure also elicited binding of **60** to DR. Complexes **61–63** having amphiphilic peptides enabled cell membrane imaging and exhibited antimicrobial and anticancer activity [68].

Barnard's group also developed an Ir(III) complex having an NHC ligand (**64**). Complex **64** was incorporated into liposomes which bound at the surface of the membrane. The binding property enabled the phosphorescence visualization of the cell membrane [69].

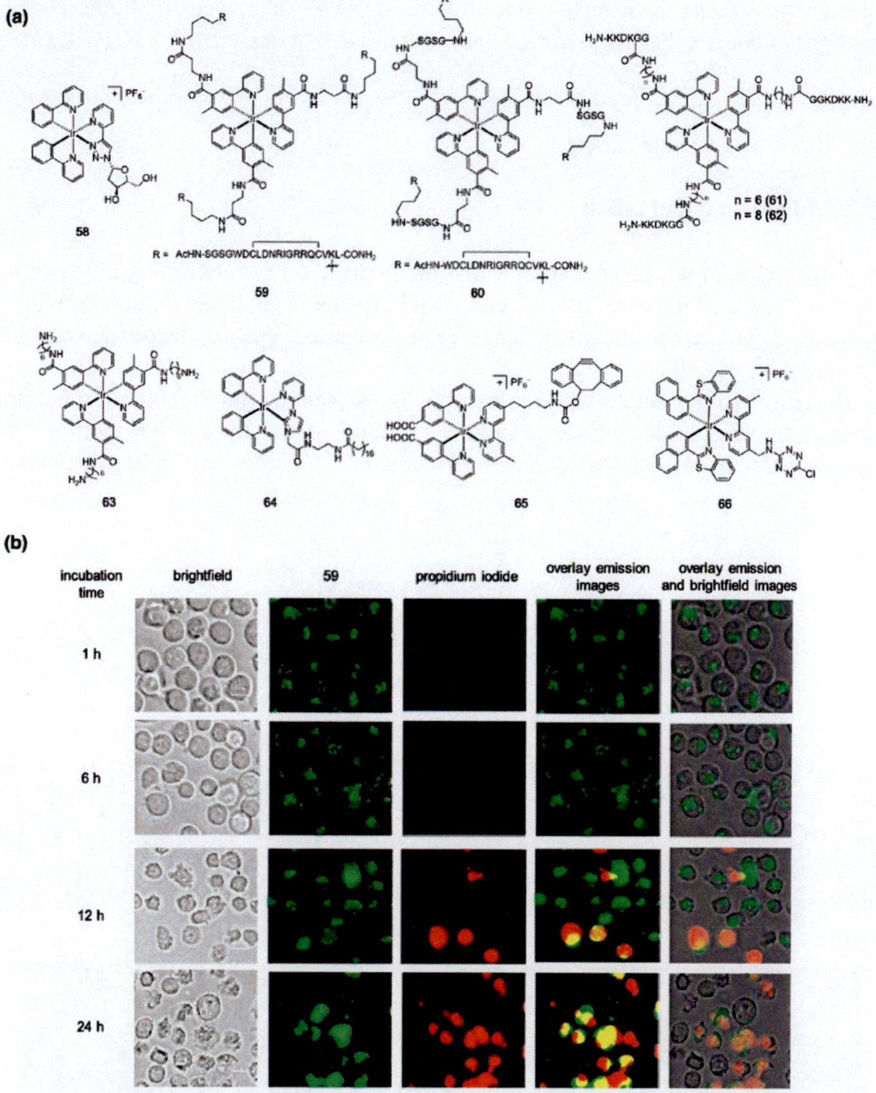

Fig. 6 **a** Phosphorescence cyclometalated Ir(III) complex cell membrane labels. **b** Time-lapse luminescence images of Jurkat cells treated with **59**. Reproduced from Ref. [66]. Copyright 2018, Hindawi Publishing Corporation

An Ir(III) complex tethering the dibenzocyclooctyne unit was prepared by Lo's group (**65**). Pretreatment of CHO cells with 1,3,4,6-tetra-*O*-acetyl-*N*-azidoacetyl-$_D$-mannosamine (Ac4ManNAz) induced the metabolic generation of *N*-azidoacetyl sialic acids which anchored at the plasma membranes. Subsequent strain-promoted alkyne-azide cycloaddition between **65** and *N*-azidoacetyl sialic acids enabled phosphorescent labeling of the cell membranes [70]. A similar approach was demonstrated using the inverse-electron-demand Diels–Alder cycloaddition between an Ir(III) complex having tetrazine (**66**) and Ac4ManNAz-treated CHO-K1 cells [71].

2.2 Tissue and Animal Labels

2.2.1 Tumor Spheroid Labels

Visualization of multicellular spheroids is prerequisite for understanding of the various pathological processes among cells. Multiphoton excitation luminescence is a powerful approach for imaging cell aggregates because it enables deeper delivery of an excitation beam [72, 73].

Several Ir(III) complex labels capable of producing multiphoton excitation phosphorescence exist in the literature. Chao's group introduced morpholine into a cyclometalated Ir(III) complex (**67** in Fig. 7). Complex **67** was capable of

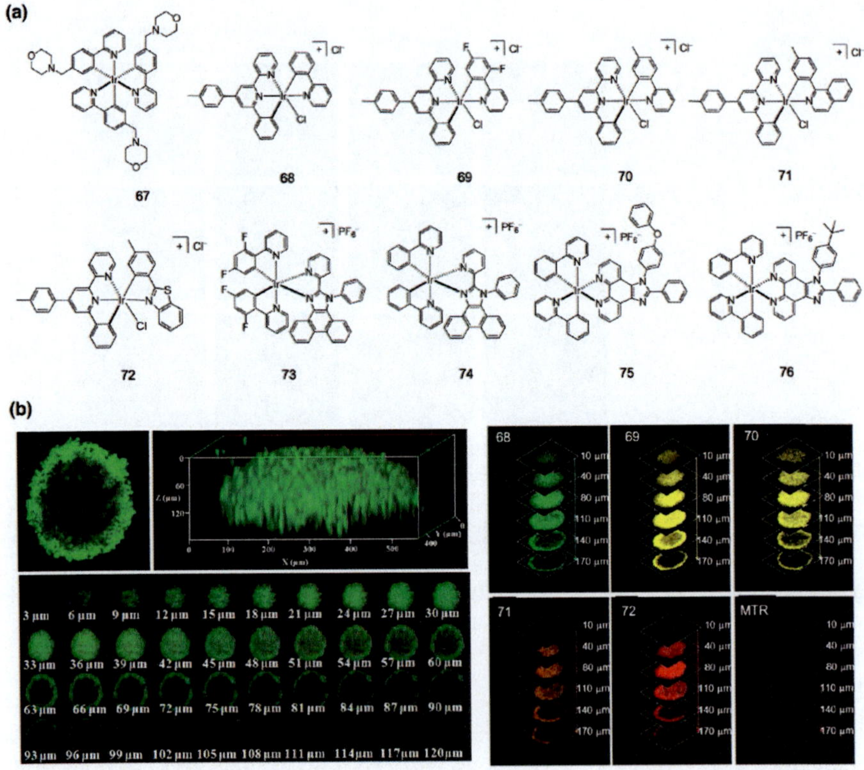

Fig. 7 **a** Multiphoton-excitable phosphorescent Ir(III) complexes for in-depth imaging of tumor spheroids. **b** Two-photon-excitation phosphorescence images of 3D HeLa multicellular spheroids treated with **67** (left image). Z-stacked images of **68–72** taken from 10, 40, 80, 110, 140, and 170 μm section. Reproduced from Refs. [11] and [74]. Copyrights 2016, American Chemical Society and 2016, Elsevier Ltd., respectively

phosphorescence imaging of lysosomes of living cells and 3D tumor spheroids. In disodium hydrogen phosphate-citric acid solutions buffered at pH 5.5, **67** displayed intense green emission at λ_{em} of 524 nm with a photoluminescence quantum yield of 0.15. Strong phosphorescence could be produced upon two-photon excitation at a wavelength of 750 nm. The two-photon-excitable phosphorescence property enabled imaging of 3D tumor spheroids because of increased penetration depths of the excitation beam (Fig. 7) [11]. Ir(III) complexes having terpyridyl ligands (**68–72**) could also produce phosphorescence emission in the green to red regions (λ_{em} = 536–626 nm) upon two-photon excitation. The phosphorescence behavior enabled real-time tracking the mitochondrial dynamics, and the visualization of 3D multicellular tumor spheroids [74]. Chao's group further expanded two-photon-excitable phosphorescence labels. Complexes **73** and **74** exhibited aggregation-induced phosphorescence emission under two-photon excitation at a wavelength of 740 nm [75]. Ir(III) complexes having imidazophenanthroline moieties (**75** and **76**) exhibited phosphorescence emission under three-photon excitation at a wavelength of 980 nm. Three-photon-excitable phosphorescence enabled ~ 1000 µm depth of tissue imaging [76].

2.2.2 Parasite Labels

Schistosomiasis is elicited by parasitic *Schistosoma*. To analyze the mechanism of the genesis of schistosomiasis, a phosphorescent label was developed based on an Ir(III) complex. Intravascular *Schistosoma* is hydrophilic and negatively charged. *Schistosoma* labeling thus requires an amphiphilic label. Fan's group synthesized Ir(III) complexes **77** and **78** (Fig. 8) [77]. In water, **77** and **78** showed phosphorescence emission at λ_{em} of 519 nm and 561 nm, respectively. These complexes could stain live adult *Schistosoma japonicum* and produced phosphorescence emission (Fig. 8).

3 Phosphorescence Probes Based on Cyclometalated Ir(III) Complexes

The phosphorescence emission of cyclometalated Ir(III) complexes is highly sensitive to environmental changes because of its charge-transfer and long-lifetime emission nature. This sensitivity enables large phosphorescence responses to interactions with analytes. Enormous efforts have been paid to devise sensing strategies [78–82]. Notable approaches include photo-induced energy and electron transfer (Fig. 9).

Interactions with analytes or environmental changes perturb the excited-state transfer, providing useful principles for creating luminescence probes. In particular, long-lived, phosphorescent triplet states are extremely sensitive to changes in rigidity, analyte binding, molecular oxygen, and hydrophobicity. This section summarizes the recent developments in phosphorescence probes based on Ir(III) complexes. Table 2 summarizes the key targets and performance of the phosphorescence probes.

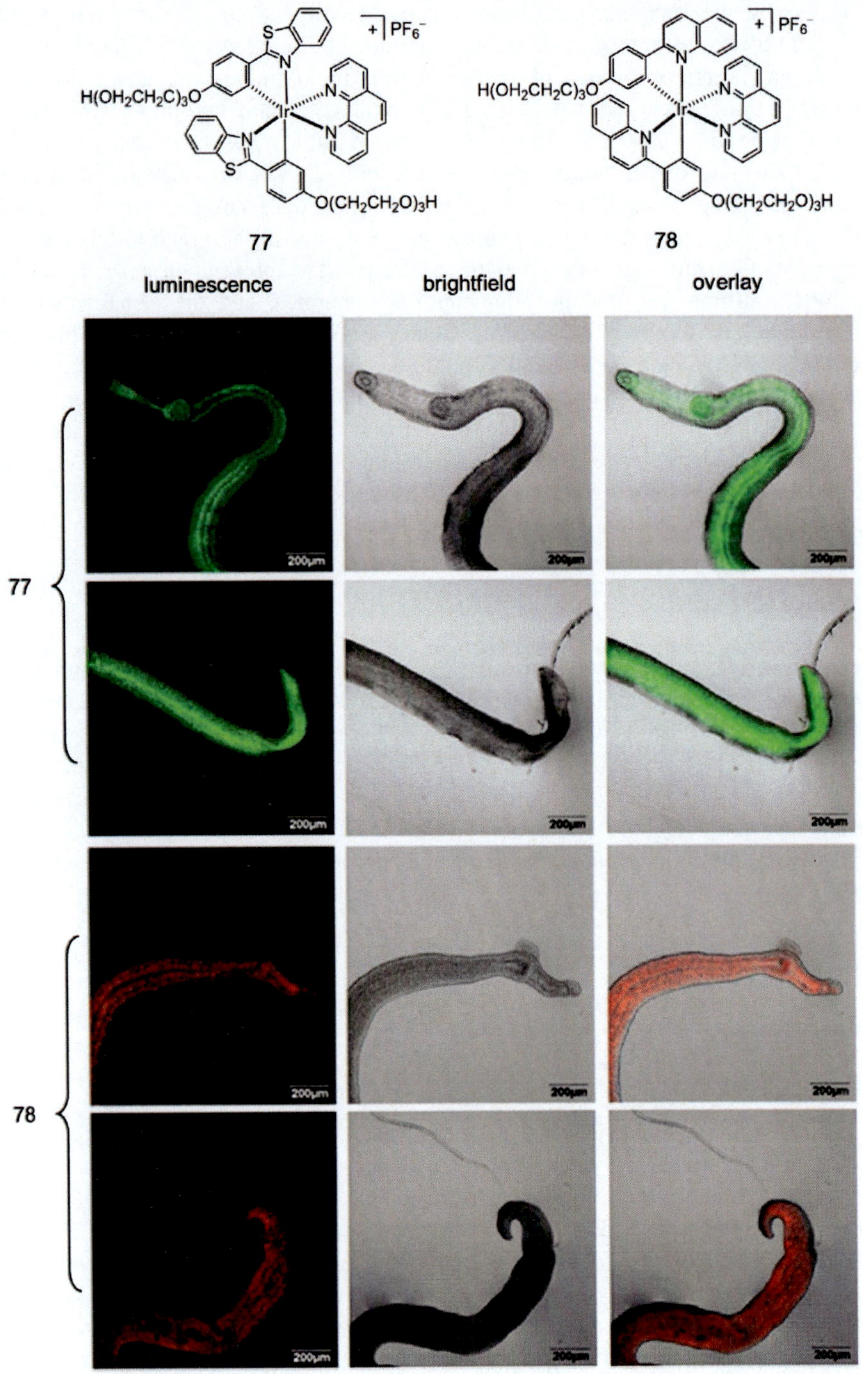

Fig. 8 Phosphorescence labels of *Schistosoma japonicum*. Reproduced from Ref. [77]. Copyright 2017, Royal Society of Chemistry

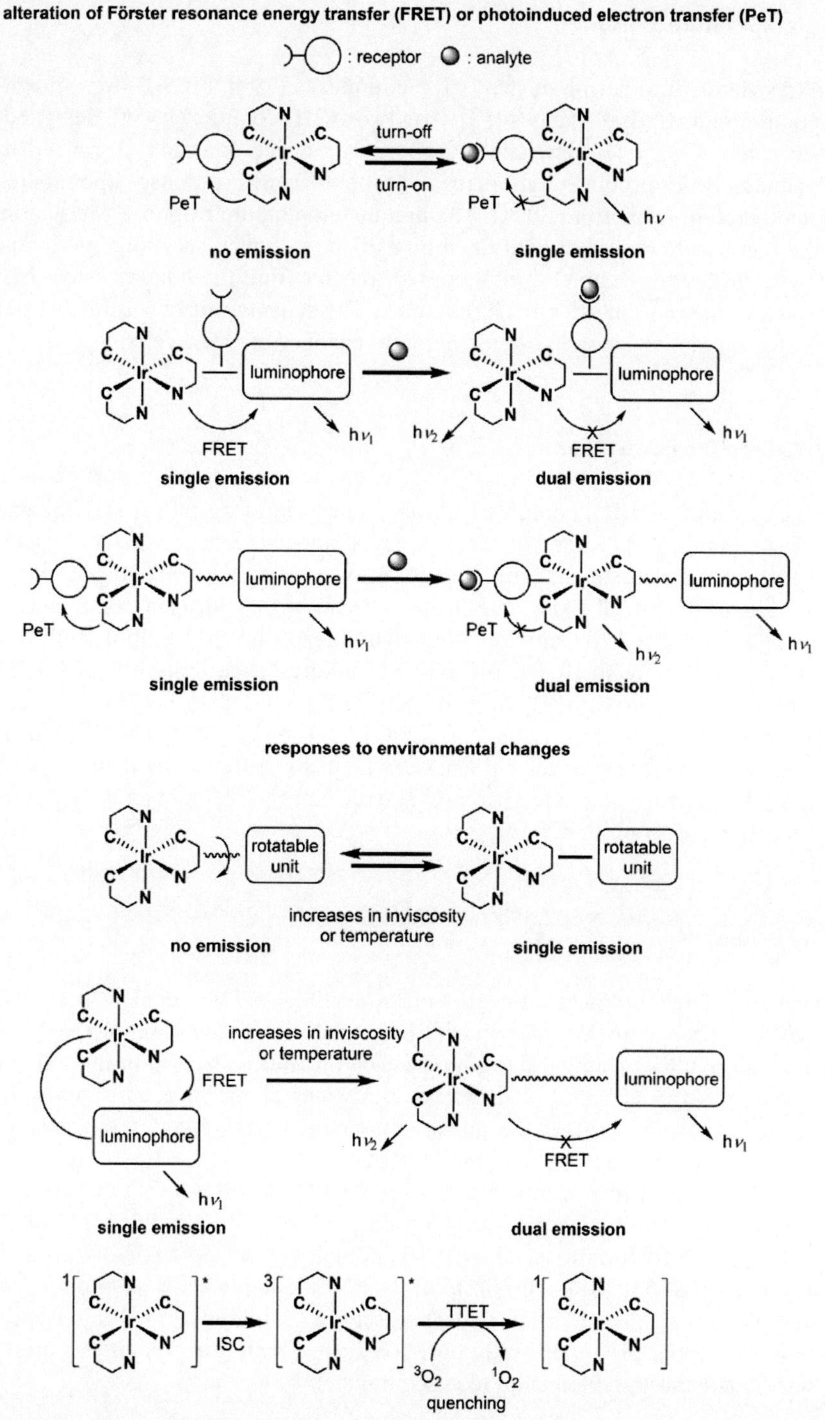

Fig. 9 Representative principles of creating luminescence probes

3.1 Temperature Sensor

Poly(acrylamide) tethering of two Ir(III) complexes (**79** in Fig. 10) was shown to respond to temperature changes [83]. The two Ir(III) complexes were designed to produce blue (λ_{em} = 470 nm) and red phosphorescence emission (λ_{em} = 590 nm) independently. **79** produced ratiometric phosphorescence responses upon increasing temperature from 16 to 36 °C. The unique temperature response was ascribed to the increased conformational flexibility of the polymer backbone, which presumably resulted in suppression of energy transfer from the blue-emissive Ir(III) complex to the red-emissive Ir(III) complex. The sensing utility was further demonstrated by the visualization of temperature changes of HeLa cells.

3.2 Carbon Dioxide Sensor

A charge-cationic Ir(III) complex bearing 2-phenylimidazo-[4,5,*f*]-[1,10]phenanthroline (**80** in Fig. 11) was found to display a phosphorescence turn-on response to CO_2 [84]. Pre-treatment with acetate anion (CH_3COO^-) led to phosphorescence quenching of **80** likely due to the formation of hydrogen bonds between CH_3COO^- and the –NH unit on 2-phenylimidazo-[4,5,*f*]-[1,10]phenanthroline. Bubbling of CO_2 restored the inherent phosphorescence emission of **80**. This phosphorescence turn-on response to CO_2 was proposed to originate from the two-step elimination of the bound CH_3COO^-: CO_2 might be attacked by a deprotonated form of **80** to produce carbamate, which was followed by hydrolysis. The dosimetric response was selective to CO_2 over other gases, including N_2, O_2 and Ar, and also electrophilic CO.

3.3 pH Sensors

Phosphorescence ratiometric imaging of intracellular pH was demonstrated using a zwitter ion consisting of two oppositely charged Ir(III) complexes (**81**) [85]. The macrocation in **81** exhibited phosphorescence turn-off responses in the red emission region (λ_{em} = 625 nm) to decreased pH, whereas the blue phosphorescence emission (λ_{em} = 451 nm) of the negatively charged counterpart remained intact (Fig. 12a). The zwitter ion thus produced phosphorescence ratiometric changes over the pH variations in the region 2.03–7.94. This pH sensing behavior was further exploited to image pH levels within HepG-2 cells. Similar pH responses could be obtained from a series of Ir(III) complexes having basic amino and or imino units (**82–85**) [86]. The basic units in the complexes were protonated at acidic pH and produced strong phosphorescence emissions (Fig. 12b). Deprotonation at elevated pH suppressed phosphorescence emission, which was presumably attributed to intramolecular electron transfer.

Table 2 Phosphorescence probes based on cyclometalated Ir(III) complexes

	Target analytes	λ_{ex} (nm)[a]	λ_{em} (nm)[b]	Signaling mode	K_d[c]	References
79	Temperature	405	470, 590	Ratiometric	–	[83]
80	CO_2	365	596	Turn-on	–	[84]
81	Proton	380	451, 625	Ratiometric	–	[85]
82	Proton	366	593	Turn-on	–	[86]
83	Proton	366	607	Turn-on	–	[86]
84	Proton	366	608	Turn-on	–	[86]
85	Proton	366	590	Turn-on	–	[86]
86	Cu^{2+}	400	593	Turn-off	–	[89]
87	Cu^{2+}	365	589	Turn-off	–	[90]
88	Cu^{2+}	322	536	Turn-off	–	[91]
89	Cu^{2+}	365	600	Turn-off	–	[92]
90	Cu^{2+}	355	560	Turn-off	–	[93]
91	Zn^{2+}	393	498	Turn-on	–	[94]
92	Zn^{2+}	372	580	Turn-on	–	[95]
93	Zn^{2+}	320	495	Turn-on	–	[96]
94	Zn^{2+}	345	572	Turn-on	82 pM	[97]
95	CN^-	380	583	Turn-on	–	[98]
96	$H_2PO_4^-$	372	560	Turn-on	–	[99]
97	HS^-	405	500	Ratiometric	–	[108]
98	HS^-	350	560	Turn-on	–	[109]
99	HS^-	350	560	Turn-on	–	[109]
100	HS^-	350	551	Turn-on	–	[109]
101	Sulfite, bisulfite	405	600	Turn-on	–	[110]
102	Sulfite, bisulfite	405	591	Turn-on	–	[111]
103	Sulfite, bisulfite	405	513, 544	Turn-on	–	[111]
104	Sulfite, bisulfite	405	529, 570	Turn-on	–	[111]
105	Cysteine	405	485, 560	Ratiometric	–	[115]
106	Cysteine, homocysteine, glutathione	365	564	Turn-on	–	[116]
107	Cysteine	365	486	Turn-on	–	[117]
108	Cysteine, homocysteine	405	564	Turn-on	–	[118]
109	Cysteine	405	632	Turn-on	–	[119]
110	Glutathione, ascorbic acid	405	614	Turn-on	–	[122]
111	O_2	–	520, 620	Ratiometric	–	[131]
112	O_2	405	630	Ratiometric	–	[132]
113	O_2	310	621	Ratiometric	–	[133]
114	O_2	310	620	Ratiometric	–	[133]
115	O_2	310	550	Ratiometric	–	[133]
116	O_2	310	619	Ratiometric	–	[133]
117	O_2	365	520, 580	Ratiometric	–	[134]
118	ClO^-	405	500, 598	Ratiometric	–	[142]
119	ClO^-	370	615	Turn-off	–	[143]
120	ClO^-	405	600	Ratiometric	–	[144]

Table 2 (continued)

	Target analytes	λ_{ex} (nm)[a]	λ_{em} (nm)[b]	Signaling mode	K_d[c]	References
121	O_2, ClO^-	405	550	Turn-off/turn-on	–	[145]
122	ClO^-	405	600	Ratiometric	–	[146]
123	ClO^-	405	534	Turn-on	–	[147]
124	ClO^-	405	560	Turn-on	–	[147]
125	ClO^-	405	580	Turn-on	–	[147]
126	ClO^-	405	598	Turn-on	–	[147]
127	ClO^-	402	590	Turn-on	–	[148]
128	$ONOO^-$	515	600	Ratiometric	–	[152]
129	$ONOO^-$	405	474	Ratiometric	–	[153]
130	$ONOO^-$	405	704	Turn-off	–	[154]
131	Phosphorylated peptides	380	475	Turn-on	–	[155]
132	Tetrapeptides	–	573	Ratiometric	–	[156]
133	Viscosity	450	521, 708	Ratiometric	–	[162]
134	Viscosity	405	479	Turn-on	–	[163]
135	Viscosity	405	567	Turn-on	–	[164]
136	Viscosity	405	585	Turn-on	–	[164]
137	Viscosity	405	551	Turn-on	–	[164]
138	Viscosity	405	530	Turn-on	–	[165]

[a]Excitation wavelength. [b]Emission wavelength. [c]Dissociation constant

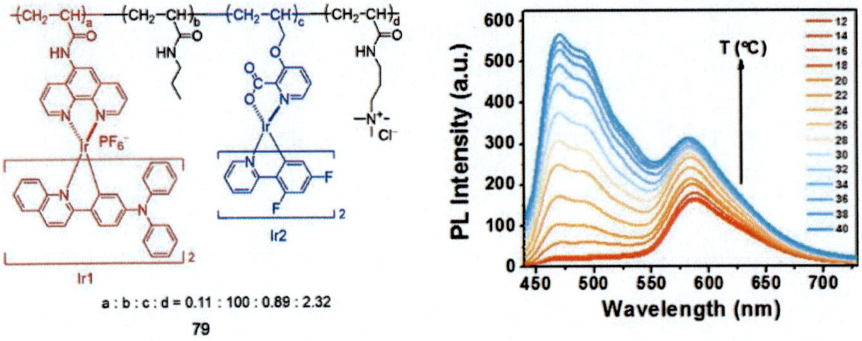

Fig. 10 Phosphorescence temperature sensor. Reproduced from Ref. [83]. Copyright 2016, WILEY-VCH Verlag GmbH & Co. KGaA, Weinheim

3.4 Metal Ion Sensors

Sensors capable of detecting biological metal ions are valuable as they can be used in the study of chemical mechanisms mediated by the metal ions. After the reports of phosphorescence probes of Cu(II) and Zn(II) [87, 88], a number of phosphorescence metal ion sensors have been established. The Cu(II) ion is a readily accessible target analyte, because it is bound most strongly to imino receptors, such as

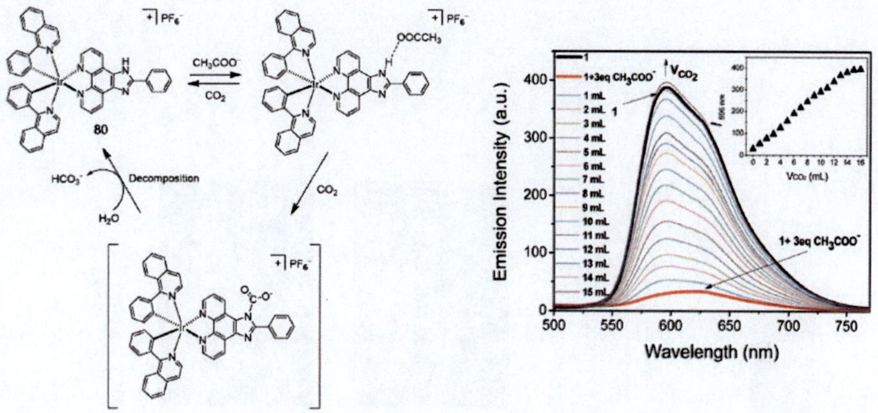

Fig. 11 Phosphorescent sensor of CO_2. Reproduced from Ref. [84]. Copyright 2015, Royal Society of Chemistry

pyridine. One representative example is the complex **86** which had a 2,2′-bipyridine pendant (Fig. 13) [89]. This pendant could bind with Cu(II) ions, which resulted in phosphorescence quenching. The emission quenching was attributed to the paramagnetic nature of the Cu(II) ion. The other metal ions caused no or slight phosphorescence responses, indicating high Cu(II) selectivity. The phosphorescence emission recovered upon the addition of iodide anions which reduced Cu(II) to Cu(I). Similar phosphorescence turn-off responses were observed for complexes **87** and **88** containing di(2-picolyl)amine receptors [90, 91], the complex **89** tethering a 2-picolylamino receptor [92], and complex **90** bearing the 5,6-bis(salicylideneimino) receptor (Fig. 13) [93].

Receptors virtually identical to those used for Cu(II) sensors are employed for creating Zn(II) sensors. The diamagnetic Zn(II) ion does not quench the emission of a luminophore. Creating zinc sensors thus requires zinc-induced perturbation of a photophysical process, such as photoinduced electron transfer. A series of phosphorescence zinc sensors, including **91**, were synthesized based on heteroleptic biscyclometalated Ir(III) complexes to examine the molecular factor that governed the photophysical perturbation exerted by zinc binding [94]. Investigations indicated that an acceleration of photoinduced electron transfer in the zinc-free state produced a large turn-on response. This mechanistic understanding provided useful guidance toward designing phosphorescence zinc sensors, such as complexes **92**, **93** and **94** [95–97].

3.5 Anion Sensors

The development of anion sensors is challenging, as exemplified by the scarcity of anion sensors in the literature. Complex **95** exhibited emission with λ_{em} of 583 nm, which was quenched upon binding with the Cu(II) ion (Fig. 14a) [98]. The non-emissive cupric complex became emissive upon an interaction with CN^-. The turn-on response was ascribed to the strong affinity of the CN^- anion with the Cu(II)

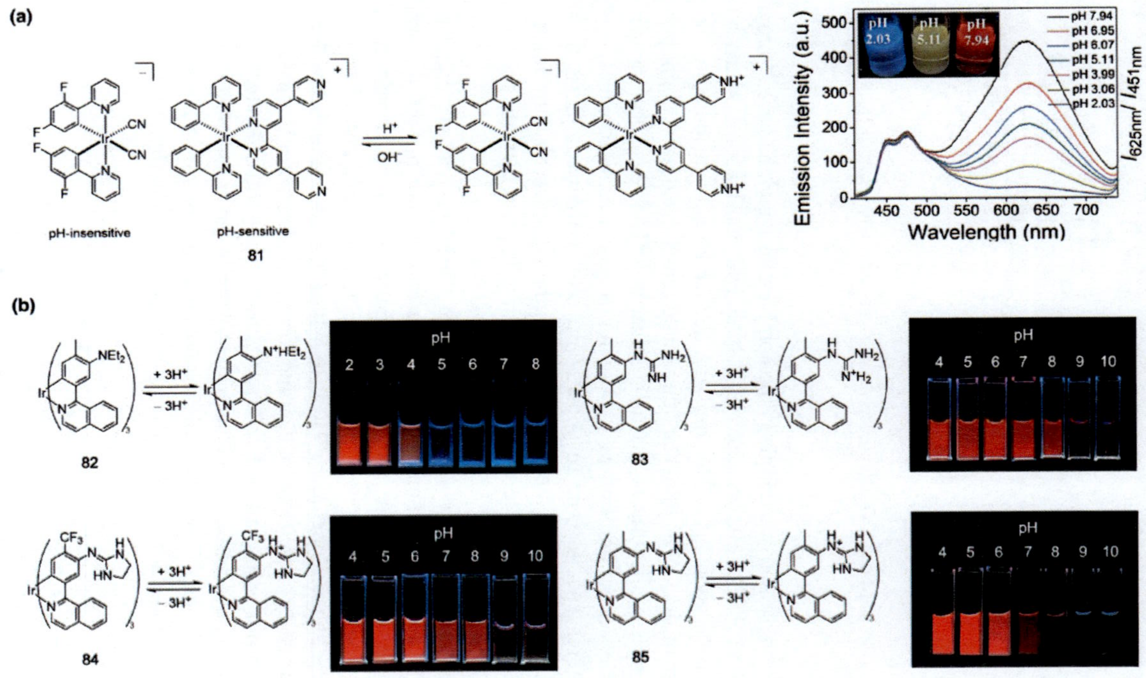

Fig. 12 **a** Phosphorescent pH sensors based on an ion pair **81** and its pH-dependent phosphorescence emission changes. **b** Ir(III) complexes having amino and imino units (**82–85**) showing phosphorescence responses to pH-changes. Reproduced from Refs. [85] and [86]. Copyright 2016, Royal Society of Chemistry and Copyright 2015, American Chemical Society

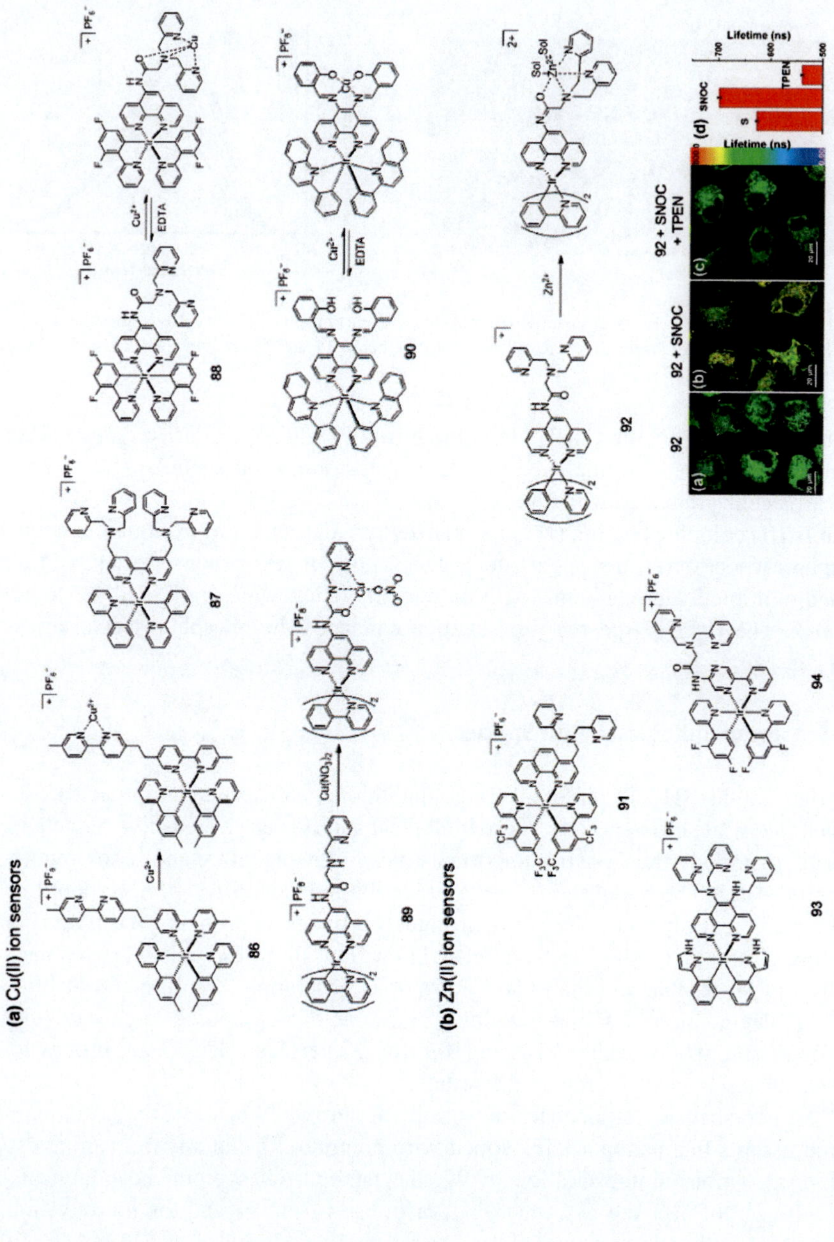

Fig. 13 Phosphorescence sensors of copper(II) (a) and zinc(II) ions (b). Reproduced from Ref. [95]. Copyright 2018, American Chemical Society

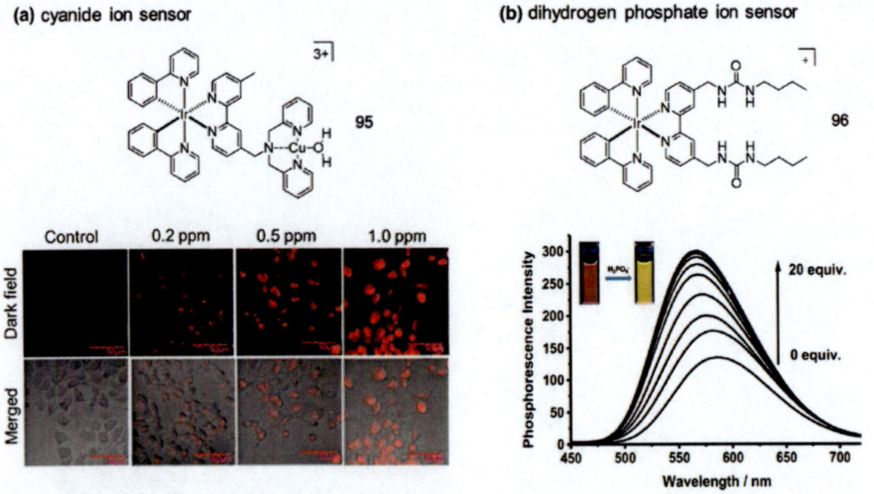

Fig. 14 Phosphorescent sensors of cyanide (**a**) and dihydrogen phosphate anions (**b**). Reproduced from Refs. [98] and [99]. Copyright 2013, Royal Society of Chemistry and Copyright 2014, Elsevier Ltd., respectively

center which liberated the Cu(II) ion in the form of $Cu(CN)_n^{-(n-2)}$ ($n = 2$ or 4). This dosimetric response was employed for the visualization of intracellular CN^- within live HeLa cells pretreated with NaCN.

An Ir(III) complex bearing two preorganized urea groups (**96**) exhibited a unique phosphorescence turn-on response to $H_2PO_4^-$ [99]. It was proposed that $H_2PO_4^-$ formed multiple hydrogen bonds with the urea moieties, which rigidified the Ir(III) complex. The $H_2PO_4^-$-specific rigidification enhanced the phosphorescence intensity (Fig. 14b).

3.6 Sensors of Biological Sulfur Species

Hydrogen sulfide (H_2S), sulfite, bisulfite, glutathione, and several amino acids constitute biological sulfur species. These biological sulfur species are active reductants in the human body thus playing important roles in the signaling and redox regulation [100–102]. The endogenous levels of the biological sulfur species should be maintained precisely, but failure in their homeostasis results in fatal conditions. For example, abnormal levels of H_2S are related to various diseases, such as Alzheimer's disease, hypertension, and diabetes. Endogenous generation of sulfite and bisulfite from the degradation of sulfur-containing amino acids is proposed to cause oxidative stress and age-associated disease [103–107]. Therefore, developing probes for biological sulfur species is highly important.

Phosphorescence ratiometric detection of intracellular H_2S was demonstrated using silica nanoparticles doped with complex **97** and a merocyanine dye (Fig. 15a). Upon photoexcitation of **97**, the nanoparticles produced dual emission with λ_{em} of 500 and 596 nm which originated from **97** and the merocyanine dye, respectively, due to partial energy transfer. An addition of NaSH selectively

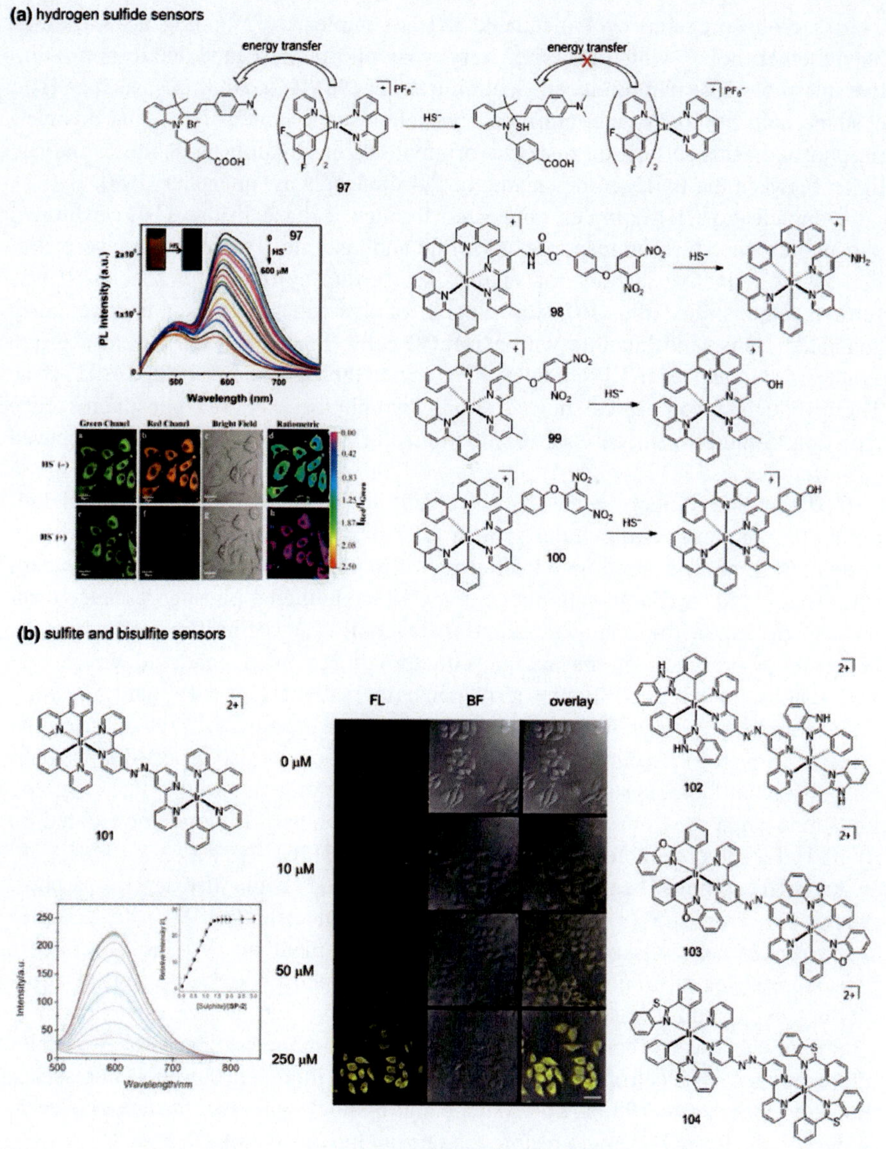

Fig. 15 Phosphorescent sensors of **a** hydrogen sulfide and **b** sulfite and bisulfite. Reproduced from Refs. [108] and [110]. Copyright 2015, American Chemical Society and Copyright 2013, Royal Society of Chemistry, respectively

quenched the fluorescence emission of the merocyanine dye. This selective quenching was ascribed to the formation of a HS^- adduct of merocyanine, which abrogated energy transfer (Fig. 15a) [108].

Lo's group prepared cyclometalated Ir(III) complexes containing a 2,4-dinitrophenyl ether moiety which exhibited very weak phosphorescence due to non-radiative intramolecular photoinduced electron transfer (**98–100**). Biothiols, such as H_2S, cysteine, and homocysteine, enhanced phosphorescence intensities. This dosimetric phosphorescence turn-on response originated from hydrolysis of the carbamate linker between the Ir(III) complex and the 2,4-dinitrophenyl quencher [109].

A dinuclear Ir(III) complex connected through a diazo bridge (**101**) exhibited phosphorescence turn-on responses to sulfite and bisulfite. The responses were elicited by the reductive degradation of the diazo group by the sulfur species. It was further demonstrated that **101** was capable of live-cell imaging of endogenously generated sulfite and bisulfite within HepG2 cells (Fig. 15b) [110]. Chao's group synthesized dinuclear Ir(III) complexes connected through a diazo bridge (**102–104**). These dinuclear complexes showed strong two-photon-excitable phosphorescence emission. Similar to the case of **101**, the diazo bridge could be reductively cleaved by sulfur species [111].

Cysteine and homocysteine are involved in physiological processes including redox regulation and cellular growth [112–114]. A phosphorescence sensor of cysteine was created based on a zwitter ion of oppositely charged Ir(III) complexes (Fig. 16a) [115]. A charge-cationic complex **105** exhibited a phosphorescence turn-on response upon the conjugate addition reaction of cysteine. In contrast, phosphorescence emission of the anionic counterpart remained unaltered which created a phosphorescence ratiometric response to cysteine (λ_{em} = 485 and 560 nm). This ratiometric phosphorescence response enabled the visualization intracellular cysteine (Fig. 16a). A dinuclear Ir(III) complex **106** could also report the presence of cysteine and homocysteine. Phosphorescence turn-on responses to thiol-containing amino acids were observed at λ_{em} = 564 nm, which was further demonstrated for living HeLa cells pretreated with the Ir(III) complex [116].

An Ir(III) complex bearing a Cu(II) ionophore (**107**) could also serve as a phosphorescent sensor for cysteine [117]. The heterodinuclear Ir(III)–Cu(II) complex was virtually non-emissive, but the Cu(II) displacement by cysteine generated a phosphorescence turn-on response. This dosimetric phosphorescence sensing property was exploited for imaging cysteine in live zebrafish. The high nucleophilicity of a thiol moiety was further employed to design a phosphorescence sensor. Complex **108** contained an electrophilic formyl group which hindered the phosphorescence transition inherent to **108** [118]. Cysteine and homocysteine can form cyclic structures with the formyl group to restore the phosphorescence emission.

Complex **109** exhibited weak phosphorescence emission with λ_{em} of 650 nm under two-photon excitation at a wavelength of 800 nm [119]. The conjugate addition reaction occurred with cysteine, which significantly enhanced a phosphorescence intensity. The unique two-photon-excitable phosphorescence activity enabled imaging intracellular cysteine within live KB cells under 750 nm excitation.

Redox potentials govern numerous intracellular processes, such as DNA synthesis [120, 121]. The redox state is precisely regulated by complex actions exerted by biological sulfur species, such as glutathione. An Ir(III) complex bearing a redox-active 2,2,6,6-tetramethyl-piperidine-1-oxyl (TEMPO) unit (**110**) exhibited phosphorescence emission at λ_{em} of 614 nm [122]. The

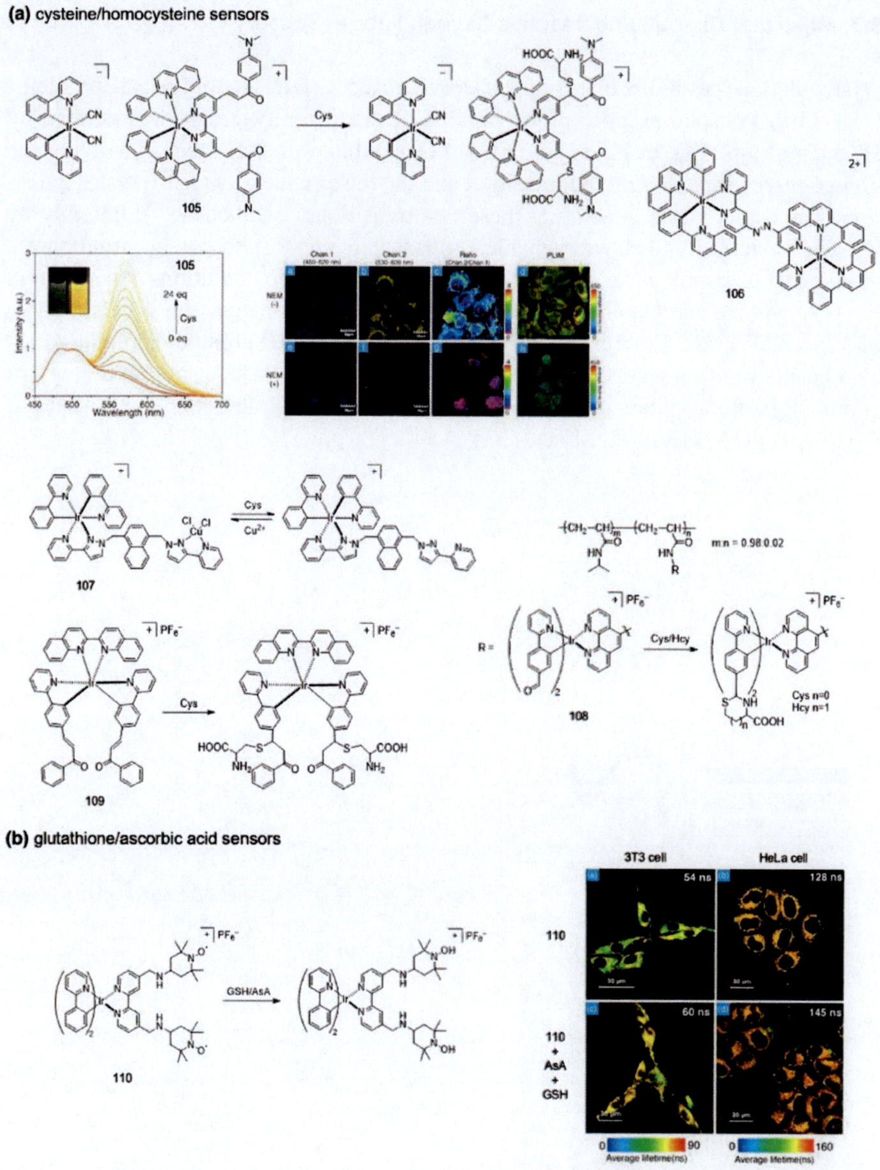

Fig. 16 Phosphorescent sensors of **a** cysteine and homocysteine, and **b** glutathione and ascorbic acid. Reproduced from Refs. [115] and [122]. Copyright 2016 and 2017, Optical Society of America

phosphorescence intensity increased upon the addition of glutathione and ascorbic acid into the solution containing **110**. The turn-on response was due to reduction of paramagnetic TEMPO into a hydroxyamino species by the biothiols. **110** was used to visualize the intracellular redox state of 3T3 cells (Fig. 16b).

3.7 Molecular Oxygen and Reactive Oxygen Species Sensors

Dysregulation of molecular oxygen levels causes fatal pathological problems [123–130]. Phosphorescence probe capable of reporting hyperoxia was devised by Huang's group (**111** in Fig. 17a) [131]. **111** exhibited dual emission involving the higher-energy ligand-centered transition and the lower-energy charge-transfer transition. The triplet states generating these two transitions are quenched differently by O_2. This interesting behavior enabled ultrasensitive phosphorescence ratiometric monitoring of hypoxic (i.e., lower O_2 levels) and hyperoxic conditions. As shown in Fig. 17, hypoxic and hyperoxic regions was easily discriminated with different phosphorescence colors. Fluorescent polymers containing Ir(III) complex moieties (**112**) also enabled luminescence ratiometric oxygen sensing [132]. A similar strategy of combining a fluorophore and a phosphore was further exploited to create a series of O_2 sensors (**113–116** in Fig. 17a) [133].

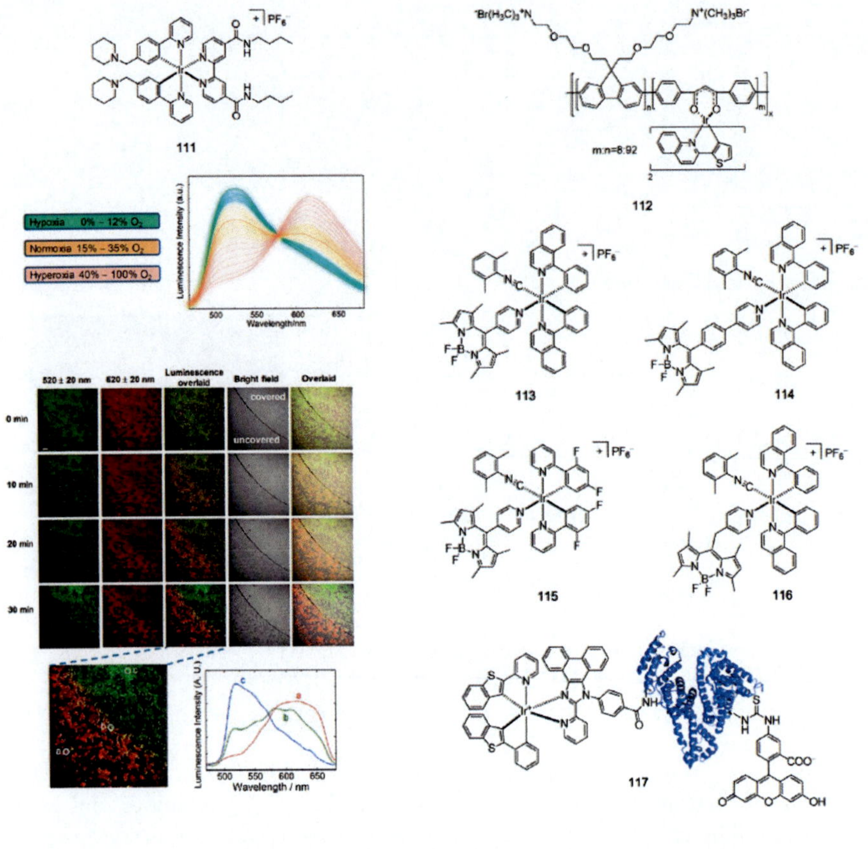

Fig. 17 Phosphorescent sensors of molecular dioxygen. The blue 3D structure in **117** denotes HSA. Reproduced from Ref. [131]. Copyright 2018, American Chemical Society

An interesting approach was based on a conjugate of an Ir(III) complex and human serum albumin (HSA) bearing fluorescein (**117**). The conjugate showed perturbations in pH and O_2 levels through phosphorescence ratiometric responses and phosphorescence lifetime changes [134].

The oxygen metabolism inevitably generates reactive oxygen species (ROSs). Studies indicated that ROSs are involved in a variety of pathological and physiological processes [135–140]. Low levels of ROSs are essential for signaling cell proliferation, whereas abnormally high levels of ROSs induce severe cell damage and even cell death. The endogenous levels of ROSs are precisely regulated, but they can rapidly increase upon stimulation, such as inhalation of toxic substances. Excessive ROSs can damage biological molecules, including DNA, lipids and proteins, which is believed to be related to the genesis of cancer. However, molecular mechanisms underlying the pathophysiology of ROSs remains insufficiently understood. Therefore, sensors capable of detecting ROSs are critically needed [141]. Core–shell silica nanoparticles consisting of an Ir(III) complex core and a hypochlorite (ClO^-)-responsive Ir(III) complex (**118**) shell were created (Fig. 18) [142]. The particles exhibited dual phosphorescence emission with λ_{em} of 500 nm and 598 nm, which originated from the core and shell parts, respectively. The latter emission increased upon a selective interaction with ClO^-. The luminescence ratiometric response was utilized to perform photoluminescence lifetime imaging of intracellular ClO^-.

An Ir(III) complex having acrylate pendants exhibited strong phosphorescence emission, but oxidative elimination of the acrylate by ClO^- turned off the phosphorescence (**119**) [143]. The oxidative cleavage by ClO^- was exploited to create a phosphorescence ratiometric sensor. A ClO^--insensitive Ir(III) complex ($\lambda_{em} = 680$ nm) and a ClO^--sensitive Ir(III) complex (**120**; $\lambda_{em} = 600$ nm) were doped into an amphiphilic DSPE-mPEG5000 polymer [144]. The 4-amino-3-nitrophenyloxy moiety in **120** was cleaved upon the reaction with ClO^-, which produced a tenfold enhancement in the phosphorescence emission. The turn-on response was elicited due to abrogation of photoinduced electron transfer from the nitroanilino group. Ir(III) complexes capable of sensing molecular dioxygen and ClO^- were developed by Huang et al. An Ir(III) complex having lipophilic alkyl chains (**121**) was found to be localized within the cell membrane and cytoplasm. The localization permitted the spatially resolved visualization of molecular dioxygen and ClO^- [145]. Polymer tethering a phenothiazine-based fluorophore and an Ir(III) complex (**122**) could also report the phosphorescence changes to ClO^- and molecular dioxygen [146].

Chao's group developed Ir(III) complexes having diaminomaleonitrile units (**123–126**). The Ir(III) complexes showed weak phosphorescence emission because of the non-radiative decay process centered at the C=N moiety. ClO^- transformed diaminomaleonitrile into carboxylate, producing phosphorescence turn-on responses [147]. The ClO^- reaction was utilized to create a two-photon-excitable probe (**127**) [148].

Drug-induced hepatotoxicity causes critical damage to the liver, but the pathophysiological mechanisms of the drug-mediated liver injury are poorly understood [149]. Recent studies revealed abnormal levels of ROSs in the injured liver, which

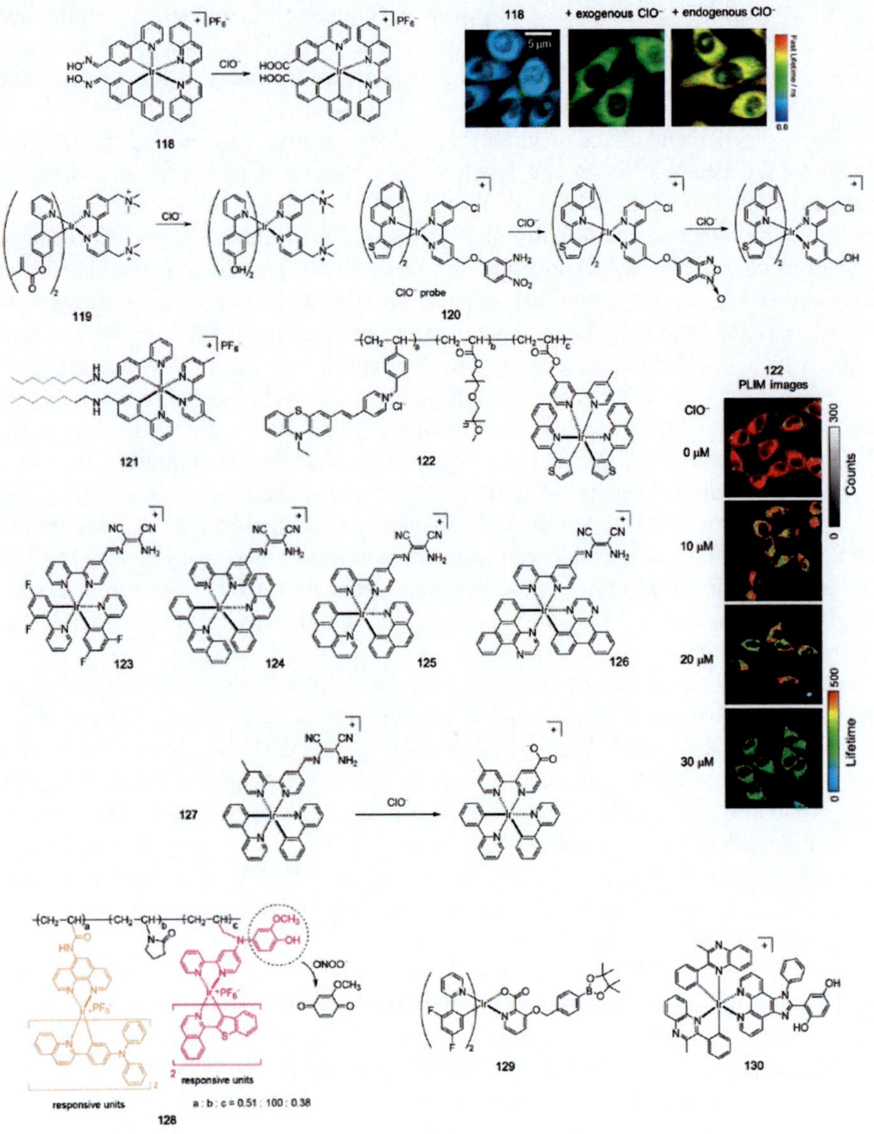

Fig. 18 Phosphorescent sensors of reactive oxygen species. Reproduced from Refs. [142] and [146]. Copyright 2015 and 2021, Royal Society of Chemistry, respectively

indicated certain activity of ROSs [150, 151]. Polymers tethering Ir(III) complexes (**128**) served as a probe for peroxynitrite ($ONOO^-$), which is a marker of drug-induced liver injury [152]. **128** produced photoluminescence ratiometric responses to $ONOO^-$: upon addition of $ONOO^-$, phosphorescence emission at 680 nm increased selectively due to removal of the *N*-(4-hydroxylphenyl)rhodol unit. The polymers were used for the direct visualization of elevated endogenous $ONOO^-$ levels after the ketoconazole-induced liver injury.

Zhao's group incorporated an $ONOO^-$-responsive Ir(III) complex (**129**) and an $ONOO^-$-inactive reference Ir(III) complex into mesoporous silica nanoparticles [153]. Peroxynitrite quenched the phosphorescence emission of **129**, generating a phosphorescence ratiometric response. The silica nanoparticles could penetrate into live RAW 264.7 cells and report elevated intracellular levels of $ONOO^-$.

Chao's group also developed an NIR-phosphorescent Ir(III) complex (**130**) for sensing ONOO. Peroxynitrite quenched the phosphorescence emission of the complex through oxidative dehydrogenation of hydroquinone moiety, while glutathione turned on the phosphorescence emission by reducing benzoquinone to hydroquinone. The reversible redox responses of **130** were utilized for the visualization of intracellular redox states [154].

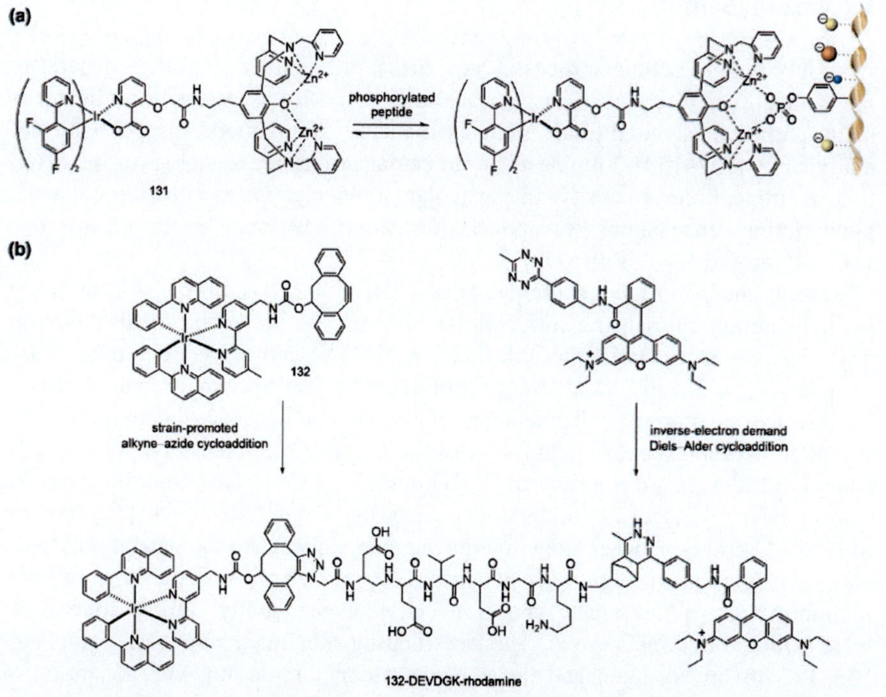

Fig. 19 Phosphorescent sensors of **a** phosphorylated peptides and **b** tetrapeptides of Asp-Glu-Val-Asp

3.8 Peptide Sensors

An Ir(III) complex bearing zinc ionophores (**131**) exhibited enhanced phosphorescence emission upon an interaction with negatively charged phosphorylated peptides (Fig. 19a). The enhanced emission originated from increased hydrophobicity provided by phosphorylated peptides. A linear correlation between the phosphorescence intensity and the net charge of phosphorylated peptides implied reduced repulsion between adjacent Zn^{2+} ions upon binding with peptides. This behavior enabled the selective detection of phosphorylated peptides without suffering from interferences caused by the other anions, including phosphate [155].

An Ir(III) complex bearing dibenzocyclooctyne (**132**) was able to label peptides through biorthogonal reactions (Fig. 19b). An Asp-Glu-Val-Asp tetrapeptide was selectively labeled by **132** through the strain-promoted alkyne–azide cycloaddition reaction, and the inverse-electron-demand Diels–Alder cycloaddition reaction with rhodamine. The resulting **132**-DEVDGK-rhodamine conjugate facilitated intra-peptide energy transfer from **132** to rhodamine. Cutting enzymes, such as caspase-3, could cleave the DEVD moiety, suppressing the energy transfer. Incubation of living HeLa cells with cisplatin cleaved the internalized **132**-DEVDGK-rhodamine, through the intracellular generation of caspase-3. The intracellular reaction enabled tracking cellular apoptosis [156].

3.9 Viscosity Sensors

A variety of intracellular processes are diffusion-mediated, and they depend on the intracellular viscosity. Thus, intracellular viscosity serves as an indicator of many pathological conditions, such as diabetes, Alzheimer's disease, and cell malignancy [157–159]. Luminescent molecular probes are valuable for the detection of intracellular viscosity. In particular, molecular rotors showing enhanced photoluminescence intensities upon rigidification have been employed to create viscosity probes [78, 160, 161].

Several phosphorescent cyclometalated Ir(III) complexes have also been tested for their ability for sensing intracellular viscosity. A molecular dyad having an NIR-phosphorescent Ir(III) complex (λ_{em} = 708 nm) and a green-emitting Ir(III) complex (λ_{em} = 521 nm) (**133**) were synthesized by Sun's group. Increased viscosity resulted in different enhancements of the phosphorescence intensities of the two Ir(III) complexes, producing a phosphorescence ratiometric response [162]. Chao et al. developed a dinuclear Ir(III) complex (**134**). This complex revealed lysosomal microviscosity. Under a low-viscosity environment, the free rotation of the C–C bridge between the two emitting cores quenched the phosphorescence emission. In contrast, high-viscosity media suppressed the non-radiative rotation to improve phosphorescence emission [163]. Conceptually similar approaches were explored by Mao's group. The free-rotating substituents in Ir(III) complexes **135–137** effectively quenched the phosphorescence emission. Internalization of the complexes into mammalian cells enabled monitoring mitochondrial viscosity

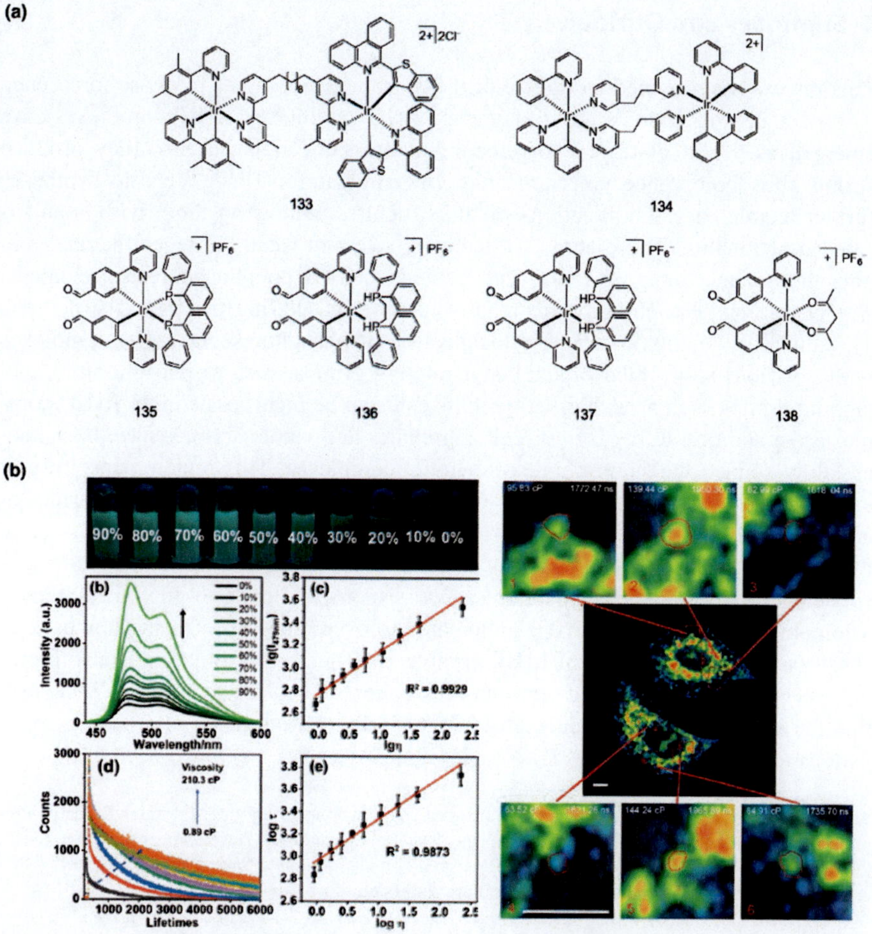

Fig. 20 **a** Phosphorescent viscosity sensors. **b** Phosphorescence responses of **134** to viscosity changes (left), and the determination of the mitochondrial viscosity within A549 cells stained with **137** (right). Reproduced from Refs. [163] and [164]. Copyright 2019, Royal Society of Chemistry, respectively

through photoluminescence lifetime-imaging microscopic techniques (Fig. 20b) [164]. The same group further demonstrated that an archetype Ir(III) complex **138** exhibited enhanced phosphorescence in high-viscosity media. Although the photophysical mechanism underlying the rigidochromic response remained elusive, the phosphorescence response enabled mapping the viscosity of endoplasmic reticulum in MCF-7 cells [165].

4 Summary and Outlook

This review has provided an update of the recent developments in phosphorescence labels and sensors based on cyclometalated Ir(III) complexes. Ir(III) complexes have emerged as useful platforms for creating bioimaging tools because they produce strong phosphorescence emission at room temperature. Their versatile syntheses further enable easy modifications of the structures, endowing them with sophisticated functionality. These attractive features have motivated the recent advances in phosphorescence imaging tools. This review focused on phosphorescence labels. The charge and lipophilicity are readily tunable by altering the ligand structures. The tunability results in selective localization of phosphorescent Ir(III) complexes within various subcellular organelles. Directing groups, such as phosphonium, and tagging groups such as a chloroethyl units can also be incorporated into Ir(III) complexes. In addition to organelles, cell aggregates and whole animals have been successfully stained with phosphorescent Ir(III) complexes. This review paper further examined phosphorescence sensors prepared from Ir(III) complexes. A variety of analytes were able to be detected with the phosphorescence sensors, and many of them were visualized under biological conditions. Notable advances include sensors capable of producing phosphorescence responses under two-photon excitation. The nonlinear optical property is anticipated to overcome the intrinsic drawback of short-wavelength excitation of Ir(III) complexes. Theranostic function is also promising because it offers unique opportunities to control biological activity. We expect that the advances will provide useful insights into creating novel bioimaging applications for cyclometalated Ir(III) complexes.

Acknowledgements Y.Y. acknowledges the financial support provided by the Midcareer Research Program (NRF-2019R1A2C2003969) and the Nano Material Technology Development Program (NRF-2021M3D1A2049323) through the National Research Foundation grants funded by the Ministry of Science, Information, and Communication Technology and Future Planning of Korea.

Declarations

Conflict of interest The authors declare that there is no conflict of interest.

References

1. You Y, Nam W (2012) Chem Soc Rev 41:7061–7084
2. Chi Y, Wang SF, Ganesan P (2019) Chem Rec 19:1644–1666
3. Buil ML, Esteruelas MA, López AM (2021) Eur J Inorg Chem 2021:4731–4761
4. Ho P-Y, Ho C-L, Wong W-Y (2020) Coord Chem Rev 413:213267
5. Shaikh S, Wang Y, Rehman F, Jiang H, Wang X (2020) Coord Chem Rev 416:213344
6. Zhang Y, Qiao J (2021) iScience 24:102858
7. Tan C-P, Zhong Y-M, Ji L-N, Mao Z-W (2021) Chem Sci 12:2357–2367
8. Chen Y, Guan R, Zhang C, Huang J, Ji L, Chao H (2016) Coord Chem Rev 310:16–40
9. He L, Tan CP, Ye RR, Zhao YZ, Liu YH, Zhao Q, Ji LN, Mao ZW (2014) Angew Chem Int Ed 53:12137–12141
10. Fan Y, Zhao J, Yan Q, Chen PR, Zhao D (2014) ACS Appl Mater Interfaces 6:3122–3131

11. Qiu K, Huang H, Liu B, Liu Y, Huang Z, Chen Y, Ji L, Chao H (2016) ACS Appl Mater Interfaces 8:12702–12710
12. Zhang KY, Yu Q, Wei H, Liu S, Zhao Q, Huang W (2018) Chem Rev 118:1770–1839
13. Zhao Q, Huang C, Li F (2011) Chem Soc Rev 40:2508–2524
14. Green DR, Reed JC (1998) Science, 1309–1312
15. Fulda S, Galluzzi L, Kroemer G (2010) Nat Rev Drug Discov 9:447–464
16. Zhu H, Fan J, Du J, Peng X (2016) Acc Chem Res 49:2115–2126
17. Yousif LF, Stewart KM, Kelley SO (2009) ChemBioChem 10:1939–1950
18. Guan R, Xie L, Rees TW, Ji L, Chao H (2020) J Inorg Biochem 204:110985
19. Li Y, Tan C-P, Zhang W, He L, Ji L-N, Mao Z-W (2015) Biomaterials 39:95–104
20. Li Y, Liu B, Lu X-R, Li M-F, Ji L-N, Mao Z-W (2017) Dalton Trans 46:11363–11371
21. Zhang KY, Liu H-W, Tang M-C, Choi AW-T, Zhu N, Wei X-G, Lau K-C, Lo KK-W (2015) Inorg Chem 54:6582–6593
22. Ye R-R, Tan C-P, Ji L-N, Mao Z-W (2016) Dalton Trans 45:13042–13051
23. Li Y, Liu B, Xu C-X, He L, Wan Y-C, Ji L-N, Mao Z-W (2020) J Biol Inorg Chem 25:597–607
24. Wu XW, Zheng Y, Wang FX, Cao JJ, Zhang H, Zhang DY, Tan CP, Ji LN, Mao Z-W (2019) Chem Eur J 25:7012–7022
25. Yi S, Lu Z, Zhang J, Wang J, Xie Z, Hou L (2019) ACS Appl Mater Interfaces 11:15276–15289
26. Cao J-J, Tan C-P, Chen M-H, Wu N, Yao D-Y, Liu X-G, Ji L-N, Mao Z-W (2017) Chem Sci 8:631–640
27. Zhang Q, Cao R, Fei H, Zhou M (2014) Dalton Trans 43:16872–16879
28. Nakagawa A, Hisamatsu Y, Moromizato S, Kohno M, Aoki S (2014) Inorg Chem 53:409–422
29. Kim I, Rodriguez-Enriquez S, Lemasters JJ (2007) Arch Biochem Biophys 462:245–253
30. Youle RJ, Narendra DP (2011) Nat Rev Mol Cell Biol 12:9–14
31. Chen M-H, Wang F-X, Cao J-J, Tan C-P, Ji L-N, Mao Z-W (2017) ACS Appl Mater Interfaces 9:13304–13314
32. Li SP-Y, Lau CT-S, Louie M-W, Lam Y-W, Cheng SH, Lo KK-W (2013) Biomaterials 34:7519–7532
33. Zhu JH, Tang BZ, Lo KK-W (2019) Chem Eur J 25:10633–10641
34. Lo KK-W, Law WH-T, Chan JC-Y, Liu H-W, Zhang KY (2013) Metallomics 5:808–812
35. Guan R, Chen Y, Zeng L, Rees TW, Jin C, Huang J, Chen Z-S, Ji L, Chao H (2018) Chem Sci 9:5183–5190
36. Jin C, Liu J, Chen Y, Zeng L, Guan R, Ouyang C, Ji L, Chao H (2015) Chem Eur J 21:12000–12010
37. Sun L, Chen Y, Kuang S, Li G, Guan R, Liu J, Ji L, Chao H (2016) Chem Eur J 22:8955–8965
38. Jin C, Guan R, Wu J, Yuan B, Wang L, Huang J, Wang H, Ji L, Chao H (2017) Chem Commun 53:10374–10377
39. Chen Q, Jin C, Shao X, Guan R, Tian Z, Wang C, Liu F, Ling P, Guan JL, Ji L (2018) Small 14:1802166
40. Lee LCC, Cheung HMH, Liu HW, Lo KK-W (2018) Chem Eur J 24:14064–14068
41. Lee LC-C, Tsang AW-Y, Liu H-W, Lo KK-W (2020) Inorg Chem 59:14796–14806
42. Qiu K, Zhu H, Rees TW, Ji L, Zhang Q, Chao H (2019) Coord Chem Rev 398:113010
43. Sansee A, Meksawangwong S, Chainok K, Franz K, Gál M, Pålsson L-O, Puniyan W, Traiphol R, Pal R, Kielar F (2016) Dalton Trans 45:17420–17430
44. Gao T-B, Yan R-Q, Metherell AJ, Cao D-K, Ye D-J, Ward MD (2017) Dalton Trans 46:16787–16791
45. Zhang DY, Zheng Y, Zhang H, Sun JH, Tan CP, He L, Zhang W, Ji LN, Mao Z-W (2018) Adv Sci 5:1800581
46. Zhang D, Wu M, Zeng Y, Wu L, Wang Q, Han X, Liu X, Liu J (2015) ACS Appl Mater Interfaces 7:8176–8187
47. Haubner R, Wester H-J, Burkhart F, Senekowitsch-Schmidtke R, Weber W, Goodman SL, Kessler H, Schwaiger M (2001) J Nucl Med 42:326–336
48. Dubey PK, Mishra V, Jain S, Mahor S, Vyas S (2004) J Drug Target 12:257–264
49. Moromizato S, Hisamatsu Y, Suzuki T, Matsuo Y, Abe R, Aoki S (2012) Inorg Chem 51:12697–12706
50. Yang Y, Guo L, Ge X, Tian Z, Gong Y, Zheng H, Du Q, Zheng X, Liu Z (2019) Dyes Pigm 161:119–129
51. Jin C, Li G, Wu X, Liu J, Wu W, Chen Y, Sasaki T, Chao H, Zhang Y (2021) Angew Chem Int Ed 60:7597–7601

52. Kuang S, Liao X, Zhang X, Rees TW, Guan R, Xiong K, Chen Y, Ji L, Chao H (2020) Angew Chem Int Ed 59:3315–3321
53. Yip AMH, Lai CKH, Yiu KSM, Lo KK-W (2022) Angew Chem Int Ed 134:e202116078
54. Shepherd VA (2006) Curr Top Dev Biol 75:171–223
55. Ma D-L, Zhong H-J, Fu W-C, Chan DS-H, Kwan H-Y, Fong W-F, Chung L-H, Wong C-Y, Leung C-H (2013) PLoS One 8:e55751
56. Li X, Yin Y, Gao P, Li W, Yan H, Lu C, Zhao Q (2017) Dalton Trans 46:13802–13810
57. Wu N, Cao J-J, Wu X-W, Tan C-P, Ji L-N, Mao Z-W (2017) Dalton Trans 46:13482–13491
58. He L, Liao S-Y, Tan C-P, Lu Y-Y, Xu C-X, Ji L-N, Mao Z-W (2014) Chem Commun 50:5611–5614
59. Sun H, Yang L, Yang H, Liu S, Xu W, Liu X, Tu Z, Su H, Zhao Q, Huang W (2013) RSC Adv 3:8766–8776
60. Leung PK-K, Lo KK-W (2020) Chem Commun 56:6074–6077
61. Leung PK-K, Lee LC-C, Yeung HH-Y, Io K-W, Lo KK-W (2021) Chem Commun 57:4914–4917
62. Liu C, Gao X, Yuan J, Zhang R (2020) Trends Anal Chem 133:116092
63. Xu W, Zeng Z, Jiang JH, Chang YT, Yuan L (2016) Angew Chem Int Ed 55:13658–13699
64. Carter KP, Young AM, Palmer AE (2014) Chem Rev 114:4564–4601
65. Maity A, Choi J-S, Teets TS, Deligonul N, Berdis AJ, Gray TG (2013) Chem Eur J 19:15924–15932
66. Masum AA, Hisamatsu Y, Yokoi K, Aoki S (2018) Bioinorg Chem Appl, 7578965
67. Masum A-A, Yokoi K, Hisamatsu Y, Naito K, Shashni B, Aoki S (2018) Bioorg Med Chem 26:4804–4816
68. Hisamatsu Y, Shibuya A, Suzuki N, Suzuki T, Abe R, Aoki S (2015) Bioconjug Chem 26:857–879
69. Quan LM, Mechler AI, Barnard PJ (2020) J Inorg Biochem 206:111047
70. Lo KK-W, Chan BT-N, Liu H-W, Zhang KY, Li SP-Y, Tang TS-M (2013) Chem Commun 49:4271–4273
71. Li SP-Y, Yip AM-H, Liu H-W, Lo KK-W (2016) Biomaterials 103:305–313
72. Helmchen F, Denk W (2005) Nat Methods 2:932–940
73. Cahalan MD, Parker I, Wei SH, Miller MJ (2002) Nat Rev Immunol 2:872–880
74. Huang H, Yang L, Zhang P, Qiu K, Huang J, Chen Y, Diao J, Liu J, Ji L, Long J (2016) Biomaterials 83:321–331
75. Qiu K, Ouyang M, Liu Y, Huang H, Liu C, Chen Y, Ji L, Chao H (2017) J Mater Chem B 5:5488–5498
76. Jin C, Liang F, Wang J, Wang L, Liu J, Liao X, Rees TW, Yuan B, Wang H, Shen Y (2020) Angew Chem Int Ed 59:15987–15991
77. Wu Y, Zeng G, Lvyue N, Wu W, Jiang T, Wu R, Guo W, Li X, Fan X (2017) J Mater Chem B 5:4973–4980
78. Yang Z, Cao J, He Y, Yang JH, Kim T, Peng X, Kim JS (2014) Chem Soc Rev 43:4563–4601
79. Sahoo SK, Kim G-D, Choi H-J (2016) J Photochem PhotoBiol C Photochem Rev 27:30–53
80. Kaur N, Kaur G, Fegade UA, Singh A, Sahoo SK, Kuwar AS, Singh N (2017) Trends Anal Chem 95:86–109
81. Chowdhury S, Rooj B, Dutta A, Mandal U (2018) J Fluoresc 28:999–1021
82. Wang X-D, Wolfbeis OS, Meier RJ (2013) Chem Soc Rev 42:7834–7869
83. Chen Z, Zhang KY, Tong X, Liu Y, Hu C, Liu S, Yu Q, Zhao Q, Huang W (2016) Adv Funct Mater 26:4386–4396
84. Ma Y, Xu H, Zeng Y, Ho C-L, Chui C-H, Zhao Q, Huang W, Wong W-Y (2015) J Mater Chem C 3:66–72
85. Ma Y, Liang H, Zeng Y, Yang H, Ho C-L, Xu W, Zhao Q, Huang W, Wong W-Y (2016) Chem Sci 7:3338–3346
86. Kando A, Hisamatsu Y, Ohwada H, Itoh T, Moromizato S, Kohno M, Aoki S (2015) Inorg Chem 54:5342–5357
87. You Y, Han Y, Lee Y-M, Park SY, Nam W, Lippard SJ (2011) J Am Chem Soc 133:11488–11491
88. You Y, Lee S, Kim T, Ohkubo K, Chae W-S, Fukuzumi S, Jhon G-J, Nam W, Lippard SJ (2011) J Am Chem Soc 133:18328–18342
89. Lu F, Yamamura M, Nabeshima T (2013) Tetrahedron Lett 54:779–782
90. Kim H-B, Li Y, Hyun MH (2013) Bull Korean Chem Soc 34:653–656
91. Shen W, Yan L, Tian W, Cui X, Qi Z, Sun Y (2016) J Lumin 177:299–305
92. Yang H, Zhu Y, Li L, Zhou Z, Yang S (2012) Inorg Chem Commun 16:1–3

93. Ma D-L, He H-Z, Chan DS-H, Wong C-Y, Leung C-H (2014) PLoS One 9:e99930
94. Woo H, Cho S, Han Y, Chae W-S, Ahn D-R, You Y, Nam W (2013) J Am Chem Soc 135:4771–4787
95. Zhang C, Liu M, Liu S, Yang H, Zhao Q, Liu Z, He W (2018) Inorg Chem 57:10625–10632
96. Ma D-L, He H-Z, Zhong H-J, Lin S, Chan DS-H, Wang L, Lee SM-Y, Leung C-H, Wong C-Y (2014) ACS Appl Mater Interfaces 6:14008–14015
97. Ryu SY, Huh M, You Y, Nam W (2015) Inorg Chem 54:9704–9714
98. Das P, Saha S, Baidya M, Ghosh SK, Das A (2013) Chem Commun 49:255–257
99. Li Y, Liu Y, Nam D, Park S, Yoon J, Hyun MH (2014) Dyes Pigm 100:241–246
100. Abe K, Kimura H (1996) J Neurosci 16:1066–1071
101. Papapetropoulos A, Pyriochou A, Altaany Z, Yang G, Marazioti A, Zhou Z, Jeschke MG, Branski LK, Herndon DN, Wang R (2009) Proc Natl Acad Sci 106:21972–21977
102. Li L, Rose P, Moore PK (2011) Annu Rev Pharmacol Toxicol 51:169–187
103. Šimić G, Leko MB, Wray S, Harrington CR, Delalle I, Jovanov-Milošević N, Bažadona D, Buée L, de Silva R, Di Giovanni G (2017) Prog Neurobiol 151:101–138
104. Yang G, Wu L, Jiang B, Yang W, Qi J, Cao K, Meng Q, Mustafa AK, Mu W, Zhang S (2008) Science 322:587–590
105. Kaneko Y, Kimura Y, Kimura H, Niki I (2006) Diabetes 55:1391–1397
106. Feng X, Chen Y, Zhao J, Tang C, Jiang Z, Geng B (2009) Biochem Biophys Res Commun 380:153–159
107. Eto K, Asada T, Arima K, Makifuchi T, Kimura H (2002) Biochem Biophys Res Commun 293:1485–1488
108. Yu Q, Zhang KY, Liang H, Zhao Q, Yang T, Liu S, Zhang C, Shi Z, Xu W, Huang W (2015) ACS Appl Mater Interfaces 7:5462–5470
109. Tso KK-S, Liu H-W, Lo KK-W (2017) J Inorg Biochem 177:412–422
110. Li G, Chen Y, Wang J, Lin Q, Zhao J, Ji L, Chao H (2013) Chem Sci 4:4426–4433
111. Li G, Chen Y, Wang J, Wu J, Gasser G, Ji L, Chao H (2015) Biomaterials 63:128–136
112. Wood ZA, Schröder E, Harris JR, Poole LB (2003) Trends Biochem Sci 28:32–40
113. Rhee SG, Chae HZ, Kim K (2005) Free Radic Biol Med 38:1543–1552
114. Circu ML, Aw TY (2010) Free Radic Biol Med 48:749–762
115. Liu S, Xu A, Chen Z, Ma Y, Yang H, Shi Z, Zhao Q (2016) Opt Express 24:28247–28255
116. Li G, Chen Y, Wu J, Ji L, Chao H (2013) Chem Commun 49:2040–2042
117. Deng P, Pei Y, Liu M, Song W, Wang M, Wang F, Wu C, Xu L (2021) RSC Adv 11:7610–7620
118. Ma Y, Liu S, Yang H, Wu Y, Sun H, Wang J, Zhao Q, Li F, Huang W (2013) J Mater Chem B 1:319–329
119. Xu W, Zhao X, Lv W, Yang H, Liu S, Liang H, Tu Z, Xu H, Qiao W, Zhao Q (2014) Adv Healthc Mater 3:658–669
120. Kamata H, Hirata H (1999) Cell Signal 11:1–14
121. Lu T, Pan Y, Kao S-Y, Li C, Kohane I, Chan J, Yankner BA (2004) Nature 429:883–891
122. Liu S, Zhou N, Chen Z, Wei H, Zhu Y, Guo S, Zhao Q (2017) Opt Lett 42:13–16
123. Harris AL (2002) Nat Rev Cancer 2:38–47
124. Murdoch C, Muthana M, Lewis CE (2005) J Immunol 175:6257–6263
125. Brahimi-Horn MC, Chiche J, Pouysségur J (2007) J Mol Med 85:1301–1307
126. Freeman BA, Crapo JD (1981) J Biol Chem 256:10986–10992
127. Reiter RJ (1998) Prog Neurobiol 56:359–384
128. Shen J, Yang X, Dong A, Petters RM, Peng YW, Wong F, Campochiaro PA (2005) J Cell Physiol 203:457–464
129. Brueckl C, Kaestle S, Kerem A, Habazettl H, Krombach F, Kuppe H, Kuebler WM (2006) Am J Respir Cell Mol Biol 34:453–463
130. Bhandari V, Choo-Wing R, Lee CG, Zhu Z, Nedrelow JH, Chupp GL, Zhang X, Matthay MA, Ware LB, Homer RJ (2006) Nat Med 12:1286–1293
131. Zhang KY, Gao P, Sun G, Zhang T, Li X, Liu S, Zhao Q, Lo KK-W, Huang W (2018) J Am Chem Soc 140:7827–7834
132. Shi H, Ma X, Zhao Q, Liu B, Qu Q, An Z, Zhao Y, Huang W (2014) Adv Funct Mater 24:4823–4830
133. Choung KS, Marroquin K, Teets TS (2019) Chem Sci 10:5124–5132
134. Solomatina AI, Chelushkin PS, Su S-H, Wu C-H, Chou P-T, Tunik SP (2022) Chem Commun 58:419–422

135. Finkel T, Holbrook NJ (2000) Nature 408:239–247
136. Droge W (2002) Physiol Rev 82:47–95
137. Barnham KJ, Masters CL, Bush AI (2004) Nat Rev Drug Discov 3:205–214
138. Valko M, Leibfritz D, Moncol J, Cronin MT, Mazur M, Telser J (2007) Int J Biochem Cell Biol 39:44–84
139. Murphy MP (2009) Biochem J 417:1–13
140. Reuter S, Gupta SC, Chaturvedi MM, Aggarwal BB (2010) Free Radic Biol Med 49:1603–1616
141. Calas-Blanchard C, Catanante G, Noguer T (2014) Electroanalysis 26:1277–1286
142. Zhang KY, Zhang J, Liu Y, Liu S, Zhang P, Zhao Q, Tang Y, Huang W (2015) Chem Sci 6:301–307
143. Lu Z, Shangguan M, Jiang X, Xu P, Hou L, Wang T (2019) Dyes Pigm 171:107715
144. Meng X, Shi Y, Chen Z, Song L, Zhao M, Zou L, Liu S, Huang W, Zhao Q (2018) ACS Appl Mater Interfaces 10:35838–35846
145. Zhang KY, Zhang T, Wei H, Wu Q, Liu S, Zhao Q, Huang W (2018) Chem Sci 9:7236–7240
146. Wu Q, Dai P, Wang Y, Zhang J, Li M, Zhang KY, Liu S, Huang W, Zhao Q (2021) Chem Sci 12:11020–11027
147. Li G, Lin Q, Ji L, Chao H (2014) J Mater Chem B 2:7918–7926
148. Li G, Lin Q, Sun L, Feng C, Zhang P, Yu B, Chen Y, Wen Y, Wang H, Ji L (2015) Biomaterials 53:285–295
149. Kramer JA, Sagartz JE, Morris DL (2007) Nat Rev Drug Discov 6:636–649
150. Cederbaum AI, Lu Y, Wu D (2009) Arch Toxicol 83:519–548
151. Du K, Ramachandran A, Jaeschke H (2016) Redox Biol 10:148–156
152. Chen Z, Meng X, Zou L, Zhao M, Liu S, Tao P, Jiang J, Zhao Q (2020) ACS Appl Mater Interfaces 12:12383–12394
153. Chen Z, Yan P, Zou L, Zhao M, Jiang J, Liu S, Zhang KY, Huang W, Zhao Q (2018) Adv Healthc Mater 7:1800309
154. Wu W, Liao X, Chen Y, Ji L, Chao H (2021) Anal Chem 93:8062–8070
155. Kang JH, Kim HJ, Kwon T-H, Hong J-I (2014) J Org Chem 79:6000–6005
156. Wu Q, Zhang KY, Dai P, Zhu H, Wang Y, Song L, Wang L, Liu S, Zhao Q, Huang W (2019) J Am Chem Soc 142:1057–1064
157. Zubenko GS, Kopp U, Seto T, Firestone LL (1999) Psychopharmacology 145:175–180
158. Shinitzky M (1984) Physiol Membr Fluid 1:1–51
159. Nadiv O, Shinitzky M, Manu H, Hecht D, Roberts CT Jr, LeROITH D, Zick Y (1994) Biochem J 298:443–450
160. Kuimova MK (2012) Phys Chem Chem Phys 14:12671–12686
161. Kuimova MK, Botchway SW, Parker AW, Balaz M, Collins HA, Anderson HL, Suhling K, Ogilby PR (2009) Nat Chem 1:69–73
162. Liu F, Wen J, Chen S-S, Sun S (2018) Chem Commun 54:1371–1374
163. Zhang P, Chen H, Huang H, Qiu K, Zhang C, Chao H, Zhang Q (2019) Dalton Trans 48:3990–3997
164. Hao L, Li Z-W, Zhang D-Y, He L, Liu W, Yang J, Tan C-P, Ji L-N, Mao Z-W (2019) Chem Sci 10:1285–1293
165. Hao L, Zhong Y-M, Tan C-P, Mao Z-W (2021) Chem Commun 57:5040–5042

Topics in Current Chemistry (2022) 380:36
https://doi.org/10.1007/s41061-022-00401-w

REVIEW

Check for updates

Post-complexation Functionalization of Cyclometalated Iridium(III) Complexes and Applications to Biomedical and Material Sciences

Shin Aoki[1,2,3] · **Kenta Yokoi**[1] · **Yosuke Hisamatsu**[1] · **Chandrasekar Balachandran**[1,3] · **Yuichi Tamura**[1] · **Tomohiro Tanaka**[1]

Received: 23 March 2022 / Accepted: 20 June 2022 / Published online: 10 August 2022

Abstract

Cyclometalated iridium(III) (Ir(III)) complexes exhibit excellent photophysical properties that include large Stokes shift, high emission quantum yields, and microsecond-order emission lifetimes, due to low-lying metal-to-ligand charge transfer (spin-forbidden singlet–triplet (^{3}MLCT) transition). As a result, analogs have been applied for research not only in the material sciences, such as the development of organic light-emitting diodes (OLEDs), but also for photocatalysts, bioimaging probes, and anticancer reagents. Although a variety of methods for the synthesis and the applications of functionalized cyclometalated iridium complexes have been reported, functional groups are generally introduced to the ligands prior to the complexation with Ir salts. Therefore, it is difficult to introduce thermally unstable functional groups such as peptides and sugars due to the harsh reaction conditions such as the high temperatures used in the complexation with Ir salts. In this review, the functionalization of Ir complexes after the formation of cyclometalated Ir complexes and their biological and material applications are described. These methods are referred to as "post-complexation functionalization (PCF)." In this review, applications of PCF to the design and synthesis of Ir(III) complexes that exhibit blue–red and white color emissions, luminescence pH probes, luminescent probes of cancer cells, compounds that induce cell death in cancer cells, and luminescent complexes that have long emission lifetimes are summarized.

Chapter 5 was originally published as Aoki, S., Yokoi, K., Hisamatsu, Y., Balachandran, C., Tamura, Y. & Tanaka, T. Topics in Current Chemistry (2022) 380: 36. https://doi.org/10.1007/s41061-022-00401-w.

✉ Shin Aoki
shinaoki@rs.tus.ac.jp

Extended author information available on the last page of the article

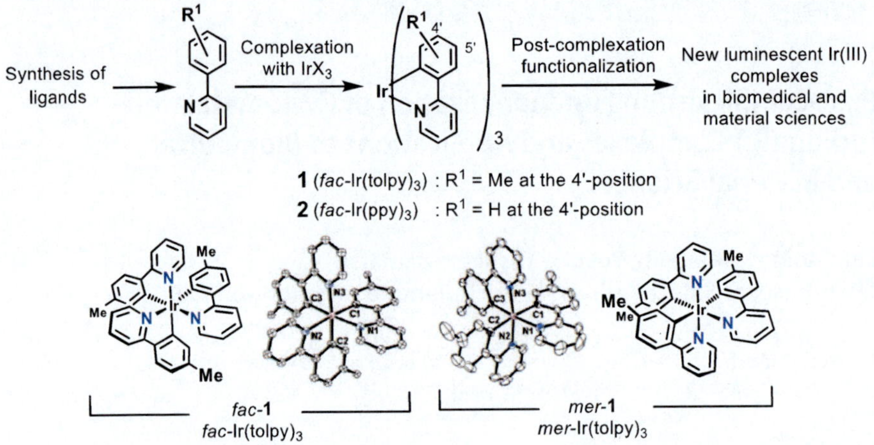

Scheme 1 Post-complexation functionalization of cyclometalated Ir(III) complexes. ORTEP drawings of *facial* form and *meridional* form of Ir(tolpy)$_3$ **1** reported by Thompson et al. [2]

Keywords Cyclometalated iridium(III) complexes · Post-complexation functionalization · Luminescence emission · pH probes · Bioimaging · Programmed cell death · Paraptosis · Reversible electronic electron transfer · Degradation reactions

1 Introduction

Cyclometalated iridium(III) (Ir(III)) complexes exhibit excellent photophysical properties that include long Stokes shifts, high emission quantum yields, and microsecond-order emission lifetimes, due to low-lying metal-to-ligand charge transfer (spin-forbidden singlet–triplet (^{3}MLCT) transition), and these analogs have been applied for research in the material sciences such as the development of organic light-emitting diodes (OLEDs) and for photocatalysts, bioimaging probes, and anticancer reagents [1–23]. Typical examples of such cyclometalated Ir(III) complexes include *fac*-Ir(tolpy)$_3$ **1** (tolpy = 2-(4′-tolyl) pyridine) and *fac*-Ir(ppy)$_3$ **2** (ppy = 2-phenylpyridine) (Scheme 1) and the emission quantum yield and emission lifetime (phosphorescence) of **1** at room temperature in CH_2Cl_2 were reported to be 0.5 and 2.0 μs, respectively [1, 2]. Although many methods for the synthesis and applications of functionalized cyclometalated iridium complexes have been reported, functional groups are usually introduced to the ligands prior to the complexation with Ir salts. This makes it difficult to introduce thermally unstable functional groups such as peptide and sugar under the harsh reaction conditions such as high temperature for the complexation with Ir salts. In this review, the functionalization of Ir complexes after the formation of cyclometalated Ir complexes and their biological and material applications for such complexes are described. We refer to these processes as "post-complexation functionalization (PCF)" in this review.

In the conventional synthesis of cyclometalated metal complexes, functionalized ligands are first synthesized and then reacted with metal salts at a high temperature, which makes the synthesis of metal complexes that contain labile functionalities not an easy task. If various modifications could be made after the complex was formed, metal complexes having various structures and functions could be obtained by using Ir(III) complex cores as luminescent building blocks that have well-defined structures.

It should be noted here that **1** and **2** are composed of nonsymmetric (described as "asymmetric" in many cases) ligands such as tolpy and ppy, which coordinate to Ir at different atoms (C and N), and as a result, there two stereoisomers for **1** and **2**, facial (*fac*) and meridional (*mer*) forms, as shown in Scheme 1. Thompson et al. reported on that *mer*-forms are kinetically favorable products and *fac*-forms are thermodynamically favorable, which allowed the selective synthesis of the *fac* and *mer* forms of **1** and **2**, whose structures were confirmed by X-ray crystal structure analysis [2]. Note that almost all homoleptic Ir(III) complexes described in this review were synthesized as *fac*-forms unless otherwise described.

2 Substitution Reactions of Cyclometalated Iridium(III) Complexes

To date, only a limited number of functionalization reactions of the ligand parts of metal complexes have been reported. In 1996, Sauvage et al. reported on the Suzuki–Miyaura coupling reaction of a brominated derivative of the cyclometalated Ru complex and nitration of Ru complex [24]. The direct bromination and iodination of the ruthenium complexes **3** to **4** was reported by Coudret and coworkers in 1998 (Scheme 2) [25]. They reported that treatment of the ruthenium(III) complex **3**, $[Ru(bpy)_2(ppy)]^+$ (bpy = 2,2′-bipyridine), with NBS in MeCN and $PhI(OAc)_2$ in CH_2Cl_2 afforded **4**, in which the 5′-position of the phenyl ring (ppy) was substituted with Br and I. In 2000, Clark et al. reported on the bromination and nitration of cyclometalated Ru(II) and Os(II) complexes **5a**,**b** to afford **6a**,**b** and **7**, respectively (Scheme 2) [26, 27].

In 2005, Leung et al. reported on the halogenation of Ir(III) complex **8** to give the dibrominated product **9**, which contains two Br atoms at the 5′-positions of its two ppy ligands, as shown in Scheme 3 [28]. Cross-coupling reactions of **9** with various boronic acids ($R^2B(OH)_2$) afforded **10a–d**, the photophysical properties of which were described.

Müllen et al. reported on the divergent synthesis of a polyphenylene dendrimer **13** using the monomeric Ir(III) unit **11**, which was prepared from the $Ir(ppy)_3$ complex **2** by the iodination at the 5′-position of **2**. The subsequent coupling reaction of **11** with the (trimethylsilyl)alkynyltin reagent and the deprotection of the three trimethylsilyl groups afforded **12**, which was used as a building block of **13** (Scheme 4) [29]. Stossel et al. reported on the regioselective halogenation and nitration of cyclometalated Ir(III) complexes in their patents [30, 31], although details of the reaction mechanisms involved in the substitutions were not reported, nor were the structures of the functionalized Ir complexes or their photochemical properties.

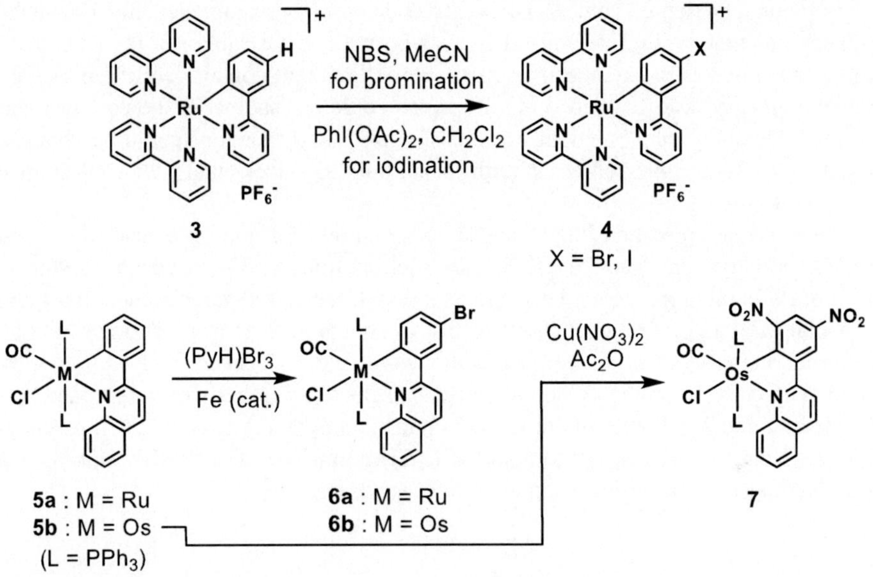

Scheme 2 Examples of substitution reactions of cyclometalated metal complexes

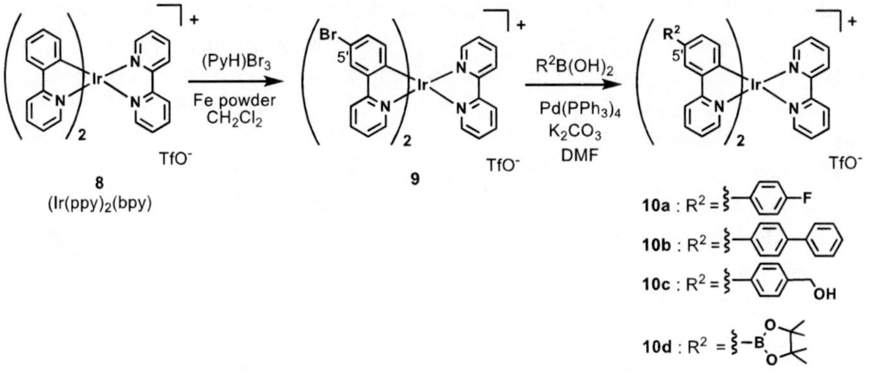

Scheme 3 Bromination and successive coupling reaction of cyclometalated Ir complexes

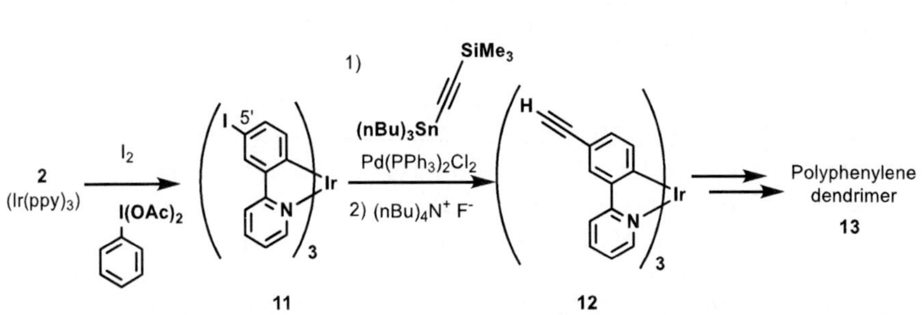

Scheme 4 Iodination of $Ir(ppy)_3$ and the introduction of alkyne unit for the synthesis of polyphenylene dendrimers

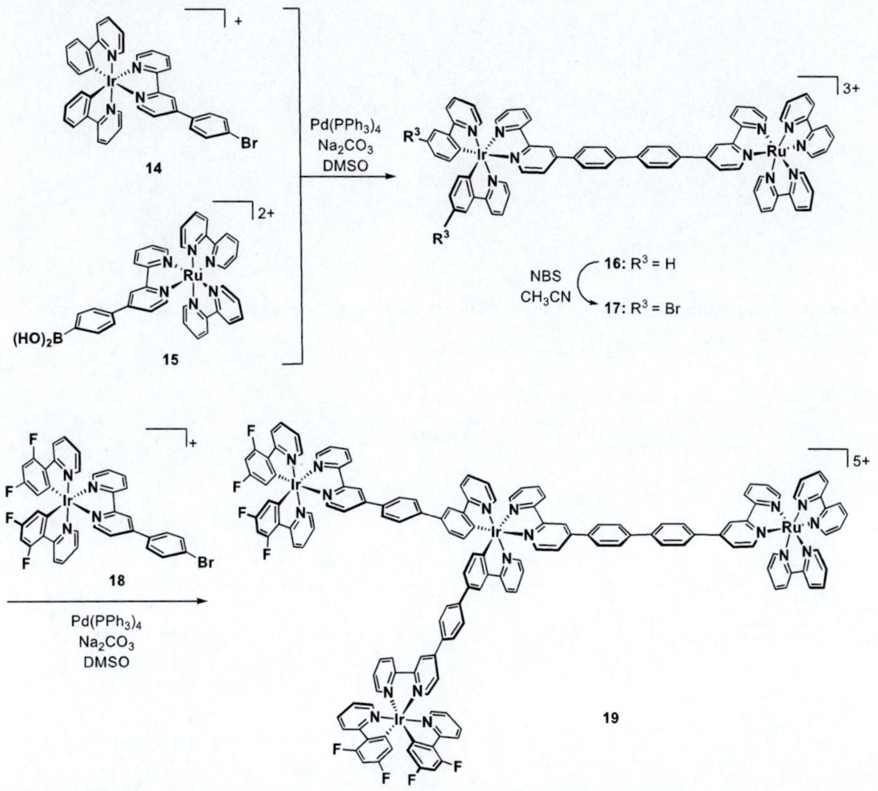

Scheme 5 The synthesis of the Y-shaped tetrameric complex **19** containing three Ir complex parts and one Ru complex unit

In 2005, Arm and Williams reported on the synthesis of the Y-shaped tetrameric metal complex **19** that contained three Ir complex cores and one Ru complex unit (Scheme 5) [32]. The heterodimer containing one Ir complex and one Ru complex **16** was prepared by the cross-coupling reaction of **14** and **15** and converted to the dibromide **17** by specific bromination by means of NBS in CH_3CN. Finally, the cross-coupling reaction of **17** with **18** afforded their target compound **19**. Linear polymers have also been prepared using similar strategies [33].

In 2015, Liang et al. reported on the bromination of **2** based on the methodology described in Stossel's patent to obtain mono-, di- and tribrominated compounds **20a–c** [34], which were then reacted with the dibenzo-24-crown-8 derivative containing the boronic acid cyclic ester moiety to give **21a–c** (Scheme 6).

In 2015, Humphrey et al. reported on the synthesis of a heterobimetallic complex **25** bearing Ir and Ru complex cores (Scheme 7) [35]. The iodo-substituted Ir complex **23** was prepared from **22** and coupled with the trimethylsilylalkyne reagent under Sonogashira coupling reaction conditions to give **24**, followed by the conversions including the coupling reaction with the Ru complex to afford the heterobimetallic complex **25**.

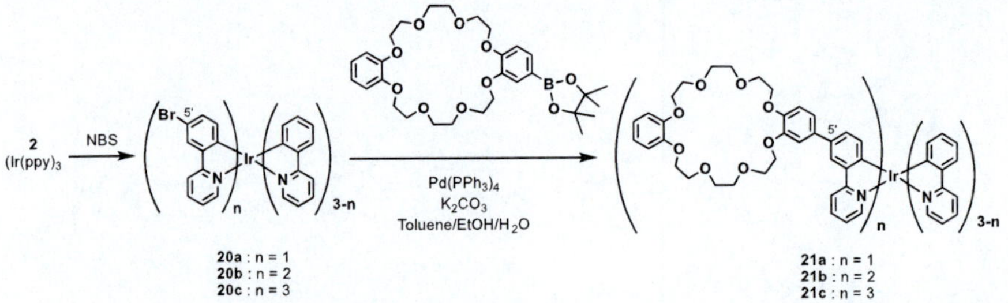

Scheme 6 Bromination of **2** ($Ir(ppy)_3$) and successive reaction with the crown ether derivative to give **21**

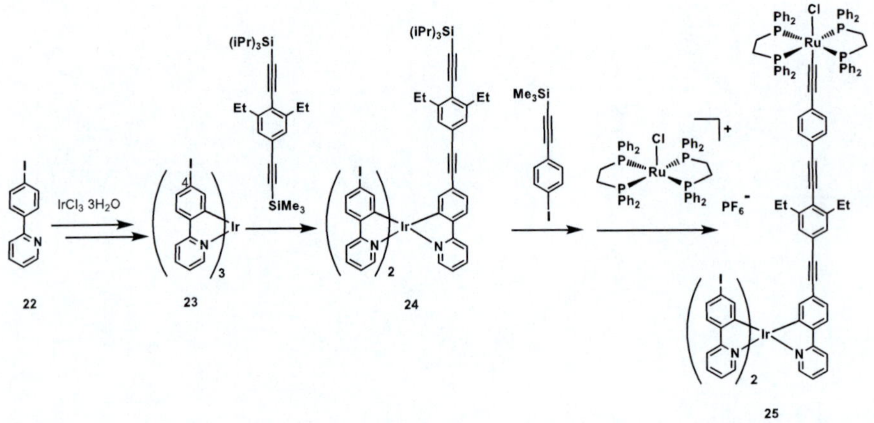

Scheme 7 Sonogashira coupling of iodo-substituted Ir complex and successive reaction with Ru complex

3 Aromatic Electrophilic Substitution Reactions of Cyclometalated Iridium(III) Complexes for the Introduction of Various Functional Groups and Control of Emission Colors

About a decade ago, we attempted to brominate three methyl groups of *fac*-$Ir(tolpy)_3$ (**1**) under typical reaction conditions (NBS, BPO in CCl_4). The product of this reaction was the Ir complex **26,** in which the 5′-position (*p*-position with respect to the Ir-C bond) of tolpy was substituted with three Br groups, as confirmed by an X-ray crystal structure analysis (Scheme 8), and not the bromomethyl derivative of **1** [36]. Analysis of the reaction mechanism for this reaction strongly suggested that these reactions proceed via an aromatic electrophilic substitution mechanism at the 5′-carbon that is activated by the Ir-C bonds and that a radical reaction mechanism is unlikely. These findings allowed us to access new Ir complexes that were equipped with labile functional groups such as peptides and related functional groups that would otherwise be very difficult to obtain, after the preparation of cyclometalated Ir complex cores such as **1**, **2**, and **27** by

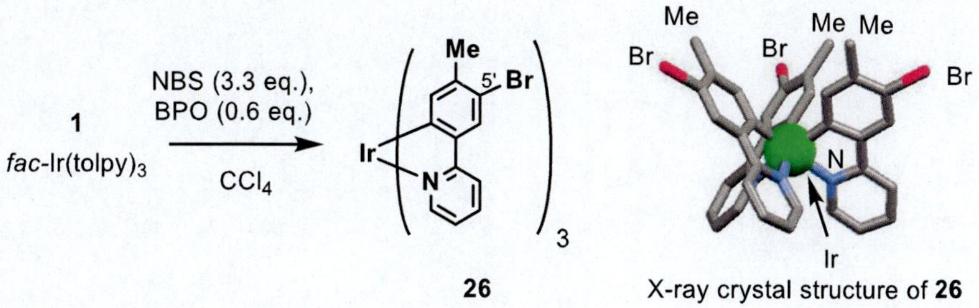

Scheme 8 Selective bromination reaction of Ir(tolpy)$_3$ **1**

the complexation of $IrCl_3$ with simple ligands (typically, the Nonoyama method [37]) as shown in Scheme 9, a process that we refer to as PCF in this review.

For example, the introduction of electron-withdrawing groups such as sulfonyl (**31**), formyl (**35**), and nitrile (**39**) groups at the 5′-position of the tolpy ligands induces a blueshift in the phosphorescence (the luminescence emissions of **1** and **2** are observed at ca. 510 nm) to blue-green (<490 nm; excitation wavelength, 366 nm) [36]. In contrast, the introduction of electron-donating groups such as an amino group afforded Ir(atolpy)$_3$ **33**, which exhibits a redshift in its luminescence emission to ca. 600 nm.

Ir complexes having basic groups such as amino and pyridyl groups at the 5′-position **40–43** were also synthesized (Scheme 10), and it was found that they function as luminescent pH probes as well (Figs. 1, 2). Figure 1a–c show the pH-dependent change in the luminescence emission spectra of Ir(deatolpy)$_3$

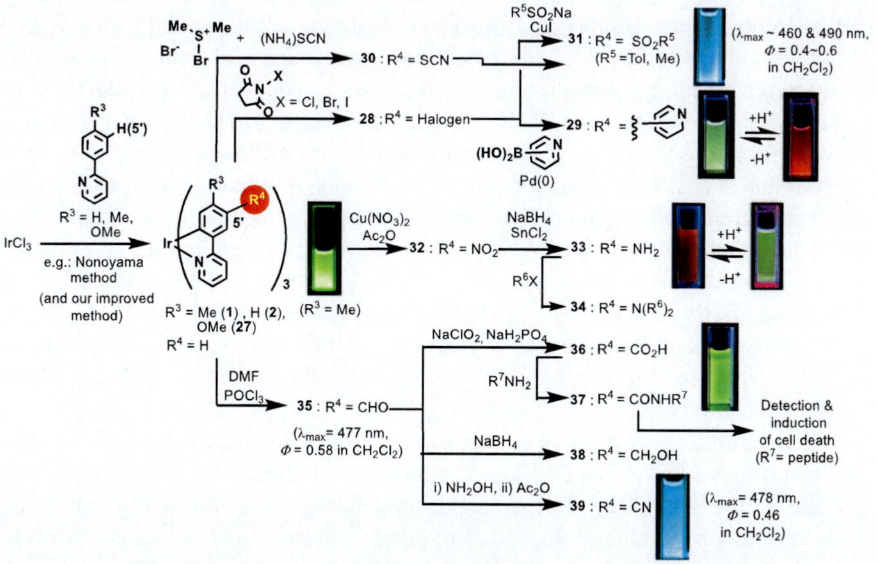

Scheme 9 Synthesis of various cyclometalated Ir complexes by post-complexation functionalization

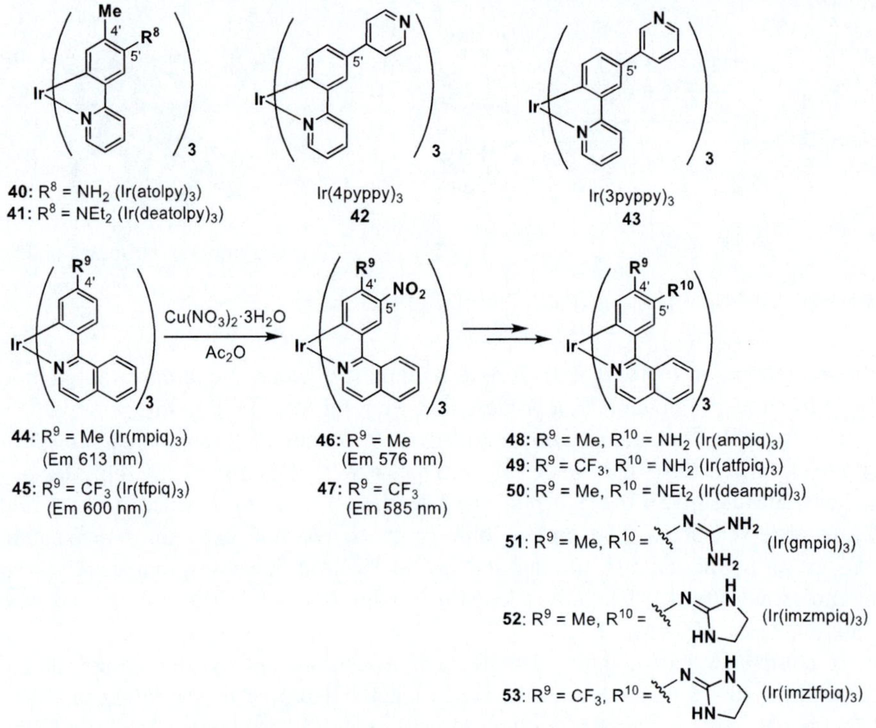

Scheme 10 Structures and synthesis of Ir complexes having basic groups

(deatolpy = 2-(5′-*N*,*N*-diethylamino-4′-tolyl)pyridine) **41**, which contains three diethylamino groups (Fig. 1a), Ir(4pyppy)$_3$ complex (4pyppy = 2-(5′-(4′-pyridyl)phenyl)pyridine) **42** (Fig. 1b), and Ir(imztfpiq)$_3$ (imztfpiq = 1-(5′-iminoimidazolidyl-4′-trifluoromethylphenyl)isoquinoline) **53** (Fig. 1c), respectively, in DMSO/100 mM buffer (pH 4–10) solutions at 25 °C (excitation at 366 nm) [38–40]. For example, **41**, which contains three diethylamino groups at the 5′-position of tolpy, shows weak red emission at an alkaline pH, and the green emission is enhanced to a considerable extent at neutral–weak acidic pH in a reversible manner (Fig. 1a) [38]. On the other hand, the Ir(4pyppy)$_3$ complex **42** exhibits a significant redshift from an emission at ca. 500 nm to around 600 nm with a decrease in the intensity of the emission upon the addition of acid (Fig. 1b) [39]. These spectra and Fig. 1d, in which the pH-dependent change in emission intensity at a given wavelength of **41** and **42** is plotted, indicate that the averaged pK_a values of the basic groups of **41** and **42** are adjusted to neutral pH.

The Ir complexes containing isoquinoline (iq)-type ligands such as **44–53** were also synthesized (Scheme 10) [40] because it was reported by Tsuboyama et al. that these types of Ir(III) complexes exhibit a strong red luminescence emission [41]. Since the pK_a value of **50** is slightly lower and that of **51** is higher than neutral pH, its methyl groups at the 4′-position and basic groups at the 5′-position of the phenyl group were changed to trifluoromethyl groups and five-membered

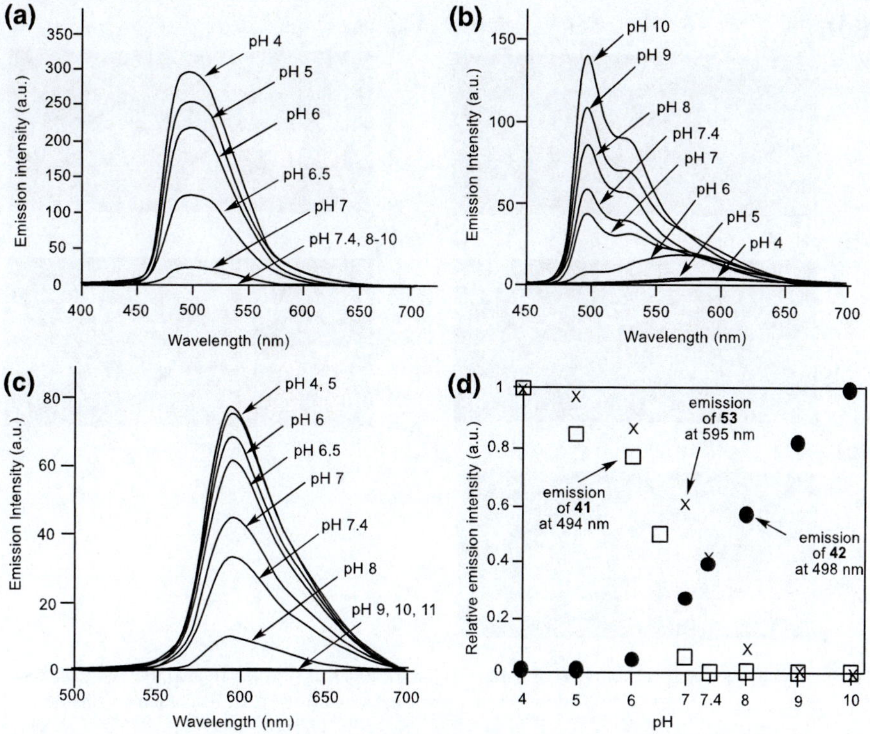

Fig. 1 **a–c** pH-dependent change in luminescence emission of Ir complexes **41** (**a**), **42** (**b**), and **53** (**c**) in DMSO/100 mM buffer (pH 4–10) at 298 K (excitation at 366 nm). **d** pH-dependent change in emission of **41** at 494 nm, **42** at 498 nm, and **53** at 595 nm, respectively

iminoimidazolidyl groups, respectively, in **52** and **53**, in order to adjust the pK_a values of these basic moieties to around pH 7 (Figs. 1c, d, 2e).

Images of DMSO/H_2O solutions of **40**, **41**, **42**, **52**, and **53** are displayed in Fig. 2. It should be noted that the emission behavior of **42** (Fig. 2b) is opposite to that of **41** (Fig. 2a) [38, 39]. The results of density functional theory (DFT) and time-dependent density functional theory (TD-DFT) calculations suggest that the HOMO–LUMO gaps of **41** and **53** are increased by the protonation of their three amino groups, which explains the blueshift in their luminescence emission at acidic pH [38, 40]. On the other hand, the HOMO–LUMO gap of **42** is decreased by the protonation of its pyridine rings [39].

The aforementioned Ir complexes were used as specific probes of intracellular organelles. For example, the Ir complex **41** stains lysosomes, which is an acidic intracellular organelle, due to the protonation of its diethylamino groups (Fig. 3) [38]. In co-staining images of HeLa-S3 cells using **41** and a mitochondrial probe (MitoTracker Red) or a lysosomal probe (LysoTracker Red) in Fig. 3, a yellow area was observed due to overlap of the green emission from **41** and the red emission from LysoTracker Red. Both **40** and **41** are taken up into HeLa-S3 cells to

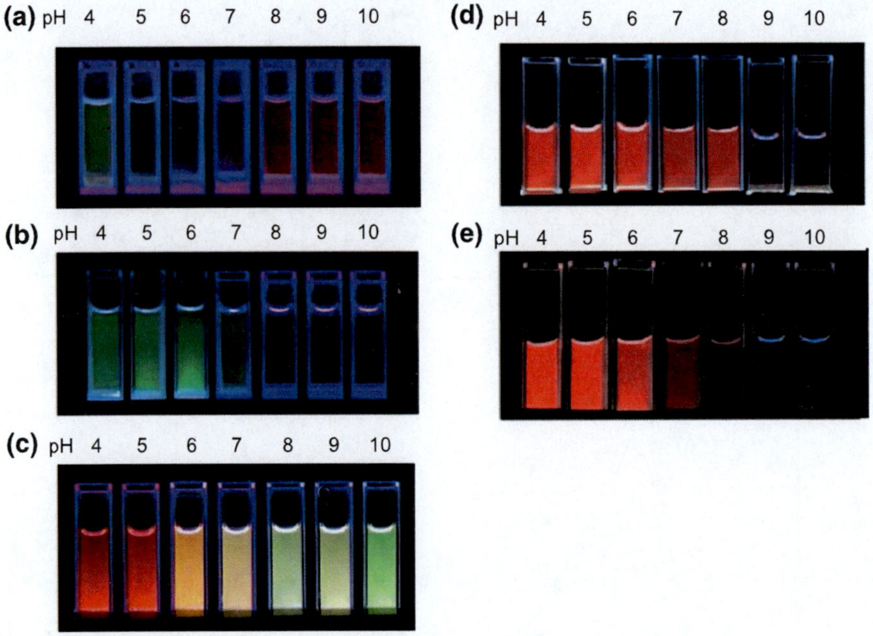

Fig. 2 Photographs showing solutions of **a** **40** (100 μM), **b** **41** (1 μM), **c** **42** (10 μM), **d** **52** (5 μM), and **e** **53** (5 μM) in degassed DMSO/100 mM pH buffer (from pH 4 to 10) at 25 °C. Excitation at 365 nm

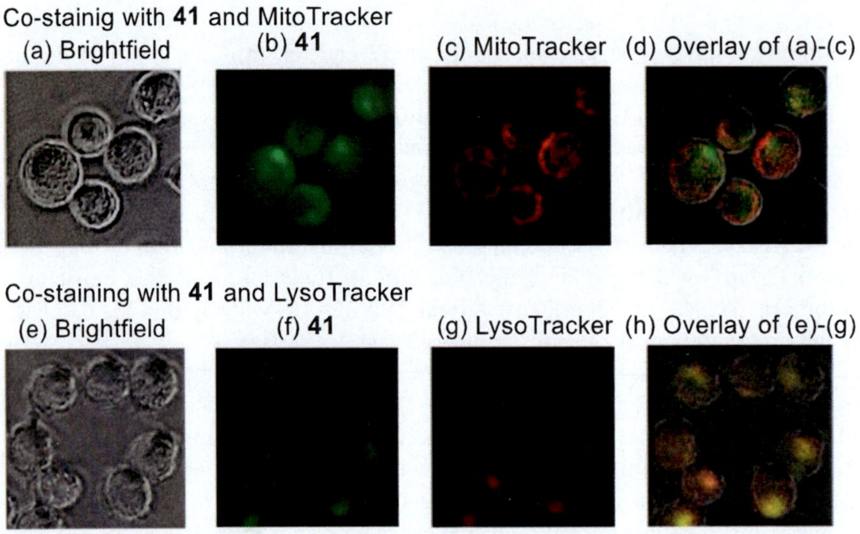

Fig. 3 Microscopic images of the luminescence emission of HeLa-S3 cells stained with Ir(deatolpy)$_3$ **41** and MitoTracker or LysoTracker

almost the same extent (0.25 and 0.20 fmol/cell, respectively), as determined by inductively coupled plasma–atomic emission spectrometry (ICP-AES), and the green emission of **41** in lysosomes is stronger than that of **40**. In addition, a green emission was observed in HeLa-S3 cells that had been treated with **41** at 4 °C. Therefore, it is very likely that **41** is taken up to living cells by passive transport and is then protonated in lysosomes, thus resulting in a strong green emission. In contrast, an overlap of the green emission from **42** and the red emission of MitoTracker Red was observed, indicating that **42** is largely deprotonated in mitochondria, the pH of which is neutral–alkaline [39].

It was also found that Ir complexes **41**, **42**, and **53** induce the cell death of HeLa-S3 cells upon photoirradiation at 377 nm (for **41** and **42**) and 465 nm (for **53**) (Fig. 4) [38–40]. We confirmed that the photoirradiation of HeLa-S3 cells in the presence of these Ir complexes causes necrosis, as evidenced by the staining experiments of HeLa-S3 cells with Annexin V-Cy3, and propidium iodide (PI), typical probes of apoptosis and necrosis, respectively. These results suggest the possible application of Ir complexes to photodynamic therapy (PDT) of cancer and related diseases.

As described above, the introduction of electron-withdrawing groups such as formyl, nitrile, or sulfonyl groups at the 5′-position of the tolpy ligands induces a blue-shift in the green phosphorescence at ca. 510 nm (excitation at 366 nm) to a blue-green color (<490 nm) (Fig. 5) [36, 42]. This can be attributed to a decrease in the HOMO level and the enhancement of its HOMO–LUMO energy gap.

4 Design and Synthesis of Ir(III) Complex-Peptide Hybrids (IPHs) and the Induction of Programmed Cell Death in Cancer Cells

Programmed cell death (PCD) plays an important role in the formation of tissues and organs, and literally in cell death (apoptosis, necroptosis, paraptosis, autophagy, etc.), as shown in Scheme 11 [43–48]. To date, many anticancer drugs have been reported to induce PCD in cancer cells. In order to overcome the resistance of cancer cells to anticancer drugs, it is necessary to develop different PCD-inducing drugs and to study their mechanisms.

Death ligands (DL) such as the tumor necrosis factor-related apoptosis-inducing ligand (TRAIL) are known to bind to the death receptor (DR) that is expressed on the membrane of cancer cells and induce their cell death. The C_3-symmetric trimeric structure of DR (DR_3) around one zinc ion [49–52] prompted us to design and synthesize some hybrid compounds based on a C_3-symmetric Ir complex with a DR-binding cyclic peptide (Ir(III) complex-peptide hybrids, IPHs) as TRAIL mimics for use in the detection of DR-expressed cancer cells and the induction of their PCD, which would contribute to the development of new cancer theranostics (a combination of diagnosis and therapy for cancer). Typically, we designed and synthesized IPHs having cyclic peptide units that had been known to interact with DR expressed on the cell membrane of human leukemic cancer cells [53]. It was expected that these compounds would bind to DR5 on the cell membrane, thus permitting the detection of cancer cells, and hopefully, activate the intracellular signaling for apoptosis or other types of PCD in cancer cells.

Fig. 4 Microscopic images of the luminescence emission of HeLa-S3 cells stained with PI (30 μM) after ▶ the photoirradiation at 465 nm and 37 °C for 30 min in the presence of **41** and **53** (10 μM)

For that purpose, the IPH-TMs (IPHs as TRAIL mimics) **58a** and **58b** (Scheme 6) were designed and synthesized via formyl and carboxylate intermediates, **35** and **36** ($R^3 = Me$) shown in Scheme 9 [54]. Hydrophilic SGSG (S: serine, G: glycine) peptide units were introduced at the *N*-terminus of cyclic peptide regions (in **58a** and **58b**) and in the linker between $Ir(tolpy)_3$ core and the cyclic peptide part (in **58c**) so as to increase their solubility in water (Scheme 12).

The dissociation constants (K_d values) of the DR5-TRAIL complex, the DR5-**58a** complex, and the DR5-**58b** complex were determined to be (4.3 ± 0.1) nM, (2.7 ± 0.1) μM, and (2.5 ± 0.1) μM, respectively, by a 27 MHz quartz crystal microbalance (27 MHz QCM) on a sensor chip [54, 55]. Figure 6 shows our proposed scheme for the binding of **58a** with DR on the cell membrane of Jurkat cells and the successive phenomena based on the luminescence images (i) before the addition of **58a**, the DR on the cell membrane is stained with an anti-DR5 antibody (Fig. 6a and red emission in Fig. 6b, d), (ii) **58a** binds to the DR5 as shown in Fig. 6f–h, which indicates a yellow area due to the overlap of the green emission from **58a** and the red emission from the anti-DR5 antibody, indicating that **58a** formed a complex with DR5 (Fig. 6e), (iii) the green emission from **58a** is transferred to the cytoplasm and the red emission from anti-DR5 antibody was decreased considerably (Fig. 6j–l), indicating that **58a** had been internalized into the cytoplasm (Fig. 6i), and (iv) Jurkat cells were incubated for 1 or 6 h after Fig. 6i and an anti-DR antibody was then added again, resulting in a strong red emission from the anti-DR5 antibody, possibly due to the re-expression of DR5 (Fig. 6n–p). We carried out this cell staining with different types of cancer cell lines including K562 cells and Molt-4 cells and found a positive relationship between the green emission from **58a** and the red emission from the anti-DR5 antibody, thus providing support for the binding of **58a** with DR5 [54].

It was also found that cell death in Jurkat cells is slowly induced during the incubation with **58a**, **58b**, and **58c** at 37 °C for 24 h (Fig. 7a) with EC_{50} values (50% growth inhibitory concentration) of 114 μM, 89 μM, and 68 μM, respectively, as determined by MTT assay [54, 55]. In contrast, TRAIL induces apoptosis strongly within 1 h. On the other hand, the peptide **59** part alone and Ir complexes containing no DR-binding peptide **60** negligibly induced cell death. A positive correlation between the cell death-inducing activity of these TM-IPHs with the DR5 expression levels of these cancer cell lines, Jurkat, K562, and Molt-4 cells, was observed. In addition, cell death induced by **58a** and **58b** was negligibly inhibited by Z-VAD-FMK (15 μM), a broad caspase inhibitor of caspase-dependent apoptosis, while cell death induced by **58c** was inhibited by Z-VAD-FMK to a considerable extent (Fig. 7b). It was therefore concluded that **58a** and **58b** induce a relatively slow necrosis-type PCD [54] and that **58c** induces a slow apoptosis [55], suggesting that the cell death pathway of cancer cells could be controlled by adjusting the position of the SGSG linker or the length of the linker. It was also confirmed that **58** binds only weakly to normal cells (HEK

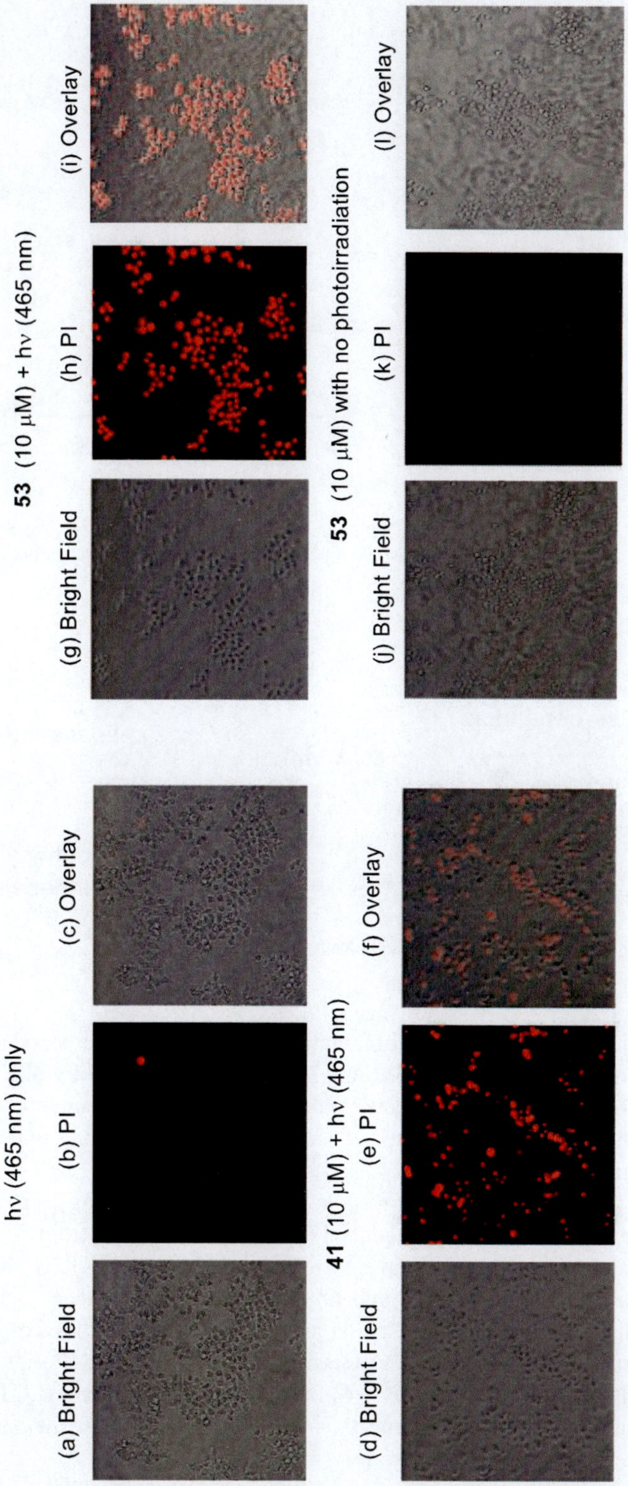
hν (465 nm) only
(a) Bright Field
(b) PI
(c) Overlay
41 (10 μM) + hν (465 nm)
(d) Bright Field
(e) PI
(f) Overlay
53 (10 μM) + hν (465 nm)
(g) Bright Field
(h) PI
(i) Overlay
53 (10 μM) with no photoirradiation
(j) Bright Field
(k) PI
(l) Overlay

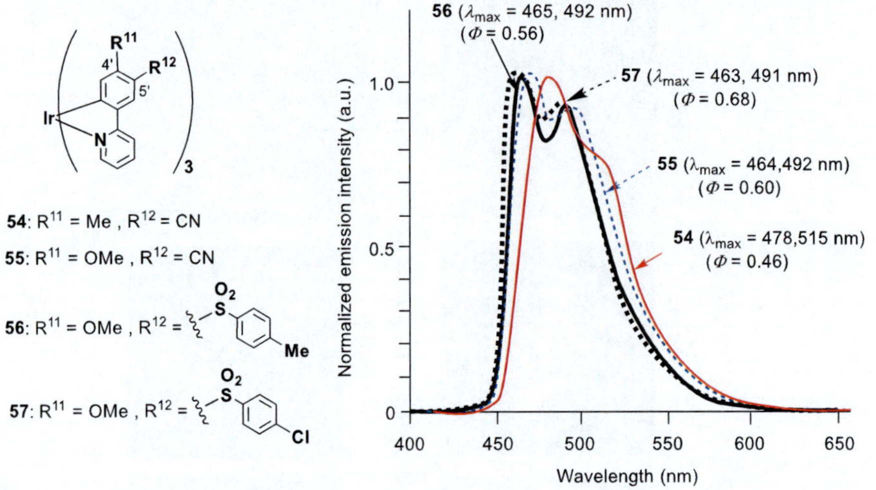

Fig. 5 Normalized emission spectra of Ir(III) complexes **54** (red plain curve), **55** (blue dashed curve), **56** (black bold curve), and **57** (bold dashed curve) (10 μM) in degassed CH_2Cl_2 at 298 K (excitation at 366 nm)

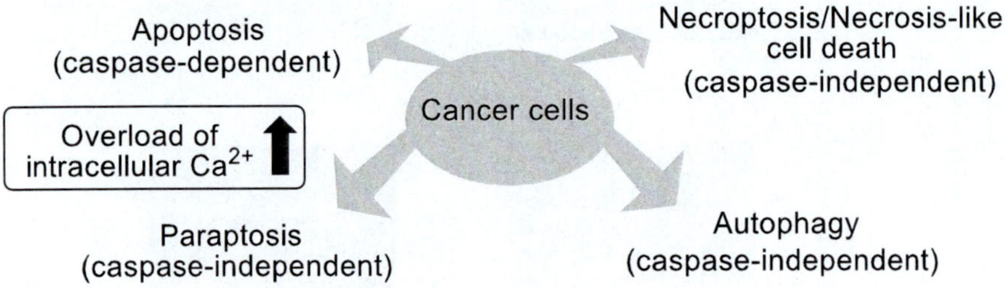

Scheme 11 Typical types of programmed cell death (PCD)

cells) and that its cell death-inducing activity is low. Jurkat cells spiked in bovine blood as a model of the cancer patients' blood were detected by **58a** [54]. These IPH-TMs are useful not only as anticancer agents but also as chemical probes for imaging cancer cells and mechanistic studies, as well as having potential for use in theranostics of cancer.

Basic peptides have been reported as candidates for serving as non-viral gene transfer agents and anticancer agents [56–58]. Therefore, $Ir(tolpy)_3$-based IPHs that contain the basic (cationic at neutral pH) peptide KK(K)GG (K: lysine, G: glycine) via alkyl linker, **61**, **62** and **63** were synthesized (Scheme 13) [59–61]. These amphiphilic IPHs (IPH-AC: IPH as an amphiphilic peptide conjugate) were confirmed to show a strong green emission at ca. 506 nm (excitation at 366 nm) in degassed 100 mM HEPES (pH 7.4) buffer at 25 °C due to ^{3}MLCT with emission quantum yields (Φ) of around 0.5–0.6, which are almost the same value as those of the unfunctionalized $Ir(tolpy)_3$ complexes.

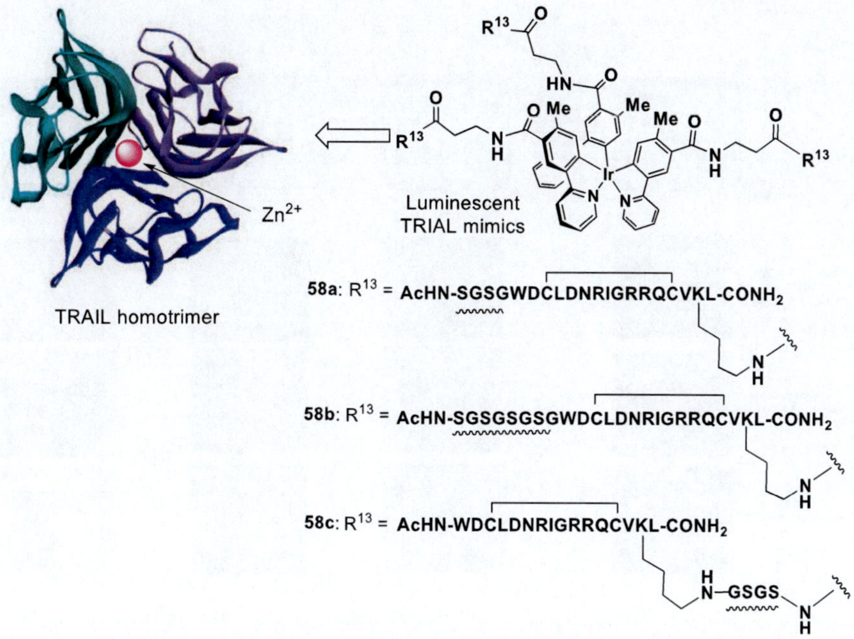

Scheme 12 Structures of IPH-TMs **58** that contain death receptor (DR)-binding peptide units

Figure 8 shows some microscopic images of Jurkat cells that had been treated with representative IPH-ACs, **61a**, **61c**, and **61f**. The findings indicated that **61a** and **61b** possessing C_2 and C_4 linkers between the KKGG peptide units are only weakly taken up into Jurkat cells after incubation for 1 h at 37 °C (Fig. 8a–c) and that **61e** and **61f** containing C_{12} and C_{16} linkers are localized on the cell membrane and have negligible cell death-inducing activity (Fig. 8j–l) [59]. On the other hand, **61c** and **61d** containing C_6 and C_8 linkers induced morphological change in Jurkat cells at 37 °C, in which a strong green emission was observed, as shown in Fig. 8d–f. Interestingly, morphological change in Jurkat cells was negligible after the incubation with **61c** at 4 °C (Fig. 8g–i). An MTT assay of **61c** and **61d** indicated their EC_{50} values of 16 μM and 7.3 μM, respectively. Moreover, the cytotoxicity of other IPHs such as **62a** containing KKKGG peptide was more potent (EC_{50} = 10 μM) than **61c** and **62b-d** including aspartic acid (D) in the peptide parts showed greater EC_{50} values (weaker cytotoxicity) than those of **61c** and **62a**, although the net cationic charges of these IPH-ACs are the same as those of **62b–d.** The IPHs **63a–c**, which possess three, two and one KKKGG peptide units, were synthesized (Scheme 13), and a good correlation between the number of these peptides (and associated charges) and anticancer activity was observed [61]. These experimental results strongly indicate that **61c** and **61d** induce cell death possibly by stimulating intracellular pathways, not only by being taken up by the cancer cells and/or the destruction of cell membrane.

It was assumed that IPH-ACs such as **61c**, **61d**, **62a**, and **63a** would bind to intracellular molecules and then stimulate intracellular pathways to induce cancer

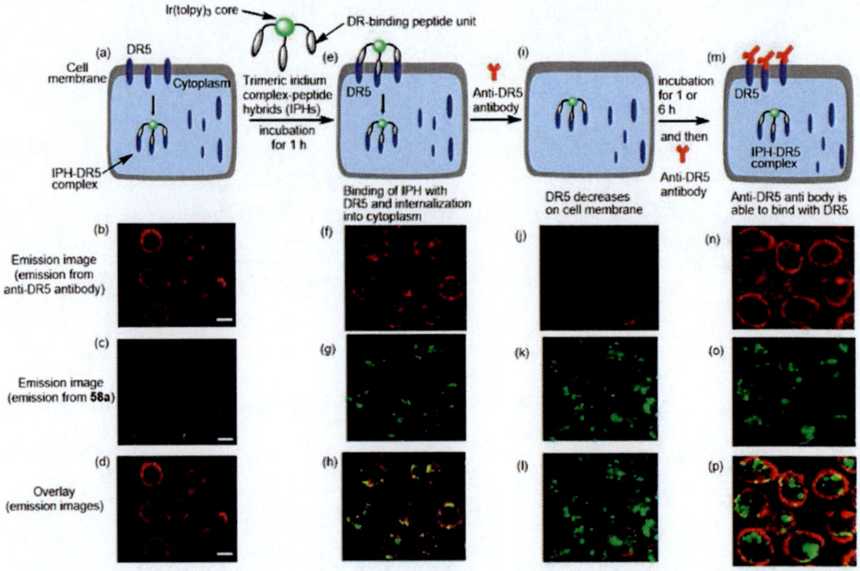

Fig. 6 Proposed scheme for the binding of IPH-TM **58a** with DR that is expressed on the cell membrane of Jurkat cells based on their luminescence microscopic images of stained with **58a** and anti-DR5 antibody. **b–d** Jurkat cells incubated with an anti-DR5 antibody (15 μg/mL) at 4 °C for 15 min. **f–h** Jurkat cells incubated with an anti-DR5 antibody (15 μg/mL) at 4 °C for 15 min and then **58a** (5 μM) at 37 °C for 1 h. **j–l** Jurkat cells incubated with **58a** (10 μM) at 37 °C for 1 h, then in fresh medium for 1 h (2 h in total) followed by treatment with an anti-DR5 antibody (15 μg/mL) at 4 °C for 15 min. **n–p** Jurkat cells were incubated with **58a** (10 μM) at 37 °C for 1 h, and incubated in fresh medium for 6 h (7 h in total), and then treated with an anti-DR5 antibody (15 μg/mL) at 4 °C for 15 min. Scale bar (white) = 10 μm

cell death, as described above. In order to identify the target molecules of the IPH, **64** having a 3-trifluoromethyl-3-phenyldiazirine (TFPD) unit (Scheme 14) was designed and synthesized as a photoaffinity probe [60]. For the photoaffinity labeling of the target molecules of **64**, Jurkat cells were photoirradiated in the presence of **64** at 4 °C to detect **64**-labeled intracellular molecules. In this case, no marker molecule is required for the detection of the target molecules in the SDS-PAGE (sodium dodecyl sulfate–polyacrylamide gel electrophoresis) experiments, because IPH-AC itself exhibits a strong luminescence emission, thus allowing **64**-bound proteins to be detected and identified.

The results of a proteomics analysis of the separated protein fraction after the aforementioned photoaffinity experiments suggested that one of the target proteins of **64** is calmodulin (CaM), which is a typical intracellular Ca^{2+}-binding protein. The interaction of **64** and related compounds with Ca^{2+}-CaM was confirmed by a 27 MHz QCM and luminescence titration experiments [60].

Since **64** has a stronger anti-cancer activity than **61c**, and **62a**, hydrophobic alkyl groups were introduced at the *N*-terminal of the peptide regions in **65** and **66**, which were found to exhibit more potent anticancer activity than those of **61c** (Scheme 14) [62]. In addition, observation of the intracellular behavior of **65** and

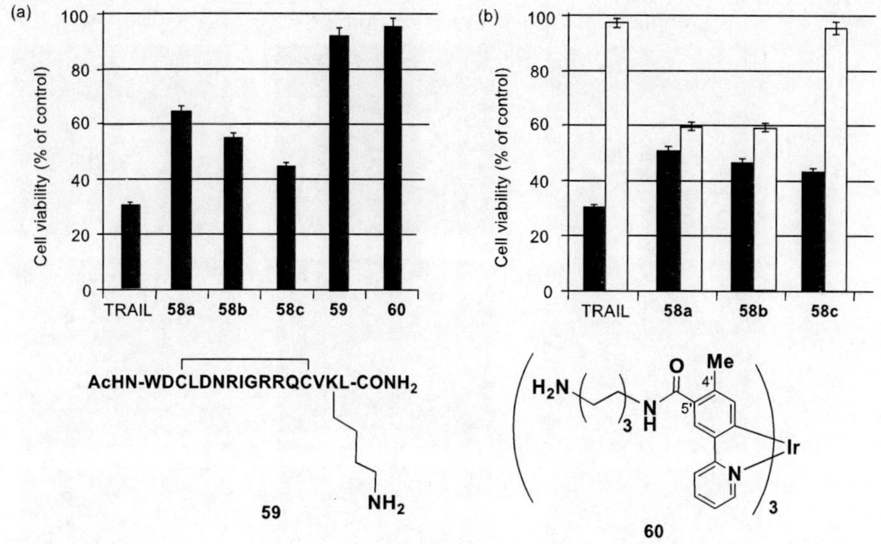

Fig. 7 **a** The results of MTT assay: cell viability of Jurkat cells (% of control at [Ir complex] = 0 μM) in the presence of TRAIL (150 ng/mL), **58a–c**, **59**, and **60** (75 μM) at 37 °C for 16 h. **b** Cell viability of Jurkat cells after incubation with TRAIL (150 ng/mL), **58a**, **58b**, and **58c** (75 μM) for 24 h in the absence of Z-VAD-FMK (closed bars) and in the presence of Z-VAD-FMK (15 μM) (open bars)

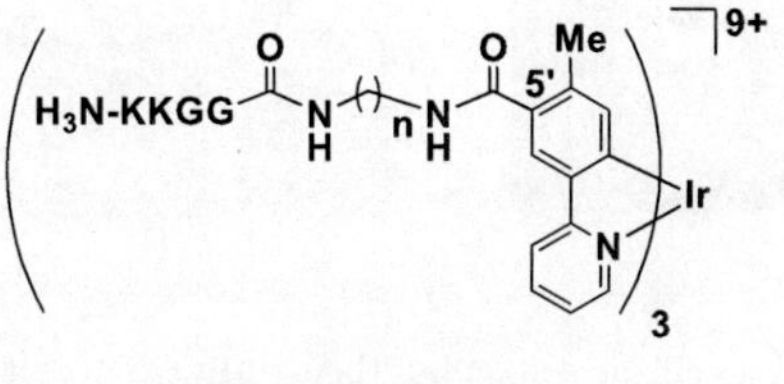

61a (n = 2): EC_{50} >50 μM against Jurkat cells
61b (n = 4): EC_{50} >50 μM
61c (n = 6): EC_{50} = 16 μM
61d (n = 8): EC_{50} = 7.3 μM
61e (n = 12): EC_{50} = 32 μM
61f (n = 16): EC_{50} >50 μM

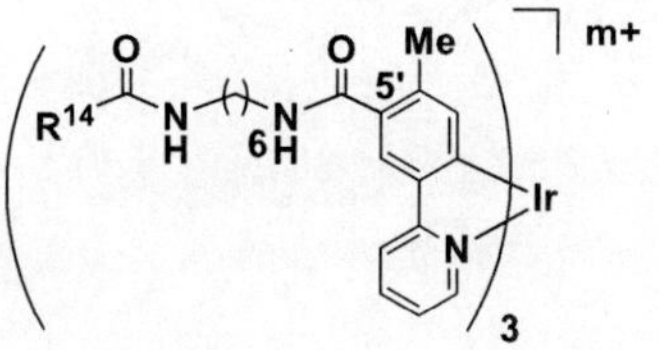

62a (R^{14} = H_3N-KKKGG- ξ, m = 12): EC_{50} = 10 μM
62b (R^{14} = H_3N-KDKKGG- ξ, m = 9): EC_{50} >50 μM
62c (R^{14} = H_3N-KKDKGG- ξ, m = 9): EC_{50} >50 μM
62d (R^{14} = H_3N-KKKDGG- ξ, m = 9): EC_{50} >50 μM

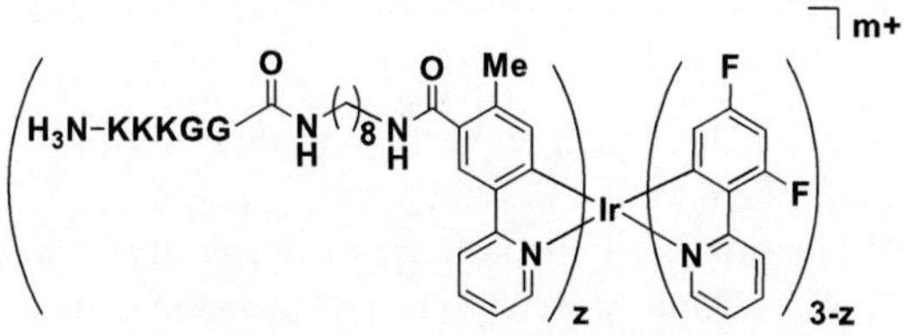

63a (z = 3, m = 12): EC_{50} =1.5 μM
63b (z = 2, m = 8): EC_{50} = 4.1 μM
63c (z= 1, m = 4): EC_{50} = 14 μM

Scheme 13 Structures of IPH-ACs **61–63** that contain basic GGKK(K) and related peptide units and their EC_{50} values against Jurkat cells

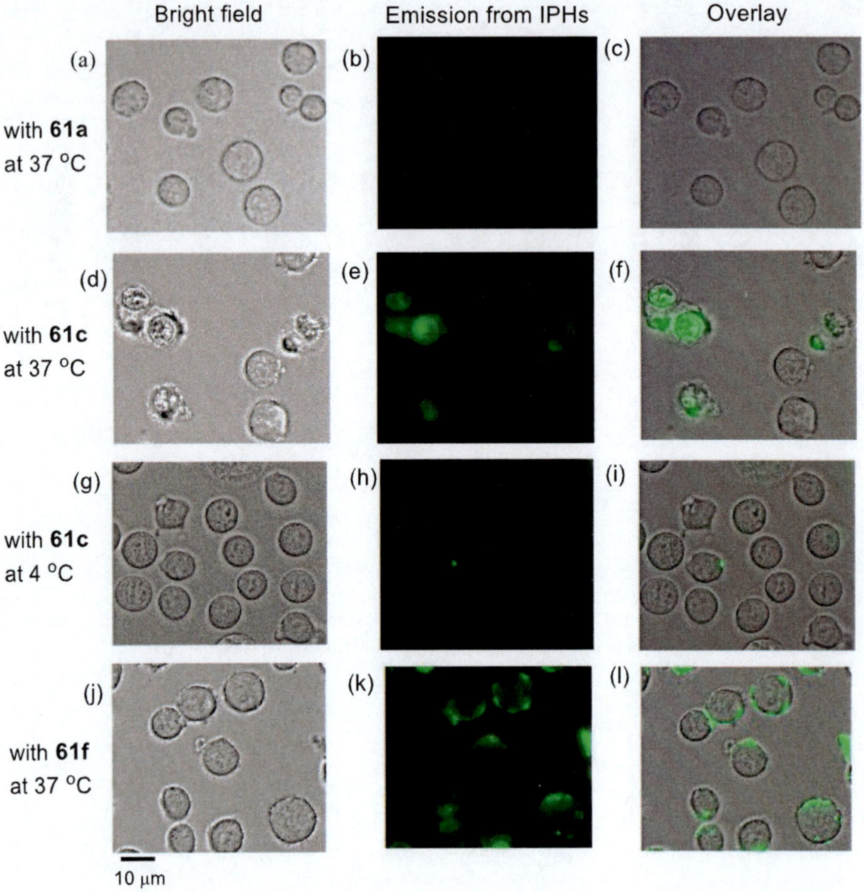

Fig. 8 Typical luminescence microscopy images of Jurkat cells treated with IPH-ACs **61a** (75 μM), **61c** (50 μM)) and **61f** (50 μM) for 1 h. **a** Bright-field image of Jurkat cells treated with **61a** at 37 °C, **b** emission image of **61a**, **c** overlay image of **a** and **b**, **d** bright-field image of Jurkat cells treated with **61c**, **e** emission image of **61c**, **f** overlay image of **d** and **e**, **g** bright-field image of **61c** at 4 °C, **h** emission image of **61c** at 4 °C, **i** overlay image of **g** and **h**, **j** bright-field image of Jurkat cells treated with **61f**, **k** emission image of **61f**, **l** overlay image of **j** and **k**. Excitation at 377 nm. Closed scale bar = 10 μm

66c suggested that these IPH-ACs enter the cell membrane and then migrate to mitochondria and then to the nuclear membrane.

In addition, **67a–d**, in which the position of the peptide on the tolpy ligand was changed from the 5′-position to the 4′-position, were synthesized and the anticancer activity of **67b,c** was even more potent than that for **61c** and **61d** [63]. However, a study of mechanisms responsive for the cell death induced by **67a–c** was hampered by their yellow luminescence emission at 500–600 nm, which overlaps with emission color of green-emitting probes such as MitoTracker Green and red-emission probes such as ER-Tracker Red.

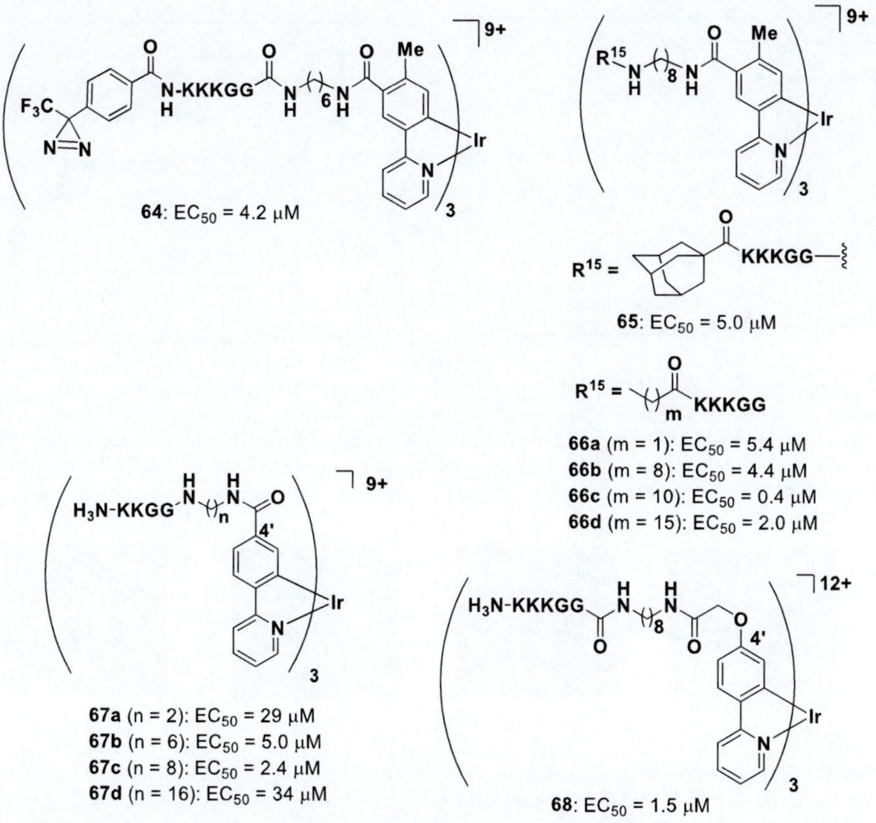

Scheme 14 Structures of IPH-ACs **64–68** and their EC_{50} values against Jurkat cells

The results of MTT assays of Jurkat cells and IMR90, which were used as a model of normal cells, after the treatment with **63a**, **67c**, and **68** are summarized in Fig. 9a, b, respectively, suggesting that these IPH-ACs act selectively against cancer cells compared to normal cells.

The EC_{50} value of **68**, that has a KKKGG peptide at the 4′-position linked via a hydroxylacetic acid unit and exhibits emission maximum at ca. 505 nm, was found to be nearly the same as that of **63a** (Scheme 14 and Fig. 9a). The intracellular uptake of **67c** and **68** in Jurkat cells (6.3–7.7 fmol/cell) is higher than that of **61c** and **66a** (0.4–1.0 fmol/cell), as determined by an inductively coupled plasma-mass spectrometer (ICP-MS), which may provide a possible explanation for the enhanced anticancer activity of **67c** and **68** [64].

Figure 10 shows the time-lapse luminescence microscopic images of Jurkat cells that had been treated with **61c**, in which a morphological change accompanied by bubble-like swelling and blebbing of the cell membrane were observed. After the incubation for 15–30 min, an enhanced green emission from **61c** was observed from dead cells (Fig. 10c, d). Our recent measurement of the

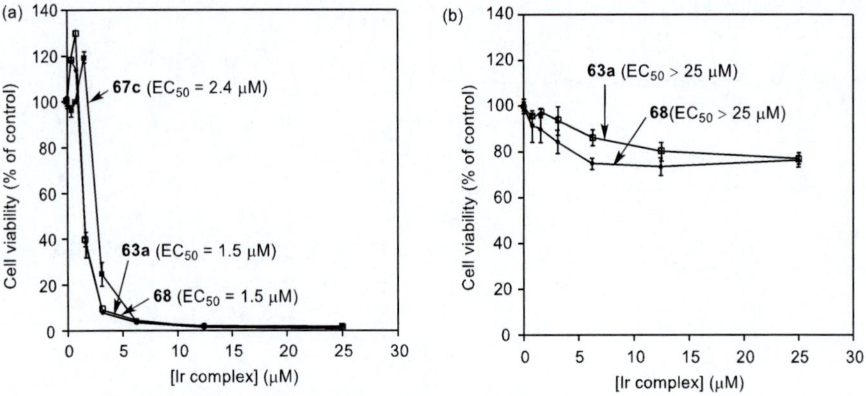

Fig. 9 The results of MTT assay of Jurkat cells with **63a** (open squares), **67c** (closed squares), and **68** (closed circles). **b** The results of MTT assay of IMR90 cells (a model of normal cells) with **63a** (open squares) and **68** (closed circles)

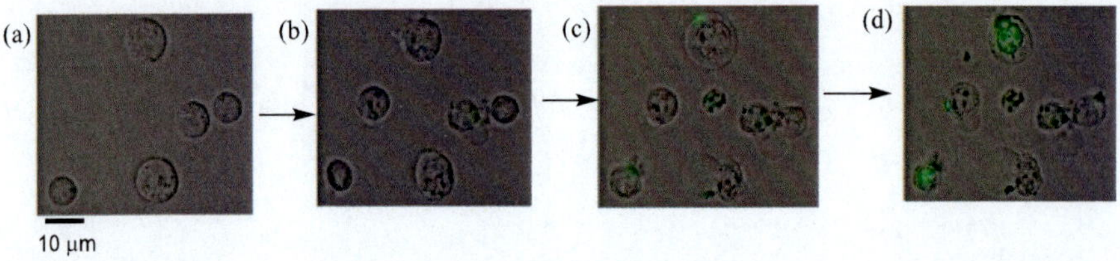

Fig. 10 Time-lapse emission images of Jurkat cells treated with **61c** (50 μM) for 0 min (**a**), for 5 min (**b**), for 15 min (**c**), and for 60 min (**d**). Black scale bar is 10 μm

intracellular amount of **68** suggested an increase from 0.64 ± 0.01 fmol/cell after the incubation for 30 min to (1.3 ± 0.1)–(1.5 ± 0.4) fmol/cell after 1–3 h, indicating the existence of a positive relationship between emission enhancement of IPH-ACs in Jurkat cells with their intracellular uptake [65]. There was negligible relationship between this emission enhancement and O_2 levels [66]. These facts allowed us to conclude that IPH-ACs that contained KK(K)GG peptide units function as inducers of cell death in cancer cells and luminescence detectors of dead cells.

In a detailed mechanistic study, the Jurkat cells were treated with **61c, 61d**, **62a**, **67b**, **67c,** and **68** in the presence of inhibitors of intracellular events, such as benzyloxycarbonyl-VAD(OMe)-fluoromethylketone (Z-VAD-FMK, a broad caspase inhibitor that stop apoptosis), necrostatin-1 (Nec-1, a specific RIPK-1 inhibitor and necroptosis inhibitor), 3-methyladenine (3-MA, an inhibitor of autophagosome formation that functions by inhibiting the action of type III phosphatidylinositol 3-kinases (PI3K), a autophagy inhibitor), and trifluoperazine (TFP, a CaM-binding molecule), and 3-chlorophenylhydrazone (CCCP, an uncoupling reagent and an inhibitor of mitochondrial Ca^{2+} uptake). Morphological changes (and cell death) in the Jurkat cells caused by **61c, 67b**, and **68** were

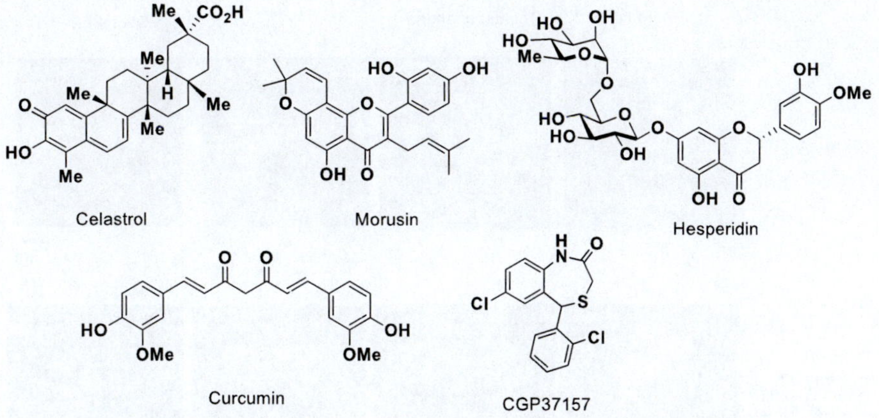

Scheme 15 Chemical structures of celastrol, morusin, hesperidin, curcumin, and CGP37157

effectively inhibited by only CCCP, indicating that the cell death induced by these IPH-ACs involves neither apoptosis, necroptosis, nor autophagy [60, 63–66].

It was reported that celastrol, a triterpenoid isolated from the roots of *Tripterygium wilfordii* [67, 68], morusin isolated from *Morus australis* [69], and hesperidin, which is a flavone glucoside contained in *Citrus* [70, 71], induce paraptosis (Scheme 15), which has been classified as one of the nonapoptotic PCD mechanisms [72, 73], in several cancer cell lines. It was found that celastrol induced cell death slowly (after incubation for 1 day) and that hesperidin induced only negligible cell death in Jurkat cells. Transmission electron microscopy (TEM) images of Jurkat cells that had been treated with **61c**, **68**, and celastrol suggest the vacuolization of the cytoplasm, indicating that these compounds induce a similar type of cell death. Consideration of these findings together with these TEM images and the experimental results presented above suggest that IPH-ACs such as **61c** and **68** induce paraptotic cell death. Moreover, it was found that CGP37157 (Scheme 15), an inhibitor of a mitochondrial Na^+/Ca^{2+} exchanger (mNCX), also induces paraptosis at somewhat high concentrations (100 μM) [65].

More detailed mechanistic studies implied that **61c** and **68** induce paraptosis via the direct transport of Ca^{2+} from the endoplasmic reticulum (ER) to mitochondria, resulting in a decrease in mitochondrial membrane potential ($\Delta\Psi_m$). Although the expression of autophagy-related molecules was observed in the case of paraptosis caused by IPH-ACs, we conclude that autophagy is the result of this paraptosis rather than the cause of cell death [64–66].

The results of co-staining experiments of mitochondria and the ER in Jurkat cells using MitoTracker Green and ER-Tracker Red in the presence of **68** and CGP37157 suggest that these two molecules induce membrane fusion (or tethering) between the ER and mitochondria (Fig. 11), the direct transport of Ca^{2+} from the ER to mitochondria and the loss of $\Delta\Psi_m$. In contrast, celastrol causes a negligible increase in mitochondrial Ca^{2+} concentrations during the induction of paraptosis, and no evidence for the ER-mitochondria membrane fusion has been observed [64–66].

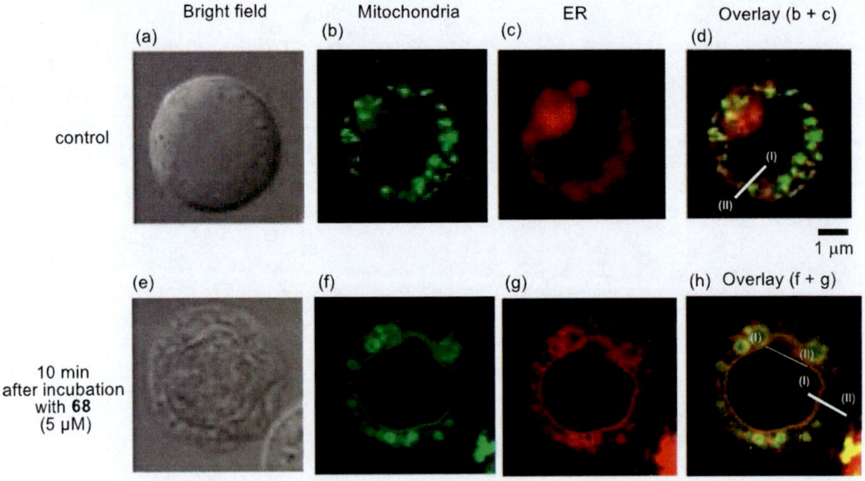

Fig. 11 Confocal microscopic images (×100) of Jurkat cells stained with MitoTracker Green (0.5 μM, 1 h) and ER-Tracker Red (1 μM, 1 h) at 37 °C under 5% CO_2, before (**a–d**) and after (**e–g**) treatment with **68** (5 μM, 10 min). Excitation at 473 nm and emission from 485 to 545 nm were used for the MitoTracker Green, and excitation at 559 nm and emission from 570 to 620 nm were used for the ER-Tracker Red. Black scale bar is 1 μm

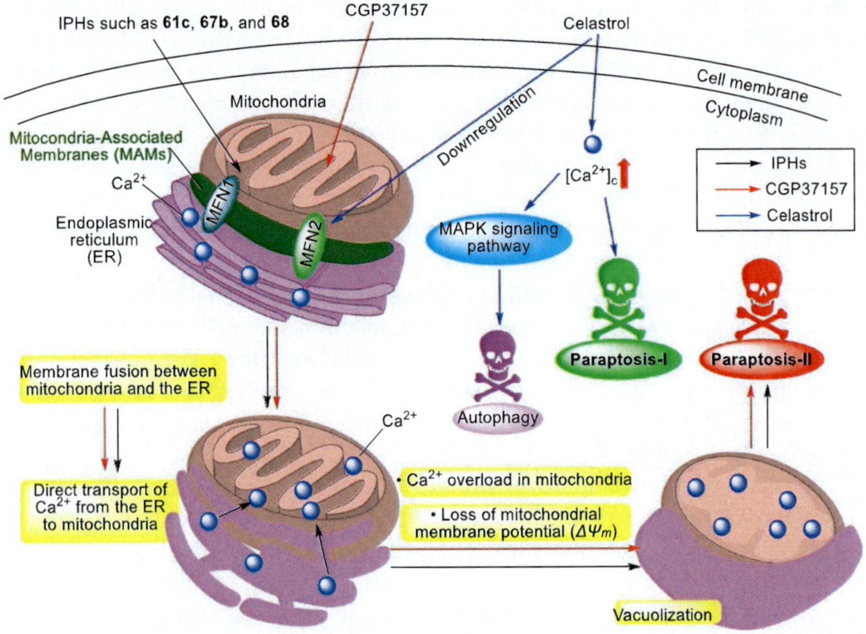

Scheme 16 Plausible mechanisms of paraptotic cell death in Jurkat cells induced by IPH-ACs having KK(K)GG peptide units (**61c**, **67b**, and **68**)

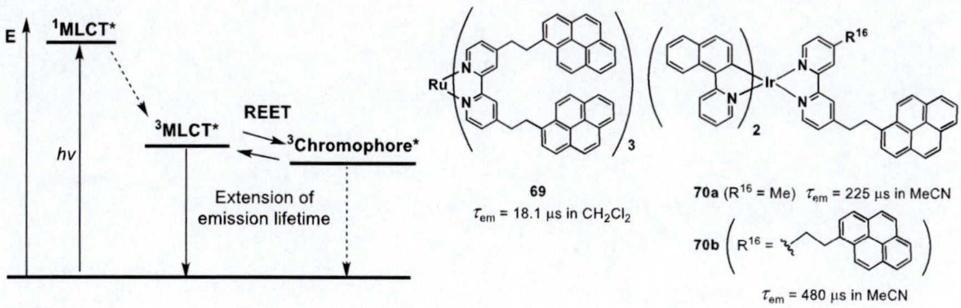

Scheme 17 Schematic representation of REET phenomena and representative metal complexes that exhibit REET

The two proposed types of paraptosis are summarized in Scheme 16. The first category is a known type of paraptosis that is induced by celastrol, which we refer to as "paraptosis I." This process involves a Ca^{2+} overload in the cytoplasm with minimal membrane fusion between the mitochondria and the ER. The second type is referred to as paraptosis II, which is induced by IPH-ACs such as **61c**, **67b**, and **68** and CGP37157, in which membrane fusion occurs between the ER and mitochondria with the induction of mitochondrial Ca^{2+} overload [65]. Note that hybrids of triptycene units (not luminescent) were developed very recently for the assembly of three KKKGG peptide units instead of Ir(III) complex cores, and it was found that these triptycene-peptide hybrids as amphiphilic peptide conjugates (TPH-ACs) induce PCD in some cancer cell lines via similar intracellular signaling pathways [74]. Moreover, it would be interesting to compare paraptosis I and II defined in this review with molecular mechanisms of paraptosis or nonapoptotic cell death induced by other metal complexes [75–79].

5 Synthesis of Iridium Complex-Dye Hybrid and Control of Emission Color and Emission Lifetime

Long emission lifetimes of cyclometalated Ir(III) complexes are one of their advantages. Many efforts have been dedicated to finding strategies for enhancing luminescence emission lifetimes in attempts to improve the luminescence properties of Ir(III) complexes by the introduction of chromophores into the side chain of the Ir complexes based on a reversible electronic energy transfer (REET) mechanism [80–86], in which an electron and/or energy exchange proceeds between the triplet excited state (^{3}MLCT*) of the Ir complex core and the excited triplet state (^{3}LC) of the dye unit, as shown in Scheme 16. For efficient REET processes, the relative triplet energy level between the central metal complexes and the chromophores, namely, a reversible process starting from a downhill energy transfer to (π–π*) followed by an uphill energy transfer from ^{3}(π–π*) to ^{3}MLCT*, is important. Typical examples of REET compounds include ruthenium (Ru(II)) complexes such as **69**, which has an emission lifetime of 18.1 μs in CH_2Cl_2 [82], and **70a**,**b**, an $Ir(npy)_3$ complex (npy: 2-(1-naphthly)pyridine) having pyrene groups, whose emission

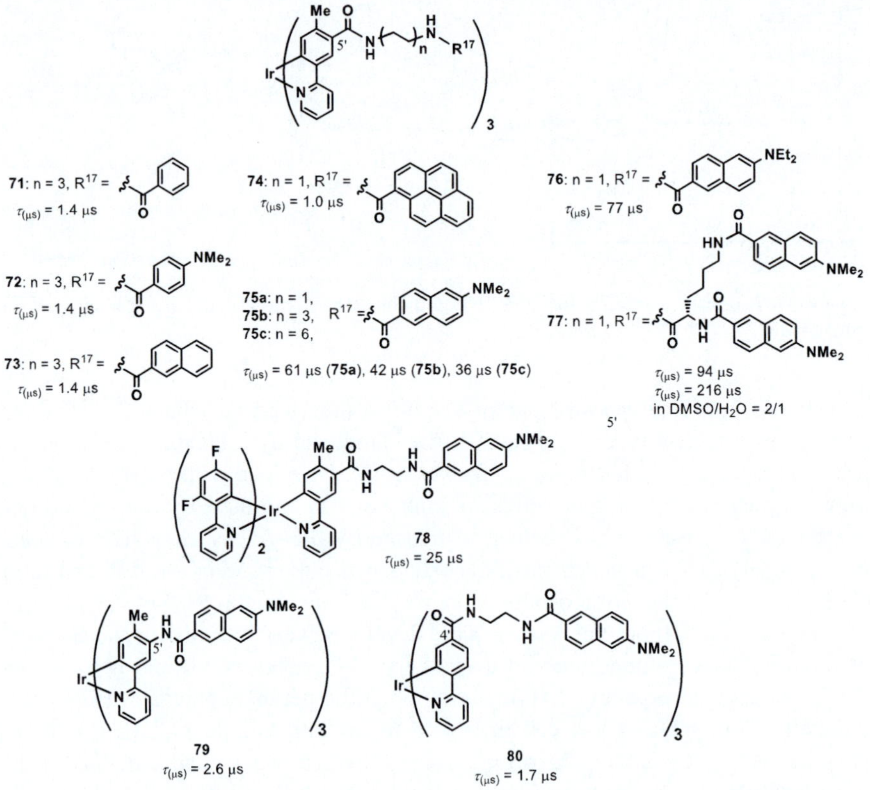

Scheme 18 The structures of Ir(III) complex-chromophore hybrids (ICHs) (*facial* forms) and their microsecond-order emission lifetimes ($\tau_{(\mu s)}$) (at > 475 nm) in DMSO at 298 K otherwise noted (excitation at 375 nm)

lifetimes are 225–448 μs in MeCN at room temperature [83, 84] (Scheme 17). However, these REET compounds are limited to red emission (emission at > 600 nm) metal complexes that contain pyrene units.

In order to explore REET compounds that exhibit other emission colors, Ir(III) complex-chromophore hybrids (ICHs) such as **71–76** were designed and synthesized via carboxylate intermediate **36** (R^3 = Me) shown in Scheme 9 [87]. As summarized in Scheme 18, the luminescence emission lifetimes of **71–74**, which possess benzoyl, 4-(*N*,*N*-dimethylamino)benzoyl, naphthoyl, and pyrenoyl groups, were 1.0–1.4 μs in degassed DMSO at 298 K, which are similar values to the lifetime of $Ir(tolpy)_3$ **1** (1.4 μs). On the other hand, the emission lifetimes ($\tau_{(\mu s)}$) of **75** and **76** containing 6-*N*,*N*-dimethylaminonaphthoyl (DMANA) and 6-*N*,*N*-diethylaminonaphthoyl (DEANA) groups were 36–77 μs (excitation at 355 nm and emission at 500–510 nm), as shown in Fig. 12. Moreover, the emission lifetime of **77** having six DMANA units was extended to 94 μs in degassed DMSO and 216 μs in a 2:1 mixture of DMSO/H_2O at 298 K, and that of **55** with one DMANA unit was 25 μs [87].

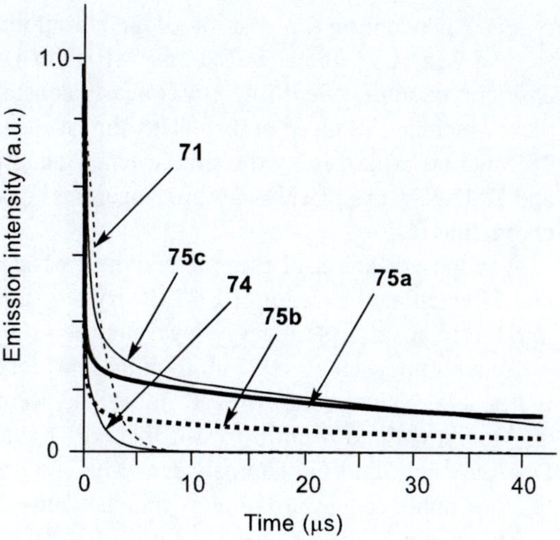

Fig. 12 Emission decay curves of **71**, **74**, and **75a–c** (10 μM) at > 475 nm in degassed DMSO at 298 K (excitation at 355 nm)

It should be noted that the nanosecond-order lifetimes ($\tau_{(ns)}$) and microsecond-order lifetimes ($\tau_{(\mu s)}$) were observed in the emission decay curves of **75–77** and microsecond-order lifetimes are mainly discussed below.

The effect of the linker length of Ir(III) complexes (**75a–c**) on emission lifetimes (closed circles) and emission quantum yields (closed squares) in degassed DMSO at 298 K is shown in Fig. 13a, and the effect of the number of DMANA groups on Ir complexes (**1**, **78**, **75a**, and **77**) on their emission lifetimes (closed circles) and emission quantum yields (closed squares) is presented in Fig. 13b [87]. These results suggest that the shorter the linker, the longer the emission lifetime, as presented in Fig. 13a. Secondly, the larger number of DMANAs contributes to the enhancement of the luminescence lifetimes (Fig. 13b). It was also found that

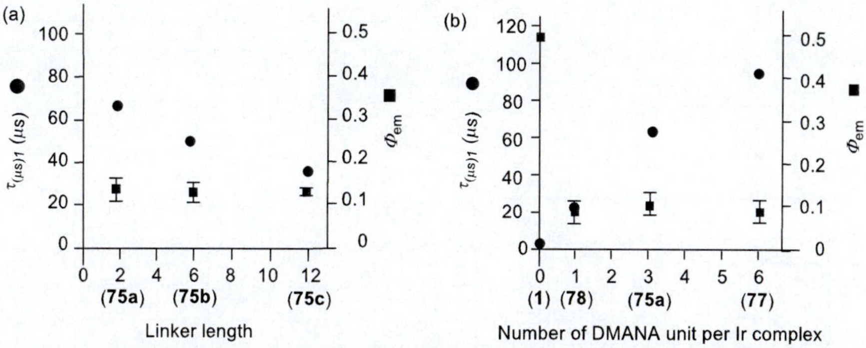

Fig. 13 Relationship of luminescence parameters about REET. **a** Effect of linker length of Ir(III) complexes (**75a**, **75b**, and **75c**) on emission lifetimes (closed circles) and emission quantum yields (closed squares) in degassed DMSO at 298 K. **b** Effect of number of DMANA groups of Ir complexes (**1, 78, 75a**, and **77**) on emission lifetimes (closed circles) and emission quantum yields (closed squares) in degassed DMSO at 298 K

$\tau_{(\mu s)}$ is dependent on the position of the chromophore and the polarity of the solvent ($\tau_{(\mu s)}$ of **75a** is 57 μs in DMSO and 216 μs in DMSO/H_2O (2/1)). Intriguingly, the emission quantum yield (Φ_{em}) was nearly constant (Φ_{em} ~ 0.1) in all of the Ir complexes included in these graphs. The almost identical Φ_{em} values of **75a–c**, **77**, and **78** could be explained by the similar emission efficiency between the Ir(tolpy)$_3$ core and DMANA (or DEANA) groups, regardless of the REET efficiency between these chromophores.

A proposed Jablonski diagram and the kinetic parameters for Ir complexes **75**, **76**, and **77** calculated based on REET theory are shown in Fig. 14 [87]. It is very likely that the relative triplet energy level between the Ir complex core and the chromophores is important for extending emission lifetimes of these molecules, as shown in Fig. 14a. Namely, the ^{3}MLCT* level is expected to be slightly higher than that of ^{3}LC to allow the downhill (forward) energy transfer from the Ir complex unit to the DMANA unit, and the gap between ^{3}MLCT* and ^{3}LC should be sufficiently small to allow uphill (backward) energy transfer from ^{3}LC to ^{3}MLCT*.

The k_f and k_b values in Fig. 14a–c are the forward and backward energy transfer rate constants and the k_{Ir} and k_{DM} values are the decay rates of the Ir complex and the DMANA chromophores, respectively [85, 87]. The K value ($=k_f/k_b$) in Fig. 14c is defined as the equilibrium constant for the reversible triplet energy transfer from the Ir complex to the attached DMANA and α value is the number (fraction) of $^3(\pi-\pi^*)$ hydrocarbon-based triplets. A good correlation between the K and α values and longer emission lifetimes was observed for these bichromophoric molecules. The greater K and α values of **75a,b** and **77** compared to those of **78** are due to the smaller k_b values of **75a,b** and **77** rather than the k_b value of **78**. As listed in Scheme 17, the lifetime of **80**, which contains DMANA groups at the 4′-position of its ppy ligand part, is 1.7 μs, suggesting a negligible REET effect. These phenomena could be explained by its emission maxima at 534 nm (yellow), meaning that the

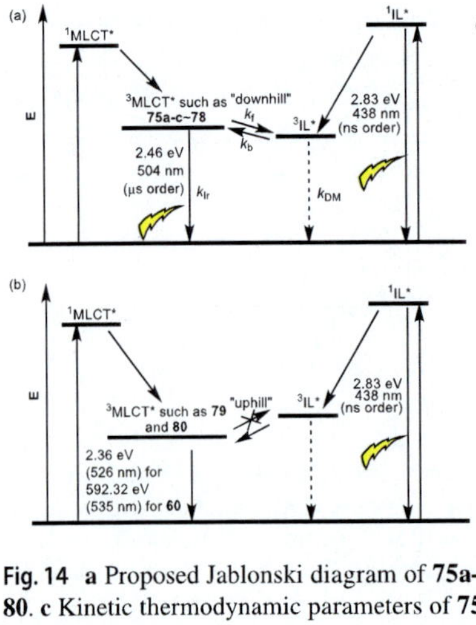

(c)

	78	75a	75b	75c	77
Number of DMANA unit	1	3	3	3	6
Linker length between Ir complex core & DMANA unit	C_2	C_2	C_4	C_6	C_2 (and L-lysine unit)
k_{Ir} (s^{-1})	5.6 x 10^5	7.1 x 10^5	7.1 x 10^5	7.1 x 10^5	7.1 x 10^5
k_f (s^{-1})	4.7 x 10^7	6.2 x 10^7	2.5 x 10^7	5.2 x 10^7	6.2 x 10^7
k_b (s^{-1})	5.7 x 10^5	2.7 x 10^5	1.2 x 10^5	6.0 x 10^5	1.6 x 10^5
k_{DM} (s^{-1})	3.4 x 10^4	1.3 x 10^4	2.1 x 10^4	2.0 x 10^4	0.9 x 10^4
K	83	229	213	86	339
α	0.987	0.996	0.995	0.988	0.997

Fig. 14 **a** Proposed Jablonski diagram of **75a–c**, **76**, **77**, and **78**. **b** Proposed Jablonski diagram of **79** and **80**. **c** Kinetic thermodynamic parameters of **75a–c**, **77**, and **78** in DMSO

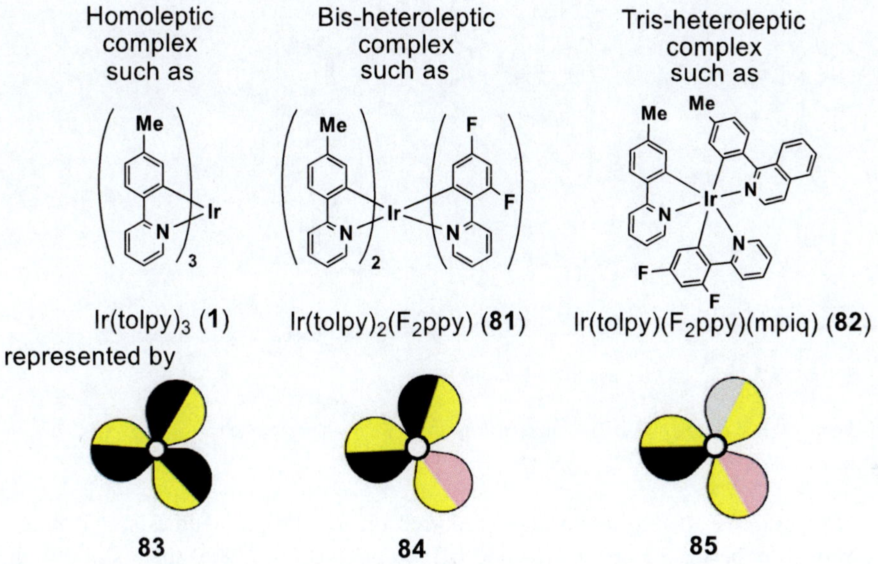

Scheme 19 Chemical structures of some representative homoleptic, bis-heteroleptic, and tris-heteroleptic tris-cyclometalated Ir complexes and their schematic representations as three-leaf clover

^{3}MLCT* of **79** is slightly lower than that of the ^{3}LC level of the DMANA moiety (Fig. 14b), which inhibits the initial (forward) energy transfer from ^{3}MLCT* to ^{3}LC.

6 Selective Synthesis of Bis- and Tris-Heteroleptic Cyclometalated Iridium(III) Complexes and Their Applications to the Fields of Material Sciences

Cyclometalated Ir(III) complexes can be mainly classified into three categories regarding the type and number of ligands that coordinate to Ir(III) ions, as shown in Scheme 19. The first class is homoleptic Ir(III) complexes such as Ir(tolpy)$_3$ **1**, Ir(ppy)$_3$ **2**, and Ir(mpiq)$_3$ **44**, which contain the three same ligands, the second is bis-heteroleptic complexes that possess two different ligands such as Ir(tolpy)$_2$(F$_2$ppy) **81** (F$_2$ppy: 2-(4′,6′-difluorophenyl)pyridine), and the third class is tris-heteroleptic complexes that contain three different ligands such as Ir(tolpy)(F$_2$ppy)(mpiq) **82**. These structures are schematically represented by three-leaf clovers (*Cryptotaenia japonica* or Japanese honeywort or "Mitsuba" in Japanese), **83**, **84**, and **85**, respectively, in which the three ligand moieties coordinate to Ir ion via C-Ir (black, pink, and gray parts) and N-Ir (yellow parts) coordination bonds (Scheme 19).

White light OLEDs usually require a combination of two or three different color emitters, blue, green, and red, with the deposition of multilayers on top of other components [85–93]. The development of metal complexes that exhibit a white emission (namely, a white emission from one compound) would be highly desirable in terms of reducing the production cost and energy consumption of such materials.

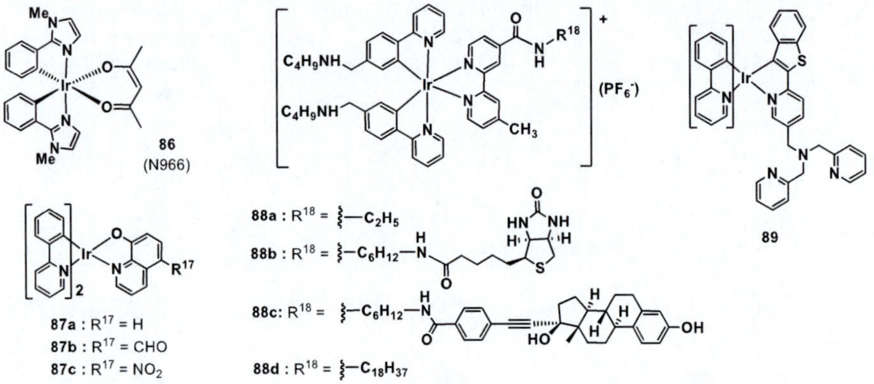

Scheme 20 Representative Ir(III) complexes that exhibit white-color emission

For example, Nazeeruddin et al. reported on the design, synthesis, and photophysical properties of bis-heteroleptic Ir(III) complex **86** (N966) that exhibit white emission due to the broad emission spectra from 440 to 800 nm from a single molecule (emission maxima at 570 nm), with emission quantum yield of 1.5×10^{-2} and emission lifetime of 1.96 μs in CH_2Cl_2 at 298 K (excitation at 355 nm) (Scheme 20) [88]. The results of TD-DFT calculations suggest that the lowest excited states of the cyclometalated complex unit are in close proximity to the $(acac)^-$ ligand part. The $Ir(ppy)_2(HQ^{-1})$ complex (HQ: 8-hydroxyquinoline) derivatives **87** were synthesized by Slugovc and coworkers [89]. Although **87a** with no functional group at 5′-position of the 8-hydroxyquinoline part is almost nonluminescent, **87b** and **87c** with electron-withdrawing groups such as formyl and nitro groups at the 5′-position emit an orange-red emission due to the dual emission maxima (emission at ca. 450 nm and a broad emission at 600–800 nm from **87c**) in solutions.

Lo and coworkers reported on some dual-emissive cyclometalated Ir(III)-polypyridine complexes **88** that exhibit a broad emission from ca. 450 to 750 nm including two emission maxima at ca. 500 nm and 600 nm with fairly good emission quantum yields in CH_2Cl_2 at 298 K [90, 91]. Interestingly, the emission maxima of **88** are observed at only 500 nm in aqueous phosphate buffer. Therefore, the emission intensity ratio of **88** functionalized with biotin (**88b**), estradiol (**88c**) and a hydrophobic alkyl (C_{18}) chain (**88d**) at 500 nm and 600 nm is altered in response to their environmental change such as the complexation with avidin at the biotin unit (for **88b**), the complexation with the estradiol receptor α (ERα) at the estradiol unit (for **88c**), and the interaction with the lipid-binding human serum albumin (HSA) at the C_{18} chain unit (for **88d**), respectively.

You et al. reported on the Ir(III) complex conjugated with di(2-picolyl)amine (DPA) unit **89** that showed dual emission maxima at 470–520 nm and 600 nm [92]. The complexation of the DPA part of **89** with the copper(II) ion (Cu^{2+}) (the K_d value for the Cu^{2+}-**89** complex was determined to be 16 μM) induced the quenching of the emission at 600 nm, while emission at 470–520 nm was negligibly quenched in a pH 7.0 buffer solution. Therefore, **89** was found to function as a ratiometric probe of

Cu^{2+} in aqueous solutions. These results indicate that the dual-emissive single molecules have potential applications in many scientific fields. In addition, an efficient white-light emission can be obtained by the encapsulation of Ir(III) complexes in metal–organic frameworks (MOFs) [93].

Bis-heteroleptic cyclometalated Ir(III) complexes, **90–92** (Fig. 15), which contain quinoline-type ligands such as 8-quinolinols, 8-mercaptoquinoline, and 8-sulfonamidoquinoline as the ancillary ligands, were synthesized [94]. It was found that Ir complex **92** which is composed of two F_2ppy ligands and an 8-benzenesulfonamidoquinoline (BSQ) ligand, exhibit two main emission maxima at about 460 nm and 620 nm, resulting in the emission of a near-white luminescence in degassed DMSO at 298 K (bold curve in Fig. 15) and 77 K (see also Fig. 21a).

The tris-heteroleptic Ir(III) complex (IrLL′A or IrLL′L″) composed of an Ir atom and three different ligands represent a class of highly functionalized Ir complexes (L, L′, and L′ denote different cyclometalating ligands, and A is a symmetric ancillary ligand such as acetylacetonate (acac), 2,2′-bipyridine (bpy), and 1,10-phenanthroline (phen) ligands). Although novel photophysical properties would be expected, there are very few examples of the selective synthesis of these types of metal complexes [95–99]. For example, when the Ir(COD)(μ-Cl)$_2$ complex (COD: 1,5-cyclooctadiene) is reacted with a mixture of different ligands, a mixture of μ-complexes having different ligands is produced [96]. The next reaction with a third ligand produces a mixture of various tris-heteroleptic Ir(III) complexes, whose purification requires tedious separation processes.

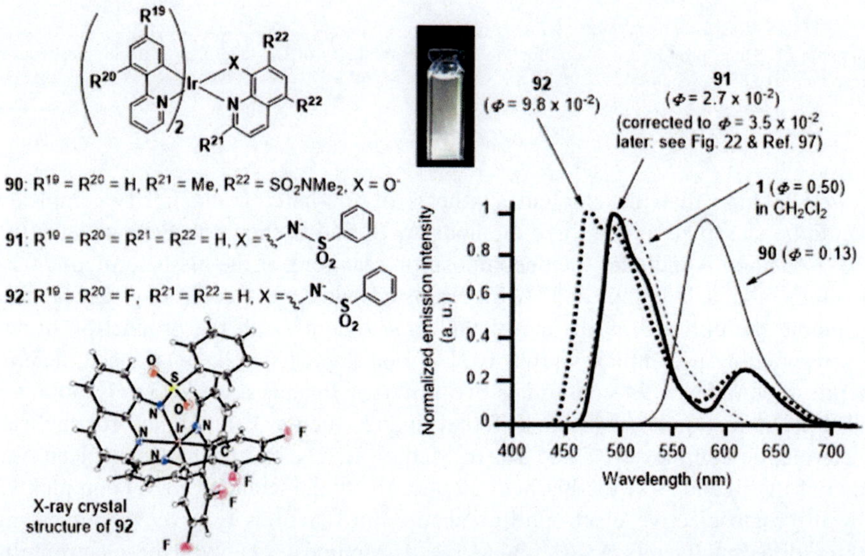

Fig. 15 Emission spectra of Ir(III) complexes **90–92** (10 μM) in degassed DMSO at 298 K (excitation at 366 nm) and X-ray crystal structure of **92** (ORTEP drawing with 50% ellipsoid). The inset shows a picture of white-emission from a solution of **92** in degassed DMSO at 298 K

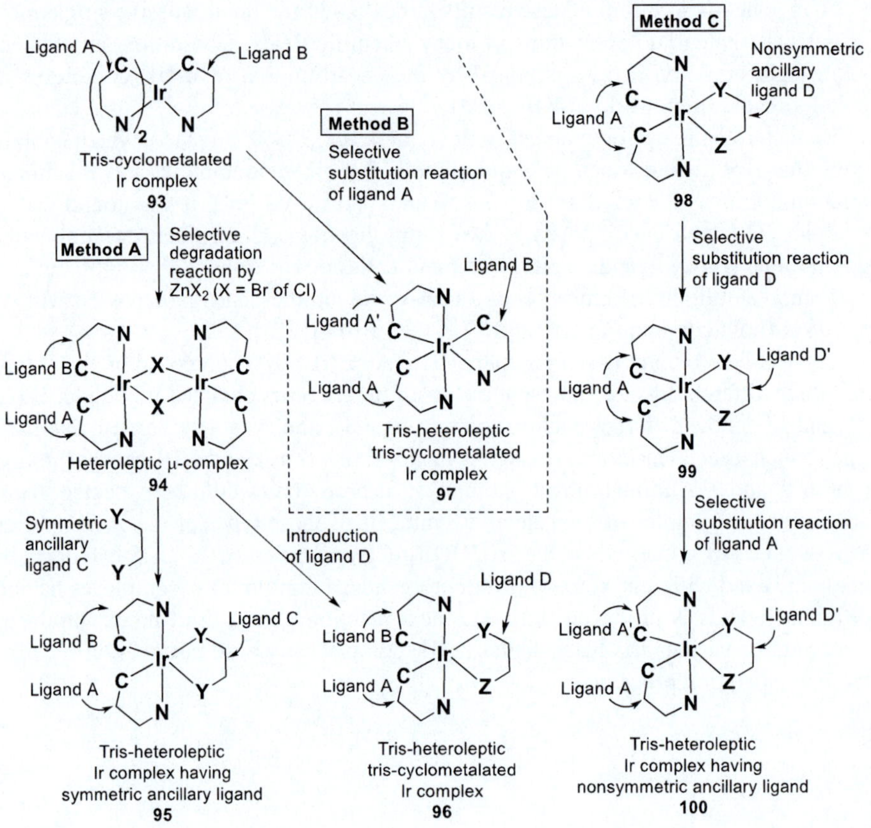

Scheme 21 Three proposed methods for the selective synthesis of tris-heteroleptic Ir complexes (corresponding to **85** in Scheme 19)

We assume that the efficient synthesis of tris-heteroleptic Ir(III) complexes could be classified as the three methods A, B, and C shown in Scheme 21 [100, 101]. Method A includes the decomposition reactions of the bis-heteroleptic tris-cyclometalated Ir complex **93** with Brønsted and Lewis acids (e.g. ZnX_2) that promote the elimination of one of the three ligands with the production of the corresponding μ-complexes [100–103]. When ZnX_2 (X = Cl, Br) is used, heteroleptic μ-complexes **94** containing two different ligands are selectively obtained [100]. The reaction of **94** with different ligands such as C and D affords the tris-heteroleptic complexes **95** and **96**. In Method B, a substituent is introduced into one of the ligand A of the bis-heteroleptic tris-cyclometalated Ir(III) complex **93** by the regioselective electrophilic substitution reactions to give **97** possessing three different ligands A, A′, and B [100]. Method C involves the electrophilic substitution reactions at the third ligand (ancillary type ligand D) of **98** with the formation of the bis-heteroleptic complex **99**, when the ligand D is most reactive (HOMO existing site) among the three ligand parts, A, A, and D in **98**. After this

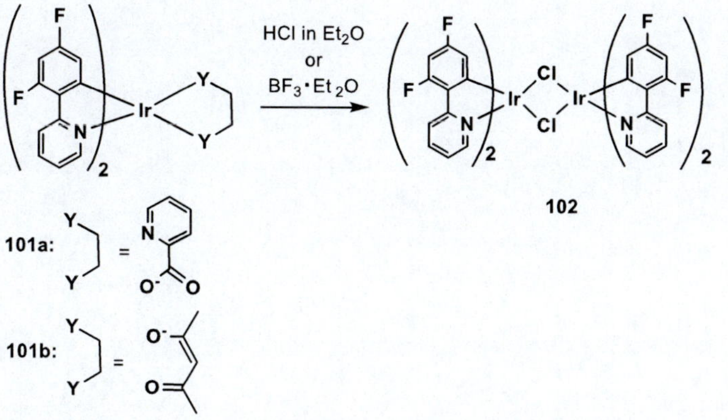

Scheme 22 Degradation reaction of **101** by HCl in Et_2O or $BF_3{\cdot}Et_2O$ to give the μ-complex **102**

manipulation, the reactive site (HOMO group) is transferred from D (D′) to one of the two ligand A's (HOMO hopping from D to A), and hence an electrophilic substitution reaction of **99** occurs at more reactive ligand A, resulting in the production of **100** having three different ligands, A, A′, and D′ [104].

The degradation of heteroleptic Ir complexes containing ancillary ligands such as $Ir(F_2ppy)_2(pic)$ **101a** and $Ir(F_2ppy)_2(acac)$ **101b** with HCl (2 M in Et_2O) and $BF_3{\bullet}Et_2O$ to give the corresponding Ir dimer $[Ir(F_2ppy)_2(\mu\text{-}Cl)]_2$ **102** was reported by Baranoff et al. (Scheme 22) [102, 103].

In Aoki's research group, it was found that decomposition reaction of cyclometalated Ir(III) complexes proceed under the typical reaction conditions for the Friedel–Crafts acylation of $Ir(tolpy)_3$ **1** using acetyl chloride in the presence of $AlCl_3$ in CH_2Cl_2, resulting in the formation of $[Ir(tolpy)_2(\mu\text{-}Cl)]_2$ **103a**, a chloro-bridged Ir(III) dimer that involves a four-membered ring consisting of two Ir and two Cl (Scheme 23) [36]. The decomposition reaction of **1** with several Brønsted acids and Lewis acids was conducted and the product from **1** after treatment with $ZnBr_2$ was confirmed to be the corresponding a bromo-bridged μ-complex **103b**, as shown in Fig. 16a [100]. Intriguingly, the reaction of the *meridional* form of the bis-heteroleptic complex $Ir(tolpy)_2(F_2ppy)$ **81** with $ZnBr_2$ in CH_2Cl_2 afforded the heteroleptic μ-complex $[Ir(tolpy)(F_2ppy)]_2(\mu\text{-}Br)]_2$ **104b** (corresponding to **94** in Scheme 21), whose structure was confirmed by an X-ray crystal structure analysis (Fig. 16b). For comparison, the reaction of **81** with TMSCl or HCl in 1,4-dioxane or $AlCl_3$ gave the homoleptic μ-complex **103a**. These collective findings suggest that the degradation reaction of heteroleptic complexes such as **81** by ZnX_2 tends to release ligands that contain electron-donating groups (namely, electron-rich ligands).

Next, **104** was converted to the tris-heteroleptic Ir complexes **105** and **107** by the reaction with ancillary ligands such as acetylacetone (acacH) and 8-benzenesulfonamidoquinoline (BSQ) (**106**) [100]. The reaction of **104b** with 4′-methylphenylisoquinoline (mpiqH) **108** in the presence of AgOTf and Et_3N in 1,2-dichloroethane afforded the tris-heteroleptic tris-cyclometalated Ir(III) complex **82**. The alternative

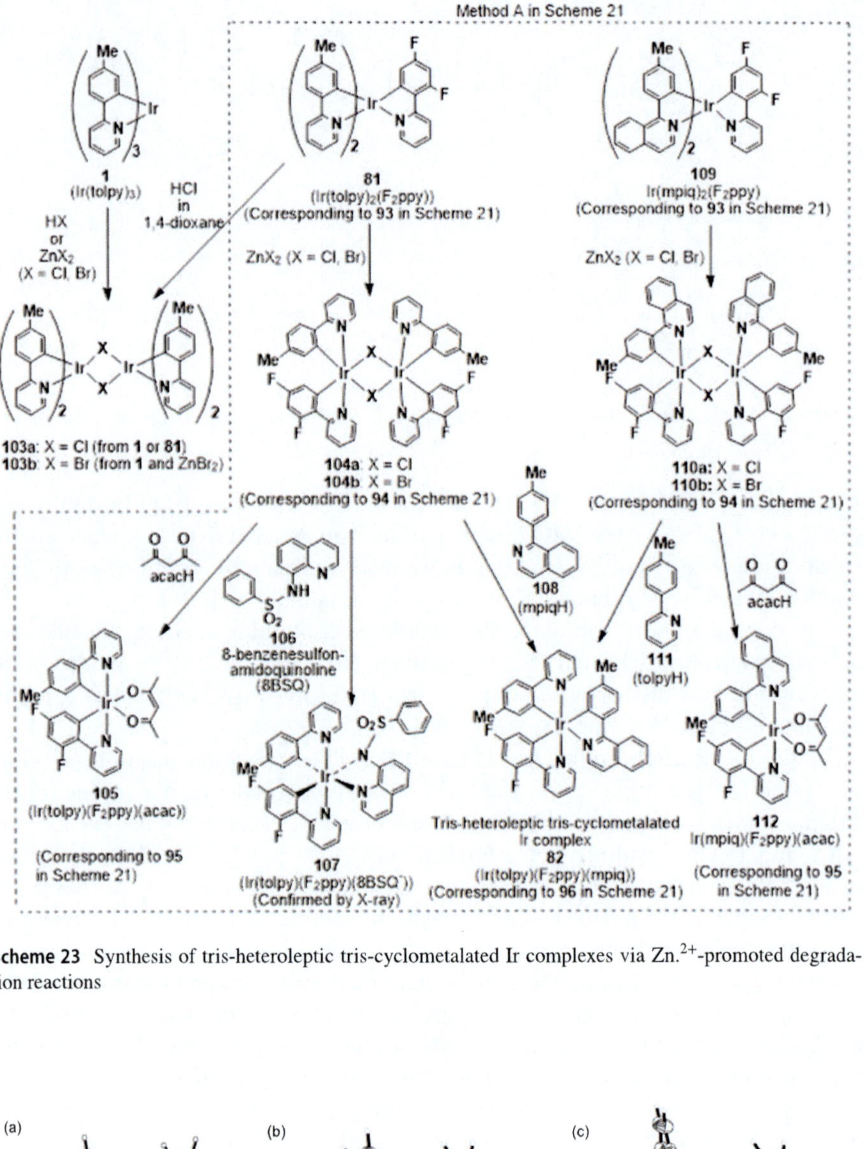

Scheme 23 Synthesis of tris-heteroleptic tris-cyclometalated Ir complexes via Zn.$^{2+}$-promoted degradation reactions

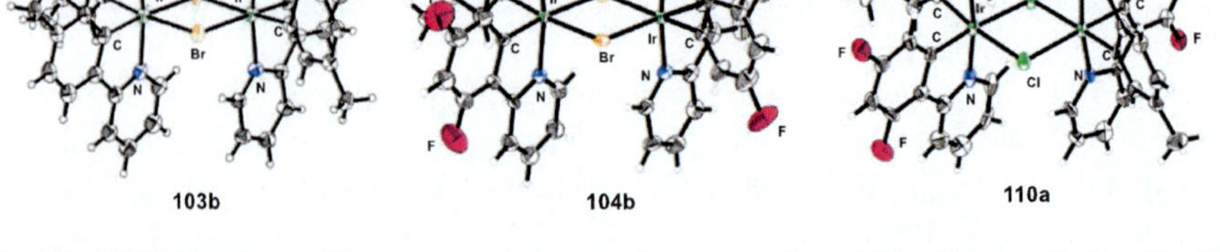

Fig. 16 ORTEP drawings of X-ray crystal structures of the μ-complexes **103b, 104b**, and **110a** with 50% probability ellipsoid. For clarity, solvent molecules contained in crystals were omitted

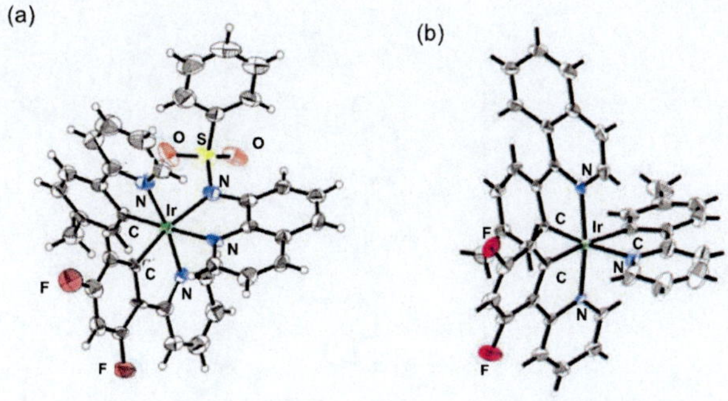

107 **82c** (meridional)

Fig. 17 ORTEP drawings of **107** and **82c** with 50% probability ellipsoid

synthesis of **82** was achieved via the heteroleptic μ-complex **110** (an X-ray crystal structure analysis of **110a** is shown in Fig. 16c), which was obtained by the Zn^{2+}-promoted decomposition reaction of the Ir(mpiq)$_2$(F$_2$ppy) complex **109**, and the following reaction with tolpyH ligand **111**. The reaction of **110** with acacH gave a heteroleptic complex **112** (Scheme 23). The X-ray crystal structure of **107** is shown in Fig. 17a and the luminescence emission spectra of **105**, **107**, and **112** are displayed in Fig. 19a, in which the dual emission of **107** and **112** is observed and **107** has a nearly single emission maxima at ca. 500 nm.

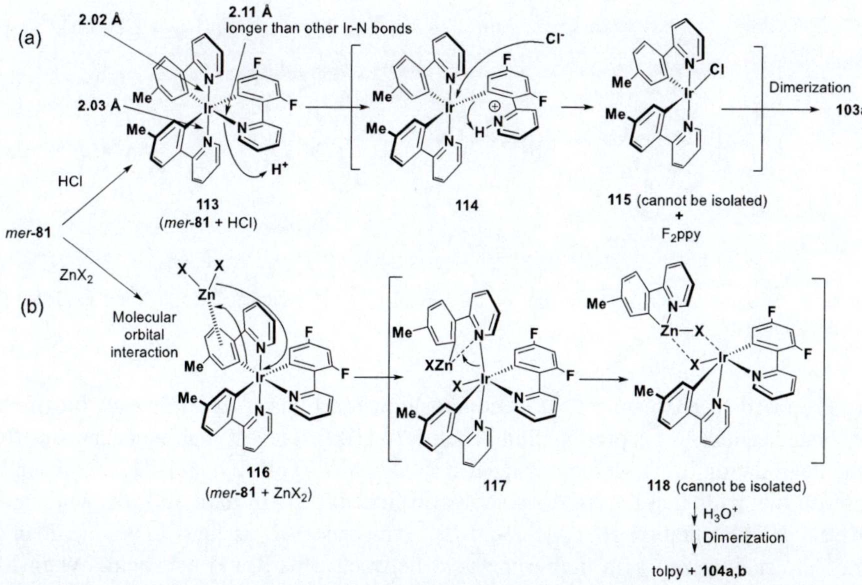

Scheme 24 Proposed scheme for the degradation of **81** promoted by Brønsted acid (H^+) and Lewis acidic Zn^{2+}

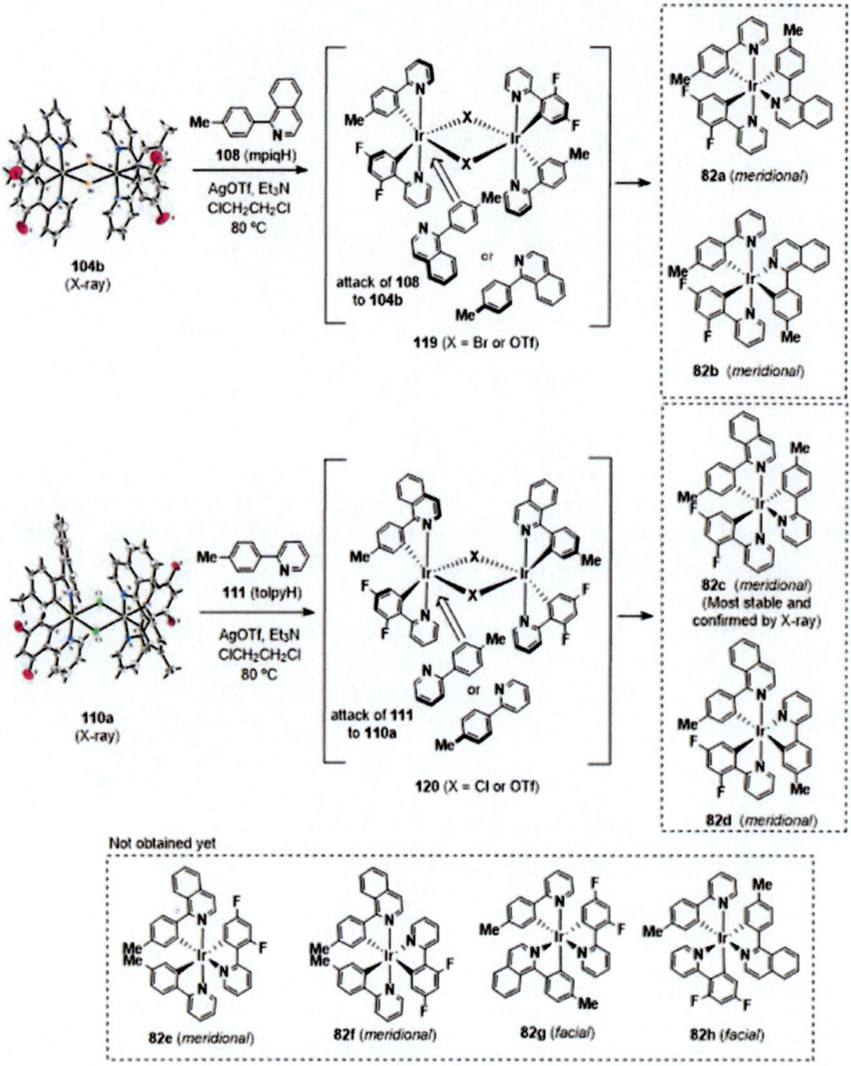

Scheme 25 Synthesis of stereoisomers of tris-heteroleptic tris-cyclometalated Ir complex Ir(tolpy)(F_2ppy)(mpiq) **82**

Proposed mechanism for the decomposition reactions of Ir complexes by Brønsted acids and ZnX_2 is presented in Scheme 24 [100]. The two main mechanisms for the degradation of tris-cyclometalated Ir complexes (typically, *mer*-**81**) are dependent on the acids that are used, as shown in Scheme 23. In hard and soft acids and bases (HSAB) theory, H^+, Al^{3+}, and Si^{4+} are classified as hard Lewis acids and Zn^{2+} is categorized as a borderline acid between a hard and soft acid. When **81** was treated with H^+, $AlCl_3$ or TMSCl, the nitrogen atom on the F_2ppy ligand in the *mer*-**81** was assumed to react with H^+, Al^{3+} (of $AlCl_3$), or Si^{4+} (of TMSCl) (**113**), resulting in the formation of the intermediate **115** via **114**. This can be explained by the longer Ir-N(F_2ppy) bond (2.11 Å) (namely a weaker bond) than those of the Ir-N(tolpy) bonds (2.02–2.03 Å), as observed in the X-ray crystal structure of *mer*-**81**. The cleavage of the Ir-C bond of **114** to give **115** and dimerization of **115** would afford **103a**. In contrast, it is assumed that *mer*-**81** is activated by ZnX_2 by the interaction at the phenyl ring to give the C-Zn intermediate **117** then undergoes the dimerization of the intermediate **118**.

It should be noted that there are eight stereoisomers for **82** (**82a–h**) based on the difference in the coordination direction of the three ligands tolpy, F_2ppy, and mpiq (in the two dashed boxes of Scheme 25), and a total of 16 stereoisomers are present when each enantiomer is taken into consideration [100, 101]. The product of the reaction of **103b** with **108** was only **82a**, because **82b** was not obtained, possibly due to its instability under the conditions of the silica gel chromatography. Only **82c** was isolated after the reaction between **110a** and **111**, as confirmed by X-ray crystallography as shown in Fig. 17b, and the isolation of **82d** was not successful. These findings suggest that different diastereomers of **82** have different stabilities and different photophysical properties, even though the combination of three ligands in **82** is the same. More interestingly, the emission spectra of **82a** and **82c** are different, as shown in Fig. 18b along with an emission spectrum of the Ir(F_2ppy)$_2$(mpiq) complex **121** for comparison [94]. The synthesis and isolation of other diastereomers **82e–h** has not yet been successful, and it is assumed that the synthesis of these stereoisomers would require other methodologies such as isomerization of either of **82a–d** and/or other reactions.

Figure 19 shows a typical example of the use of Method C for the preparation of tris-heteroleptic Ir(III) complexes via the successive ligand-selective electrophilic substitution reactions of bis-heteroleptic Ir(III) complexes, a process that is referred

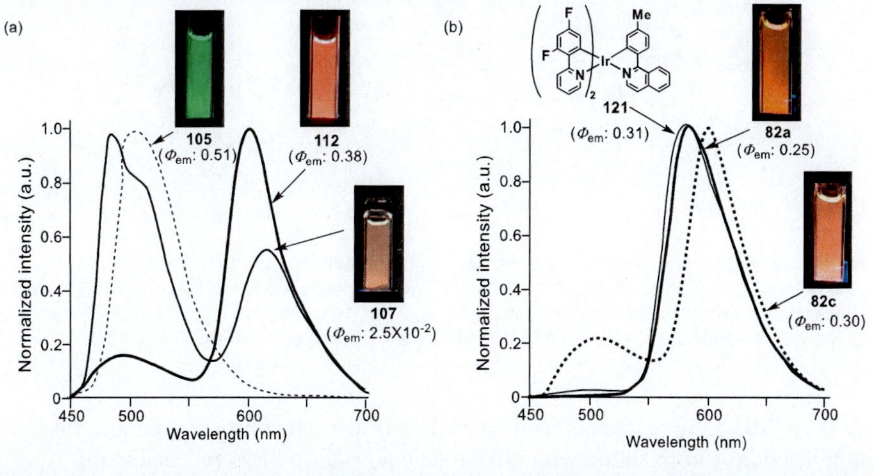

Fig. 18 **a** Normalized luminescence emission spectra of **105** (dashed curve), **107** (plain curve) and **112** (bold curve) (10 μM) in degassed DMSO at 298 K (excitation at 366 nm). **b** Normalized luminescence emission spectra of **82a** (bold curve), **82c** (dashed curve), and **121** (plain curve) (10 μM) in degassed DMSO at 298 K (excitation at 366 nm)

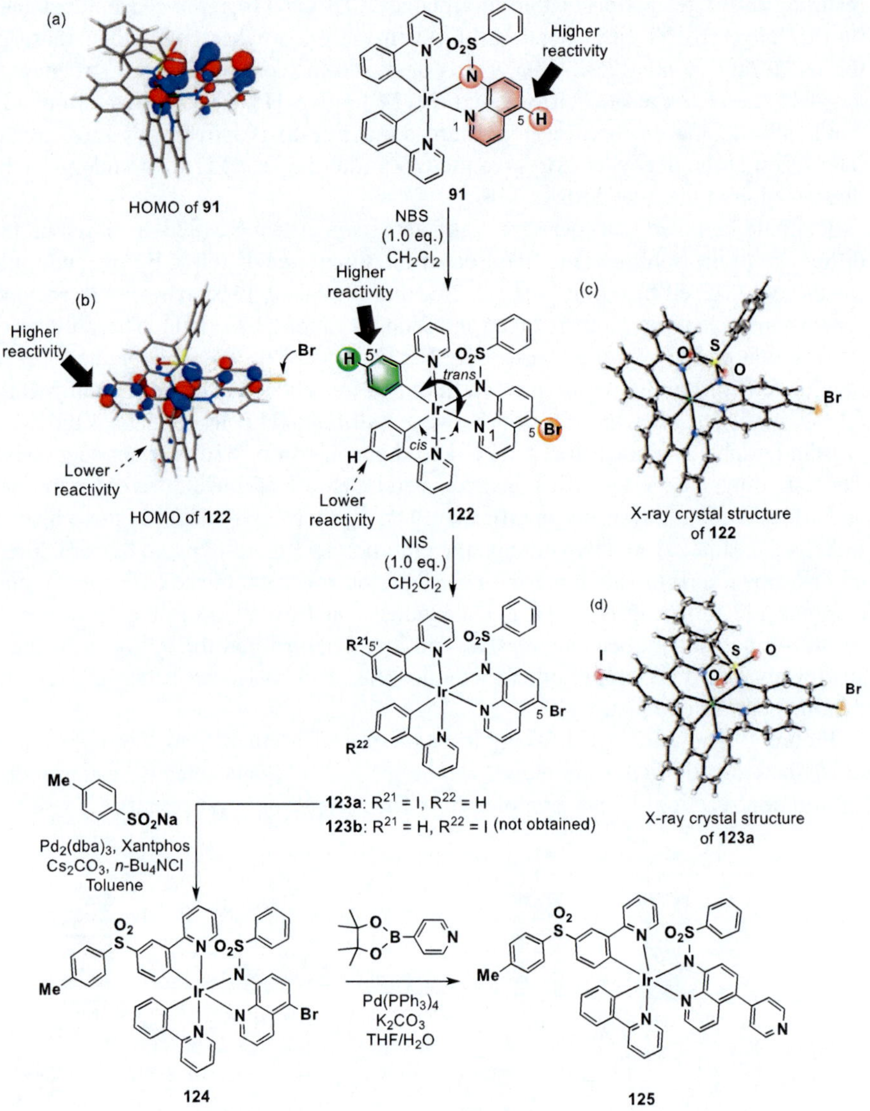

Fig. 19 The synthesis of functionalized tris-heteloreptic Ir complex **125** from **91** via the intermediates **122**, **123**, and **124** based on the HOMO transfer (HOMO hopping) phenomena, as indicated as Method C in Scheme 20. **a** HOMO of the starting material **91**, **b** HOMO of the intermediate **122**, **c** ORTEP drawing of X-ray crystal structure of **122**, and **d** ORTEP drawing of the X-ray crystal structure of **123a**

to as HOMO hopping phenomena [104]. It was found that the electrophilic substitution reactions (bromination with NBS (one equivalent)) of **91** proceed at the 5-position of its quinoline ring to afford **122**, whose structure was confirmed by an X-ray crystal structure analysis (Fig. 19c), due to the localization of the HOMO on its quinoline ligand (Fig. 19a). In the brominated product **122**, HOMO is transferred to one

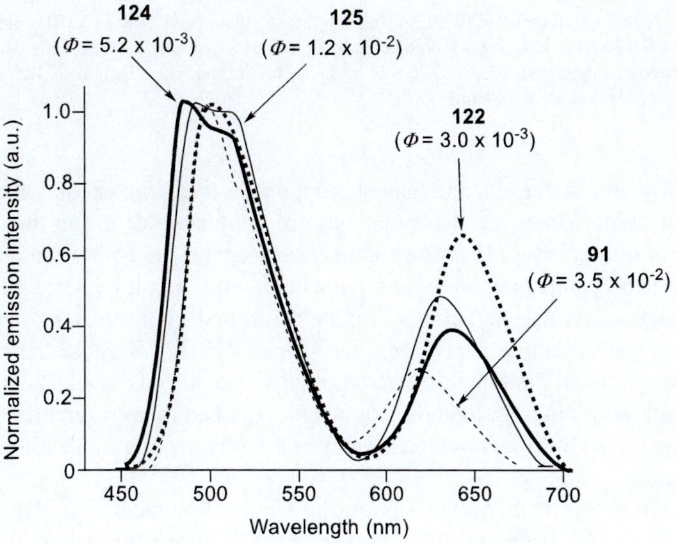

Fig. 20 Luminescence emission spectra of **91**, **122, 124**, and **125** (10 μM) in degassed DMSO at 298 K (excitation at 366 nm)

of two ppy ligands as shown in Fig. 19b, and hence the iodination of **122** with one equivalent of NIS takes place at the 5′-position of the ppy ligand (trans position with respect to the Ir-N (quinoline) bond) to give the tris-heteroleptic Ir complex **123a**, whose structure was also confirmed by an X-ray crystal structure analysis (Fig. 19d). The palladium-catalyzed cross-coupling reaction of **123a** with *p*-toluenesulfinate to give **124** followed by a Suzuki–Miyaura coupling with 4-pyridineboronic acid pinacol ester afforded **125**. As shown in Fig. 20, 91, 122, 124, and 125 exhibit dual-emission spectra with somewhat different emission intensity ratios at around 500 nm and 620–650 nm. As shown in Fig. 21b, the emission spectra of **125** is changed upon the protonation of its pyridine part, suggesting that the fine tuning of the ratio of green/red emission intensity in dual-emissive molecules would be possible by the introduction of appropriate functional groups by PCF and external stimuli such as protonation/deprotonation and so on.

Figure 20 shows luminescence emission spectra of **91**, **122**, **124**, and **125** in degassed DMSO at 298 K [104]. Figure 21a shows the luminescence emission color of Ir(III) complexes **2, 44, 46, 57**, **91**, **92** (at 298 K and 77 K), **126** (Ir(F_2ppy)$_3$, and **127** in a CID chromaticity diagram (Commission International de I'Eclairage coordinates) [40, 94] and Fig. 21b displays emission colors of **125** in DMSO/buffer (pH 4–8) [104].

7 Conclusion and Prospects

In this review, the synthesis and applications of cyclometalated Ir(III) complexes based on the "post-complexation functionalization (PCF)" principle are presented. The emission color of Ir complexes can be controlled by the introduction of halogen

Fig. 21 **a** Chromaticity diagram (CIE 1931) showing the xy color coordinates of Ir(III) complexes **2**, **44**, **46**, **57**, **90**, **91**, **126**, and **127**. The xy color coordinates of these complexes were calculated from their emission spectra in degassed DMSO at 298 K or 77 K. **b** Chromaticity diagram of **125** in DMSO/pH buffer (pH 4–8) (excitation at 366 nm)

atoms, nitro, amino, formyl, and related groups after the complexation of the ligands with the Ir atom. Moreover, these methods are promising for use in the design and synthesis of luminescent pH sensors, anticancer reagents, and other functional Ir(III) and other metal complexes, whose emission color, emission lifetimes, and biological activity (such as anticancer activity) can be controlled by introducing peptide units or chromophores (such as IPH-TMs, IPH-ACs and ICHs). Because there are many possible reactions of the functionalized or nonfunctionalized cyclometalated Ir complexes containing different ligands (homoleptic, bis-heteroleptic, and tris-heteroleptic complexes) with different reagents, it would be interesting to try all the possible combinations.

The main focus of this review is on the tris-cyclometalated Ir(III) complexes rather than the heteroleptic complexes such as $[Ir(ppy)_2(bpy)]^+$ **8**, because the emission of **8** and its analogs is dependent on the polarity of solvents, and these complexes exhibit broad emission spectra from 500 to 700 nm in polar solvents such as acetonitrile [105–107], which would overlap with the many specific fluorescent probes of intracellular organelles described in this review. In addition, **8** and related analogs have been reported as powerful homogeneous catalysts in C–C bond formation reactions and water-reducing systems [14, 15, 105–107]. Therefore, we assume that applications of PCF to various functionalization of heteroleptic and cationic or anionic metal complexes would be useful for the improvement of their catalytic activities. The second advantage of tris-cyclometalated metal complexes is their triangular (C_3-symmetric) structures. Because there are many biomolecules that possess trimeric structures such as proteins (proteins modified with oligosaccharides) and polynucleotides, the tris-cyclometalated Ir complexes could be useful as the scaffolds to mimic these trimeric biomolecules.

Note that the Ir(III) complexes described in this review were synthesized as mixtures of two enantiomers, Δ (delta) and Λ (lambda) forms, with respect to Ir(III) complex cores. Attempts at the optical resolution of the Ir(III) complexes are currently underway.

By combination with the PCF introduced in this paper, more precise and appropriate complex design and synthesis are expected. We believe that this review provides useful information for the design and synthesis of new types of cyclometalated complexes not only of Ir complexes but also of other platinum metal groups [108–111] and their applications to biomedical and material sciences and related fields.

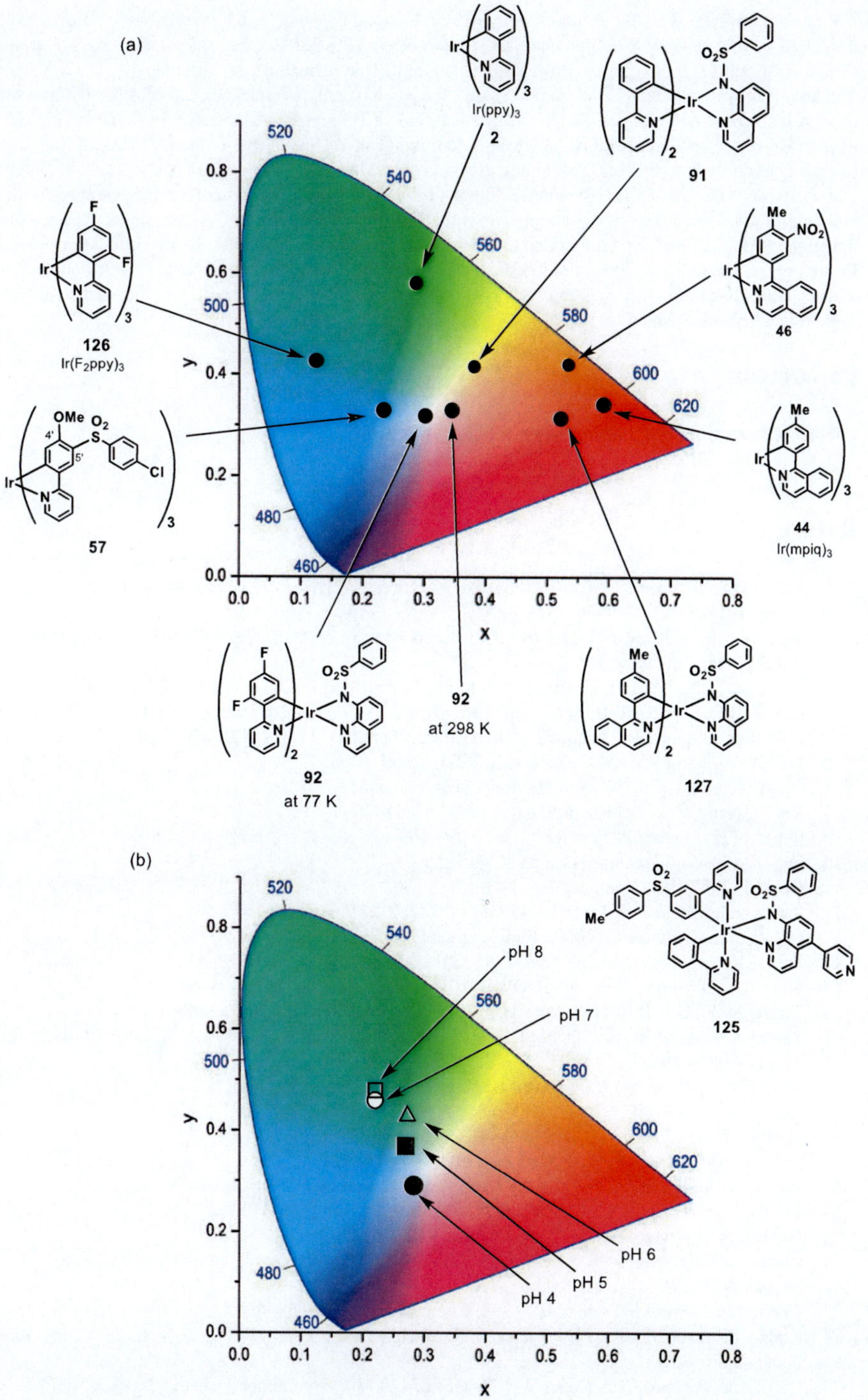

(a)
Ir(ppy)3
2
91
126
Ir(F2ppy)3
46
57
44
Ir(mpiq)3
92
at 77 K
92
at 298 K
127
(b)
125
pH 8
pH 7
pH 6
pH 5
pH 4
x
y

Acknowledgements We wish to thank our collaborators and co-workers for their contributions to work described in this review. We appreciate Dr. Motoo Shiro (Rigaku Co. Ltd.), Prof. Masahiro Kawano (Tokyo Institute of Technology), Prof. Osamu Ishitani (Tokyo Institute of Technology), Dr. Yasuyuki Yamada (Nagoya University), Dr. Junpei Yuasa (Tokyo University of Science), Dr. Masakazu Umezawa (Tokyo University of Science), Dr. Koji Tsuchiya (Tokyo University of Science), Prof. Toshiyuki Kaji (Tokyo University of Science), Prof. Takeshi Nakamura (Tokyo University of Science), Prof. Kohei Soga (Tokyo University of Science), Prof. Hideki Sakai (Tokyo University of Science), Dr. Rikio Niki (Tokyo University of Science), and Dr. Toshinari Ichihara (Tokyo University of Science) for their great assistance and helpful discussion. Financial supports from the Ministry of Education, Culture, Sports, Science and Technology (MEXT) of Japan, the Uehara Memorial Foundation, the Tokyo Ohka Foundation for the Promotion of Science and Technology, Kanagawa, Japan, the Tokyo Biochemical Research Foundation, Tokyo, Japan, Japan Society for the Promotion of Science (JSPS), and Tokyo University of Science are gratefully acknowledged.

Declarations

Conflict of interest The authors declare no conflict of interest.

References

1. Baldo MA, O'Brien DF, You Y, Shoustikov A, Sibley S, Thompson ME, Forrest SR (1998) Nature 395:151–154
2. Tamayo AB, Alleyne BD, Djurovich PI, Lamansky S, Tsyba I, Ho NN, Bau R, Thompson ME (2003) J Am Chem Soc 125(24):7377–7387
3. Flamigni L, Barbieri A, Sabatini C, Ventura B, Barigelletti F (2007) Top Cur Chem 281:143–203
4. You Y, Park SY (2009) Dalton Trans. 1267–1282.
5. Lo KKW, Louie MW, Zhang KY (2010) Coord Chem Rev 254(21–22):2603–2622
6. Lo KKW, Zhang KY (2012) RSC Adv 2(32):12069–12083
7. Chi Y, Chou PT (2010) Chem Soc Rev 39(2):638–655
8. You Y, Nam W (2012) Chem Soc Rev 41(21):7061–7084
9. Omae I (2016) Coord Chem Rev 310(1):154–169
10. Omae I (2016) J Organomet Chem 823(15):50–75
11. Omae I (2017) J Organomet Chem 841(15):12–30
12. Kappaun S, Slugovc C, List EJW (2008) Int J Mol Sci 9(8):1527–1547
13. Liu Z, Bian Z, Huang C (2009) Topics. Organomet Chem 28:113–142
14. Prier CK, Rankic DA, MacMillan DWC (2013) Chem Rev 113(7):5322–5363
15. Shaw MH, Twilton J, MacMillan DWC (2016) J Org Chem 81(16):6898–6926
16. Teegardin K, Day JI, Chan J, Weaver J (2016) Org Proc Res Dev 20(7):1156–1163
17. Zhang S, Hosaka M, Yoshihara T, Negishi K, Iida Y, Tobita S, Takeuchi T (2010) Cancer Res 70(11):4490–4498
18. Yoshihara T, Hirakawa Y, Hosaka M, Nangaku M, Tobita S (2017) J Photochem Photobiol C Photochem Rev 30:71–95
19. Ho PY, Ho CL, Wong WY (2020) Coord Chem Rev 412:213267
20. Lord RM, McGowan PC (2019) Chem Lett 48(8):916–924
21. Abbas S, Din ID, Raheel A, Din AT (2019) Appl Organomet Chem 34:e5413
22. Shi H, Wang Y, Lin S, Lou J, Zhang Q (2021) Dalton Trans 50:6410
23. Guerchais V, Fillaut JL (2011) Coord Chem Rev 255(21–22):2448–2457
24. Chodorowski-Kimmes S, Beley M, Collins JP, Sauvage JP (1996) Tetrahedron Lett 37(17):2963–2966
25. Coudret C, Fraysse S, Launay JP (1998) Chem Commun 10:663–664
26. Clark GR, Headford CEL, Roper WR, Wright LL, Yap VPD (1994) Inorg Chim Acta 220(1–2):261–272
27. Clark AM, Rickard CEF, Roper WR, Wright LJ (2000) J Organometal Chem 598(2):262–275
28. Cheung KM, Zhang QF, Chan KW, Lam MHW, Williams ID, Leung WH (2005) J Organomet Chem 690:2913–2921

29. Qin T, Ding J, Wang L, Baumgarten M, Zhou G, Müllen K (2009) J Am Chem Soc 131(40):14329
30. Stossel P, Spreitzer H, Bach I. U.S. Patent. US 2005/0253135 A1
31. Stossel P, Spreitzer H, Bach I. U.S. Patent. US 2006/7125998 B2
32. Arm NJ, Gareth Williams JA (2005) Chem Commun 20:230–232
33. Whittle VL, Gareth Williams JA (2008) Inorg Chem 47(15):6596–6607
34. Liang A, Huang G, Chen S, Hou H (2015) RSC Adv 5:49466–49470
35. Zhao H, Simpson PV, Barlow A, Moxey GJ, Morshedi M, Roy N, Philip R, Cifuentes MP, Humphrey MG (2015) Chem Eur J 21(33):11843–11854
36. Aoki S, Matsuo Y, Ogura S, Ohwada H, Hisamatsu Y, Moromizato S, Shiro M, Kitamura M (2011) Inorg Chem 50(3):806–818
37. Nonoyama M (1974) Bull Chem Soc Jpn 43(3):767–768
38. Moromizato S, Hisamatsu Y, Suzuki T, Matsuo Y, Abe R, Aoki S (2012) Inorg Chem 51(4):12697–12706
39. Nakagawa A, Hisamatsu Y, Moromizato S, Kohono M, Aoki S (2014) Inorg Chem 53(1):409–422
40. Kando A, Hisamatsu Y, Ohwada H, Mormonization S, Kohno M, Aoki S (2015) Inorg Chem 54(11):5342–5357
41. Tsuboyama A, Iwawaki H, Furugori M, Mukaide T, Kamatani J, Igawa S, Moriyama T, Miura S, Takiguchi T, Okada S, Hoshino M, Ueno K (2003) J Am Chem Soc 125(42):12971–12979
42. Hisamatsu Y, Aoki S (2011) Eur J Inorg Chem 20:5360–5269
43. Yin XM, Dong Z (eds) (2003) Essentials of apoptosis. Humana Press, Totowa
44. Zhivotovsky B, Orrenius S (2011) Cell Calcium 50:211–221
45. Orrenius S, Gogvadze V, Zhivotovsky B (2015) Biochem Biophys Res Commun 460(1):72–81
46. Har MW, Distelhorst CW (2010) Cold Spring Harbor Perspect Biol 2:a005579
47. Lee D, Kim IY, Saha S, Choi KS (2016) Pharmacol Ther 162:120–133
48. Kim E, Lee DM, Seo MJ (2021) Front Cell Dev Biol 8:607844
49. Hymowitz SG, Christinger HW, Fuh G, Ultsch MH, O'Connell M, Kelley RF, Ashkenazi AA, de Vos AM (1999) Mol Cell 4(4):563–571
50. Hymowitz SG, O'Connell M, Ultsch MH, Hurst A, Totpal K, Ashkenazi AA, de Vos AM, Kelley RF (2000) Biochemistry 39(4):633–640
51. Blanc HN, Ashkenazi AA (2003) Cell Death Diff 10:66–75
52. Wiezorek J, Holland P, Graves J (2011) Clin Cancer Res 16:1701–1708
53. Angell YM, Bhandari M, De Francisco MN, Frederick BT, Green JM, Leu K, Leuther K, Sana R, Schatz PJ, Whitehorn EA, Wright K, Holmes CP (2009) Adv Exp Med Biol 611:101–103
54. Masum AA, Hisamatsu Y, Yokoi K, Aoki S (2018) Bioinorg Chem Appl Article ID 20:7578965
55. Masum AA, Yokoi K, Hisamatsu Y, Naito K, Shashni B, Aoki S (2018) Bioorg Med Chem 26(17):4804–4816
56. Lv H, Zhang S, Wang B, Cui S, Yan J (2006) Controlled Release 114(1):100–109
57. Nakase I, Tanaka G, Futaki S (2013) Mol BioSyst 9(5):855–861
58. Gasper D, Veiga AS, Castanho ARB (2013) Front Microbiol 4:Article 294
59. Hisamatsu Y, Shibuya A, Suzuki N, Suzuki T, Abe R, Aoki S (2015) Bioconjug Chem 26(5):857–879
60. Hisamatsu Y, Suzuki N, Suzuki T, Abe R, Aoki S (2017) Bioconjug Chem 28(2):507–523
61. Haribabu J, Tamura Y, Yokoi K, Balachandran C, Umezawa M, Tsuchiya K, Yamada Y, Kavrvembu R, Aoki S (2021) Eur J Inorg Chem 2021(18):1796–1814
62. Naito K, Yokoi K, Balachandran C, Hisamatsu Y, Aoki S (2019) J Inorg Biochem 199:Article 110785
63. Yokoi K, Hisamatsu Y, Naito K, Aoki S (2017) Eur J Inorg Chem 20:5295–5309
64. Yokoi K, Balachandran C, Umezawa M, Tsuchiya K, Mitrić A, Aoki S (2020) ACS Omega 5(12):6983–7001
65. Yokoi K, Yamaguchi K, Umezawa M, Tsuchiya K, Aoki S (2022) Biochemistry 61(8):639–655
66. Balachandran C, Yokoi K, Naito K, Tamura Y, Jebiti H, Umezawa M, Tsuchiya K, Yoshihara T, Tobita S, Aoki S (2021) Molecules 26:7028
67. Yoon MJ, Lee AR, Jeong SA, Kim Y, Kim JY, Kwon Y, Choi KS (2014) Oncotarget 5(16):6816–6830
68. Wang WB, Feng LX, Yue QX, Wu WY, Guan SH, Jiang BH, Yang M, Liu X, Guo DA (2012) J Cell Physiol 227(5):2196–2206
69. Xue J, Li R, Zhao X, Ma C, Lv X, Liu L, Liu P (2018) Chem-Biol Interact 283:59–74
70. Garg A, Garg S, Zaneveld LJ, Singla AK (2001) Phytother Res 15(8):655–669

71. Yumnam S, Hong GE, Raha S, Saralamma GVV, Lee HJ, Lee WS, Kim EH, Kim GS (2016) J Cell Physiol 231(6):1261–1268
72. Spernandio S, de Belle I, Bredesen DE (2000) Proc Natl Acad Sci USA 97(26):14376–14381
73. Wang Y, Wen X, Zhang N, Wang L, Hao D, Jiang X, He G (2019) Biomed Pharmacother 118:109203
74. Yamaguchi K, Yokoi K, Umezawa M, Tsuchiya K, Aoki S (2022) Bioconjug Chem 33(4):691–717
75. Ye R-R, Tan CP, Chen MH, Hao L, Ji LN, Mao ZW (2016) Chem Eur J 22(23):7800–7809
76. Li C, Ip K-W, Man W-L, Song D, He M-L, Yiu S-M, Lau T-C, Zhu G (2017) Chem Sci 8(10):6865–6870
77. Cini M, Williams H, Fay MW, Searle MS, Woodward S, Bradshaw TD (2016) Metallomics 8(3):286–297
78. Tardito S, Bassanetti I, Bignardi C, Elviri L, Tegoni M, Mucchino C, Bussolati O, Franchi-Gazzola R, Marchiò L (2011) J Am Chem Soc 133(16):6235–6242
79. Liu Z, Romero-Canelón I, Habtemariam A, Clarkson GJ, Sadler PJ (2014) Organomet 33(19):5324–5333
80. Ford WE, Rodgers MA (1992) J Phys Chem 96(7):2917–2920
81. Howarth AJ, Davies DL, Lelj F, Wolf MO, Patrick BO (2014) Inorg Chem 53(22):11882–11889
82. McClenaghan ND, Barigelletti F, Maubert B, Campagna S (2002) Chem Commun 20:602–603
83. Denisov SA, Cudré Y, Verwilst P, Jonusauskas G, Marin-Suárez M, Fernández-Sánchez JF, Baranoff E, MaClenaghan ND (2014) Inorg Chem 53(5):2677–2682
84. Medina-Rodríguez S, Denisov SA, Cudré Y, Male L, Marin-Suárez M, Fernández-Guitiérrez JF, Tron A, Jonusauskas G, McClenaghan ND, Baranoff E (2016) Analyst 141:3090–3097
85. Jiang X, Peng J, Wang J, Guo X, Zhao D, Ma Y (2016) ACS Appl Mater Interfaces 8(6):3591–3600
86. Lavie-Combot A, Lincheneau C, Cantuel M, Leydet Y, McClenaghan ND (2010) Chem Soc Rev 39(2):506–515
87. Kazama A, Imai Y, Okayasu Y, Yamada Y, Yuasa J, Aoki S (2020) Inorg Chem 59(10):6905–6922
88. Bolink HJ, De Angelis F, Baranoff E, Kline C, Fantacci S, Coronado E, Sessolo M, Kalyanasundaram K, Gratzel M, Nazeeruddin MK (2009) Chem Commun 20:4672–5467
89. Kappaun S, Sax S, Eder S, Moller KC, Waich K, Niedermair F, Saf R, Mereiter K, Jacob J, Mullen K, List EJW, Slugovc C (2007) Chem Mater 19(6):1209–1211
90. Lo KKW, Zhang KY, Leung SK, Tang MC (2008) Angew Chem Int Eng 47(12):2213–2216
91. Lo KKW, Chan JSW, Lui LH, Chung CK (2004) Organomet 23(13):3108–3116
92. You Y, Han Y, Lee YM, Par SY, Nam W, Lippard SJ (2011) J Am Chem Soc 133(30):11488–11491
93. Sun CY, Wang WL, Zhang X, Qin C, Li P, Su ZM, Zhu DX, Shan GG, Shao KZ, Wu H, Li J (2013) Nat Commun 4:2717
94. Kumar S, Hisamatsu Y, Tamaki Y, Ishitani O, Aoki S (2016) Inorg Chem 55(8):3829–3843
95. Edkins RM, Wriglesworth A, Fucke K, Bettington SL, Beeby A (2011) Dalton Trans 40:9672–9678
96. Cudré Y, de Carvalho FF, Burgess GR, Male L, Pope SJA, Tavernelli I, Baranoff E (2017) Inorg Chem 56(19):11565–11576
97. Park GY, Kim Y, Ha Y (2007) Mol Cryst Liq Crsyt 462:179–188
98. Lepeltier M, Dumur F, Graff B, Xiao P, Gigmes D, Lalevée J, Mayer CR (2014) Helv Chim Acta 97(7):939–956
99. Lepeltier M, Graff B, Lalevée J, Wantz G, Ibrahim-Ouali M, Gigmes D, Dumur F (2016) Org Elect 37:24–34
100. Tamura Y, Hisamatsu Y, Kumar S, Itoh T, Sato K, Kuroda R, Aoki S (2016) Inorg Chem 56(2):812–833
101. Tamura Y, Hisamatsu Y, Kazama A, Yoza K, Sato K, Kuroda R, Aoki S (2018) Inorg Chem 57(8):4571–4589
102. Baranoff E, Curchod BFE, Frey J, Scopelliti S, Kessler F, Tavernelli I, Rothlisberger U, Grätzel M, Nazeeruddin MD (2011) Inorg Chem 51(1):215–224
103. Tordera D, Delgado M, Orti E, Bolink HJ, Frey J, Nazzeruddin MK, Baranoff E (2012) Chem Mater 24(10):1896–1903
104. Hisamatsu Y, Kumar S, Aoki S (2017) Inorg Chem 56(2):886–899
105. Lowry MS, William R, Hudson WR, Robert A, Pascal RA Jr, Bernhard S (2004) J Am Chem Soc 126(43):14129–14135
106. Curtin PN, Tinker LL, Burgess CM, Cline ED, Bernhard S (2009) Inorg Chem 48(22):10498–10506
107. Yen HY, Huang MJ, Chen IC (2016) J Photochem Photobiol A Chem 318:33–41

108. Cutillas N, de Yellol GS, Haro C, Vincente C, Rodriguez V, Ruiz J (2013) Coord Chem Rev 257(19–20):2784–2797
109. Garg JA, Blacque O, Fox T, Vankatesan K (2010) Inorg Chem 49(24):11463–11472
110. Omae I (2016) Curr Org Chem 20(27):2848–2864
111. Cerón-Camacho R, Roque-Ramires MA, Raybov AD, Legadec RL (2021) Molecules 26(6):1563

Authors and Affiliations

Shin Aoki[1,2,3] · Kenta Yokoi[1] · Yosuke Hisamatsu[1] · Chandrasekar Balachandran[1,3] · Yuichi Tamura[1] · Tomohiro Tanaka[1]

1 Faculty of Pharmaceutical Sciences, Tokyo University of Science, Tokyo, Japan

2 Research Institute for Science and Technology, Tokyo University of Science, Tokyo, Japan

3 Research Institute for Biomedical Sciences, Tokyo University of Science, Tokyo, Japan

Topics in Current Chemistry (2022) 380:29
https://doi.org/10.1007/s41061-022-00392-8

REVIEW

Determination and Imaging of Small Biomolecules and Ions Using Ruthenium(II) Complex-Based Chemosensors

Miaomiao Wu[1] · Zexi Zhang[1] · Jiaxi Yong[1] · Peer M. Schenk[2] · Dihua Tian[1] · Zhi Ping Xu[1] · Run Zhang[1]

Received: 30 September 2021 / Accepted: 27 May 2022 / Published online: 13 June 2022

Abstract

Luminescence chemosensors are one of the most useful tools for the determination and imaging of small biomolecules and ions in situ in real time. Based on the unique photo-physical/-chemical properties of ruthenium(II) (Ru(II)) complexes, the development of Ru(II) complex-based chemosensors has attracted increasing attention in recent years, and thus many Ru(II) complexes have been designed and synthesized for the detection of ions and small biomolecules in biological and environmental samples. In this work, we summarize the research advances in the development of Ru(II) complex-based chemosensors for the determination of ions and small biomolecules, including anions, metal ions, reactive biomolecules and amino acids, with a particular focus on binding/reaction-based chemosensors for the investigation of intracellular analytes' evolution through luminescence analysis and imaging. The advances, challenges and future research directions in the development of Ru(II) complex-based chemosensors are also discussed.

Keywords Ru(II) complexes · Chemosensors · Luminescent imaging · Ions · Small biomolecules detection

Miaomiao Wu, Zexi Zhang and Jiaxi Yong contributed equally to this work and should be regarded as joint first authors.

Chapter 6 was originally published as Wu, M., Zhang, Z., Yong, J., Schenk, P. M., Tian, D., Xu, Z. P. & Zhang, R. Topics in Current Chemistry (2022) 380:29. https://doi.org/10.1007/s41061-022-00392-8.

✉ Run Zhang
r.zhang@uq.edu.au

1 Australian Institute for Bioengineering and Nanotechnology, The University of Queensland, Brisbane, QLD 4072, Australia

2 School of Agriculture and Food Sciences, The University of Queensland, Brisbane, QLD 4072, Australia

Abbreviations

Ru(II)	Ruthenium(II)
HPLC	High-performance liquid chromatography
ICP-OES/MS	Inductively coupled plasma-optical emission spectroscopy/mass spectrometry
bpy	2,2′-Bipyridine
CT	Computerized tomography
MRI	Magnetic resonance imaging
PET	Positron emission tomography
CA	Contrast agent
LoD	Limit of detection
ϕ	Quantum yields
TGL	Time-gated luminescence
Ir(III)	Iridium(III)
Pt(II)	Platinum(II)
Au(I)	Gold(I)
Re(I)	Rhenium(I)
Os(II)	Osmium(II)
MLCT	Metal-to-ligand charge transfer
ILCT	Intraligand charge transfer
LLCT	Ligand-to-ligand charge transfer
MMLCT	Metal–metal-to-ligand charge transfer
LMCT	Ligand-to-metal charge transfer
MLLCT	Metal-to-ligand-ligand charge transfer
LMMCT	Ligand-to-metal–metal charge transfer
MC	Metal centered
LC	Ligand centered
FRET	Förster resonance energy transfer
PeT	Photo-induced electron transfer
dppz	Dipyridophenazine
ROS	Reactive oxygen species
RNS	Reactive nitrogen species
RCS	Reactive carbonyl species
RSS	Reactive sulfur species
F^-	Fluoride
CH_3COO^-	Acetate
CN^-	Cyanide
$H_2PO_4^-$	Phosphate
Cl^-	Chloride
Br^-	Bromide
CH_3CN	Acetonitrile
DMSO	Dimethyl sulfoxide
SCN^-	Thiocyanate
Cys	Cysteine
Hcy	Homocysteine
GSH	Glutathione

HSO_3^-	Hydrogen sulfite
S^{2-}	Sulfide
PPi	Pyrophosphate
DPA	Di(2-picolyl)amine
ATP	Adenosine triphosphate
UCNP	Upconversion nanoparticle
UCL	Upconversion luminescence
NO	Nitric oxide
$ONOO^-$	Peroxynitrite
HNO	Nitroxyl
$O_2^{\bullet-}$	Superoxide
Cy5	Cyanine 5
•OH	Hydroxyl radicals
H_2O_2	Hydrogen peroxide
$^1O_2$	Singlet oxygen
HOCl	Hypochlorous acid
DNP	2,4-Dinitrophenyl
BFO	Benzofurazan-1-oxide
GO	Glyoxal
CO	Carbon monoxide
MGO	Methylglyoxal
FA	Formaldehyde
CBS	Cystathionine β-synthase
CSE	Cystathionine γ-lyase
H_2S	Hydrogen sulfide
NBD	7-Nitro-2,1,3-benzoxadiazoles
His	Histidine
S/N	Signal-to-noise

1 Introduction

Small biomolecules and ions are building blocks of living organisms and play indispensable roles in all biological processes, such as enzymatic reactions, metabolism, growth, adaptation, and various disease developments and progressions [1, 2]. Determination and monitoring of the levels of these biomolecules and ions in situ are essential for better understanding their biological roles in biomedical systems, thus contributing to early diagnosis and treatment assessment of various diseases [3–5]. The most common and typical methods, such as high-performance liquid chromatography (HPLC) and inductively coupled plasma-optical emission spectroscopy/mass spectrometry (ICP-OES/MS), have been developed for the determination of small biomolecules and ions in vitro [6–8]. Onsite determination of these analytes in situ and in vivo, using these techniques, is not possible because the sample preparation in solution is an essential step to ensure performing successful analysis [9]. Imaging technologies, such as computerized tomography (CT) [10], magnetic resonance imaging (MRI) [11] and positron emission tomography (PET), have been widely used in clinical diagnosis [12], while these technologies cannot be directly used for the determination of the concentration and/or activity of these analytes in biological samples [13, 14]. This is mainly because the contrast agents (CAs) used in these technologies are generally nonspecific; more importantly, these CAs hardly respond to small biomolecules and ions at molecular level because of their resolution and sensitivity limitations [15]. Other approaches to optical detection, such as fluorescence and phosphorescence measurements, have also been successfully developed and adopted in biomedical research and clinical diagnosis [16]. In contrast to conventional bioassay and imaging technologies, luminescence bioassay and imaging using advanced optical spectroscopic and imaging instruments are featured with high sensitivity and selectivity, fast response time and low cost, enabling their use in biological and biomedical investigations involving in vitro bioassay and in vivo luminescence bioimaging [17–20].

Chemosensors are one of the most important tools for luminescence bioassay and imaging of small biomolecules and ions in situ in real time [21–23]. Luminescent chemosensors are normally designed as chemical compounds that can respond to targeted analytes through a unique binding/reaction (Fig. 1) [24, 25]. Generally, the chemosensors with reaction-based sensing mechanisms have higher selectivity, and the chemosensors with binding-based sensing mechanisms feature excellent reversibility for monitoring the targeted analyte in situ. As a result of these response processes, the luminescence signals can be switched "ON" (Fig. 1A) or "OFF" (Fig. 1B). Of the luminescence switch "OFF" and "ON" responses, the emission switch "ON" chemosensors are preferable for imaging analysis because

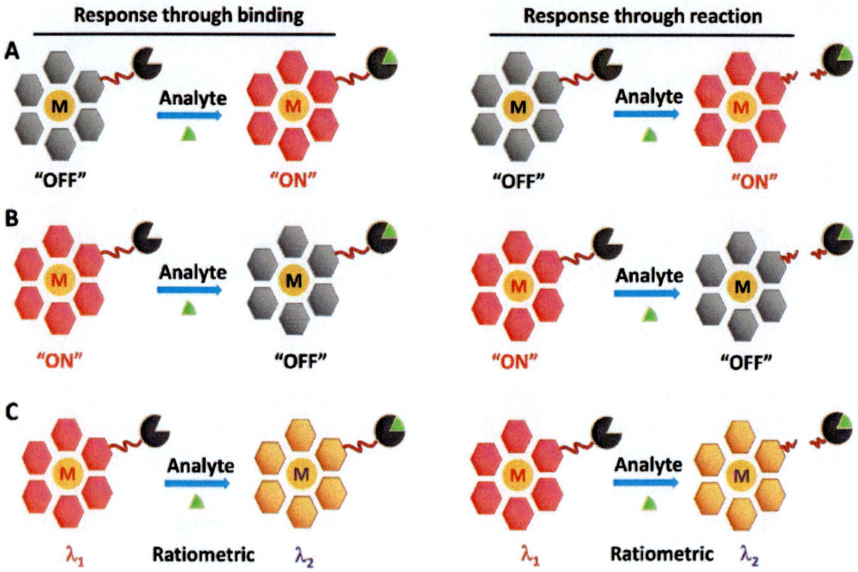

Fig. 1 Design of chemosensors for the determination of analytes through the response mechanisms of binding and reaction, resulting in "OFF–ON" (**A**), "ON–OFF" (**B**), and ratiometric (**C**) luminescence response to analytes

the enhancement of the luminescence intensity can be easily observed by microscopy. The emission wavelengths of the chemosensors can also be shifted after the response processes (Fig. 1C) [25–27], allowing ratiometric luminescence detection and imaging of targeted analytes with the potential for precise and quantitative analyses. The changes of emission signals generally correspond to the concentrations of the targeted analyte and thus can be recorded for the analyte's determination of abundance by luminescence spectroscopes and/or microscopes. Because of the unique advantages of luminescence bioassays and imaging, enormous efforts have been devoted to the development of luminescent chemosensors for the detection of a variety of analytes in complicated biological and environmental systems in the past few decades.

As shown in Fig. 1, luminescence chemosensors generally consist of three parts, including the luminophore, response unit and spacer, which links the luminophore and response unit. Of all luminophores that are being used for the development of chemosensors, fluorescent organic dyes are most widely investigated because of their high quantum yields (ϕ) and easy modification of chemical structures [28]. Lanthanide chelates are another family of luminophores that have been successfully employed in the development of chemosensors [29, 30]. Compared with fluorescent organic dyes, lanthanide chelate luminescence has high photostability, large Stokes shift and unique line-like emissions [29]. The prolonged lifetime of lanthanide chelates (microseconds to milliseconds) enables a background-free bioassay and imaging of targeted analytes through time-gated luminescence (TGL) measurement [28, 31]. Transition metal complexes, particularly the luminescent ruthenium(II) (Ru(II)) [32], iridium(III) (Ir(III)) [33], platinum(II) (Pt(II)), gold(I) (Au(I)) [34], rhenium(I) (Re(I)) [35] and osmium(II) (Os(II)) complexes with d^6, d^8 and d^{10} electron structures, have also been studied when developing chemosensors for biomolecule and ion detection and imaging [36–40]. Different from the fluorescent organic dyes that emit from excited singlet state, phosphorescence of transition metal complexes is derived from excited triplet states [41]. The excited states of these transition metal complexes are more complicated than those of fluorescent dyes and mainly include metal-to-ligand charge transfer (MLCT), intraligand charge transfer (ILCT), ligand-to-ligand charge transfer (LLCT), metal–metal-to-ligand charge transfer (MMLCT), ligand-to-metal charge transfer (LMCT), metal-to-ligand-ligand charge transfer (MLLCT) and ligand-to-metal–metal charge transfer (LMMCT) [41–43]. The excited state-mediated emission properties of transition metal complexes are varied upon the changes of the metal center, local environment and particularly chemical structure of ligands, enabling transition metal complexes to be designed as the chemosensors through modulating these parameters [43].

Of all transition metal complex-based luminophores, Ru(II) polypyridine complex, particularly the prototype of the Ru complex ($[Ru(bpy)_3]^{2+}$ (bpy: 2,2′-bipyridine)) (Fig. 2A), has been one of the most popular molecules and widely investigated in the past few decades [44]. Ru(II) polypyridine complexes have octahedral symmetry with three kinds of electronic transitions, including metal centered (MC), ligand centered (LC) and MLCT [41]. As shown in Fig. 2B, in this octahedral symmetry of Ru(II) complex, MC excited states are obtained for an electron transition from π_M to σ^*_M orbitals, LC excited states are formed through an

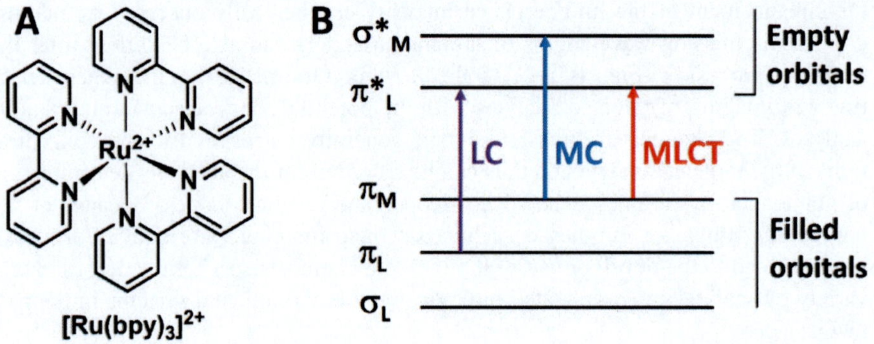

Fig. 2 Molecular structure of $[Ru(bpy)_3]^{2+}$ prototype complex (**A**). Ru(II) polypyridine complexes' molecular orbital diagram and the corresponding LC, MC and MLCT electronic transitions (**B**)

electron transition from π_L to π^*_L, and MLCT excited states are produced by promotion of an electron from π_M metal orbital to π^*_L ligand orbitals [45]. The lowest excited state MC can decay to the ground state through a fast radiationless process. In contrast, the lowest excited states LC and MLCT undergo radiative deactivation to the ground state, thus exhibiting intense luminescence at room temperature in a rigid matrix and fluid solution, respectively [41]. Consequently, the lowest excited state is ^{3}MLCT for most luminescent Ru(II) polypyridine complexes in solution. Upon the excitation at about 450 nm (spin-allowed ^{1}MLCT), the lowest spin-forbidden ^{3}MLCT excited state is obtained after a fast intersystem crossing process and then emits orange to near infrared emission [46]. The ^{3}MLCT-based emission of Ru(II) polypyridine complexes displays unique photochemical and photophysical properties, including large Stokes shift (about 150 nm), prolonged luminescence lifetime (hundreds of nanoseconds to microseconds level), high photostability and brightness by visible light excitation [41, 47].

The emission of Ru(II) complexes, including luminescence intensity and lifetime, can be fine-modulated by modifying the chemical structure of ligands, allowing the Ru(II) complexes to be used for the development of chemosensors for biomolecule and ion detection. As described in early review articles [42, 43, 48, 49], the response mechanisms of luminescent Ru(II) complex chemosensors for target analytes mainly include (1) photo-induced electron transfer (PeT), in which the Ru(II) polypyridine complexes are ideal electron donors and acceptors, (2) Förster resonance energy transfer (FRET), in which the Ru(II) complexes can serve as the energy donor and acceptor [50–53] for the energy transfer (ET), (3) distortion of ligand and the octahedral symmetry after binding/reaction with the target analyte and (4) changes of the local environment upon the analyte's binding. In addition to the luminescence response of Ru(II) complex-based chemosensors to the analyte [40, 54, 55], other prerequisites for this family of chemosensors to be applied in bioassay and imaging include (1) capability of analyte determination in aqueous solution, (2) high sensitivity and selectivity, which allow the chemosensors to be used for targeted analyte detection even at extremely low concentration without non-specific binding of interference species, (3) low cytotoxicity, enabling targeted analyte determination and imaging with minimum perturbation to the native micro-environment and (4) high cell membrane permeability to ensure the chemosensors are easily internalized into biological tissues.

The last few years have witnessed a huge leap forward in the development of Ru(II) complexes as the chemosensors for the detection of various analytes. Based on the above-described response mechanisms, hundreds of Ru(II) complex chemosensors for colorimetric and luminescent determination of anions, metal ions, small biomolecules and biomacromolecule have been available [42, 43]. For example, the triplet nature of the emission state with long lifetime of Ru(II) complexes allows them to be used for oxygen sensing by monitoring the changes of their luminescence intensity and lifetime [56–58]. Some of the Ru(II) complexes have also been revealed to be nucleic acid sensitive [59, 60], thus serving as a "light switch" for DNA detection [61]. The Ru(II) complexes with dipyridophenazine (dppz) ligands (e.g., complex **1** ($[Ru(phen)_2(dppz)]^{2+}$) are particularly interesting (Fig. 3A) [61]. These complexes are non-luminescent (undetectably small quantum yield) in water, while the emission intensity is significantly increased when bound to DNA [62, 63], which allows the Ru(II) complexes (e.g., complex **2**) to be used for imaging of DNA structure and related mitosis progression (Fig. 3A). Subsequent research has also revealed that Ru(II) complexes (e.g., complex **3**) can be used for the determination of DNA mismatches and RNA (Fig. 3B) [64–67]. Since the investigation of cellular uptake and imaging of the Ru(II) complex by Barton's group [68, 69], the application of Ru(II) complex chemosensors for sensing and imaging of intracellular biomolecules and ions has increasingly attracted interest in recent years.

In this chapter, we wish to summarize recent examples of Ru(II) complex chemosensors for the detection of small biomolecules and ions in aqueous solution, with a particular focus on binding/reaction-based chemosensors for the investigation of intracellular analytes' evolution through luminescence imaging. Specifically, the chemosensors for the determination of reactive oxygen/nitrogen/carbonyl species (ROS/RNS/RCS), biothiols, amino acids, pH, metal ions and anions are summarized, followed by the discussion of these chemosensors for luminescence bioimaging. The advances, challenges and future research directions in the development of Ru(II) complex-based chemosensors will also be discussed.

2 Ru(II) Complex Chemosensors for Anions

In various chemical and biological processes, anions play important roles in the body, such as blood pressure stabilization, blood purification, sugar level reduction, respiration and fatigue recovery. For anion sensing, the development of Ru(II) complex-based chemosensors has attracted enormous attention in the past few decades [70–72]. By virtue of their abundant photo-physical and chemical properties, hundreds of Ru(II) complexes have been designed and synthesized for the detection of various anions, such as fluoride (F^-) [73–76], acetate (CH_3COO^-) [77], cyanide (CN^-)[78], phosphate ($H_2PO_4^-$) [79–81], chloride (Cl^-) and bromide (Br^-). Similar to the design of other anion receptors [82], most Ru(II) complex-based

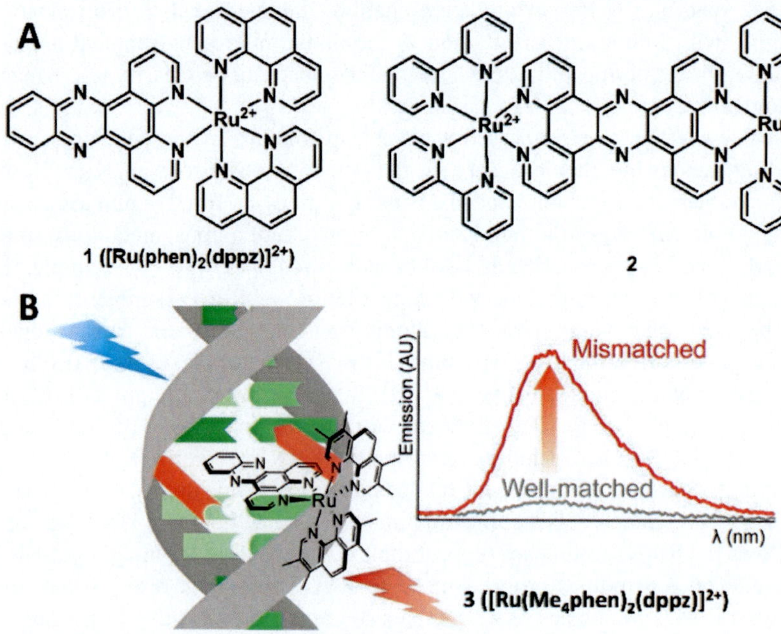

Fig. 3 Molecular structures of examples of Ru(II) complexes **1** ([Ru(phen)$_2$(dppz)]$^{2+}$) and **2** for DNA determination and imaging (**A**) and complex **3** ([Ru(Me$_4$phen)$_2$(dppz)]$^{2+}$) for detection of DNA mismatches. The luminescence spectra represent the emission of complex **3** with well-matched and mismatched DNA. Adapted with permission from Ref. [67]. Copyright 2016 American Chemical Society

chemosensors are designed using the following three response mechanisms, including (1) the binding of the Ru(II) complex's recognition unit with anions via hydrogen bonding and deprotonation [83], electrostatic and Lewis acid–base interactions [84], (2) specific reactions of the Ru(II) complex's recognition unit with anions [85] and (3) displacement of metal ions from heterobimetallic Ru(II) complex [86]. In the following section, the Ru(II) complex-based chemosensors for anions will be discussed according to their different response mechanisms.

2.1 Response Based on Hydrogen Bonding, Electrostatic and Lewis Acid-Base Interactions

Although most Ru(II) complex-based anion chemosensors are developed through the mechanism of hydrogen bonding and electrostatic and Lewis acid-base interactions, the colorimetric and luminescent response of these Ru(II) complexes to anions can only be obtained in organic solvents, including acetonitrile (CH_3CN) and dimethyl sulfoxide (DMSO) [87, 88]. This is mainly because the hydrogen bonding, electrostatic and Lewis acid-base interactions are significantly inhibited by the water molecules and other anions in the buffer solution [89]. This sub-section will discuss some examples of Ru(II) complex chemosensors for determination of anions in aqueous solution [90–92].

Through modification of the $[Ru(bpy)_3]^{2+}$ with amide containing a calixarene moiety (binding site), Maity et al. reported a Ru(II) complex (**4**) for CN^- and CH_3COO^- determination (Fig. 4) [93]. Titration of complex **4** with CN^- and CH_3COO^- in H_2O–CH_3CN (95:5) resulted in a remarkable luminescence quenching and enhancement, respectively. The different response mechanisms of complex **4** to CN^- and CH_3COO^- were investigated by ^1H NMR spectra. The CN^- can bind to each amide N–H and leads to the deprotonation to form HCN. The electron density on the bpy ligand increased, and thus the intramolecular quenching was enhanced. In contrast, the weak interaction of bidentate CH_3COO^- with two N–H protons led to the formation of electron delocalization, in which the CH_3COO^- pulls the electron density to itself and thus decreases the intramolecular quenching. The luminescence response of complex **4** to CN^- showed a LoD of 70 ppb.

Mardanya and co-workers described a Ru(II) complex (**5**) with pyrene-biimidazole ligand as a chemosensor for highly selective CN^- detection in both CH_3CN and aqueous media (Fig. 5) [94]. The imidazole N–H protons of the coordinated ligand were found to be highly acidic with $pKa_1=5.09$ and $pKa_2=8.95$. Deprotonation of these two N–Hs was found through hydrogen bonding interaction with CN^-, leading to the increase of electron density of the metal center. As a result, red shift of absorption and quenching of emission were obtained for complex **5** after CN^- binding. Detection limit of complex **5** to CN^- was determined to be 5.24×10^{-9} and 4.67×10^{-9} M for colorimetric and luminescent analyses, respectively. A similar hydrogen bonding-based interaction has been employed for the development of Ru(II) complex **6** for selective detection of thiocyanate (SCN^-) (Fig. 5) [95]. In complex **6**, the SCN^- interacts with N–H through a 1:1 binding mode, which hinders the photo-induced electron transfer (PeT) between the long pair electron of the

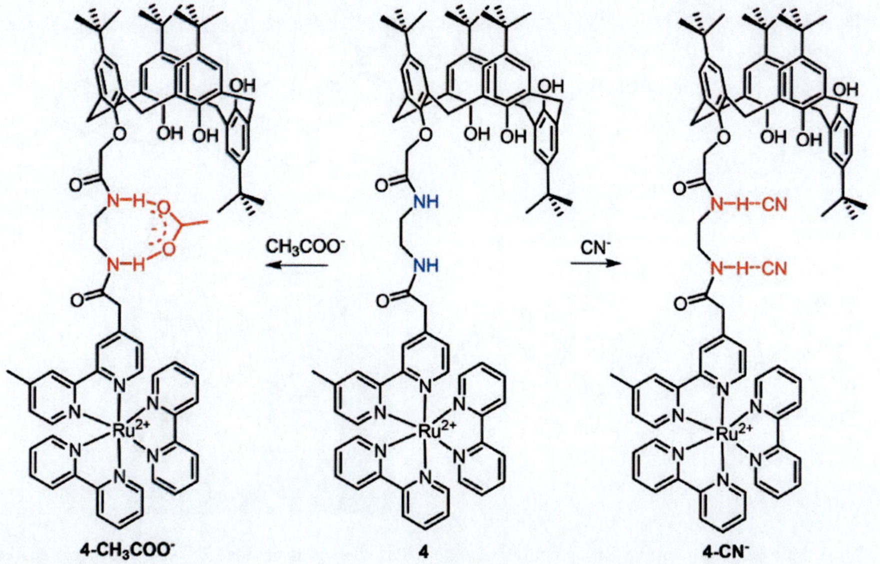

Fig. 4 Molecular structure of Ru(II) complex **4** and its response mechanism to CN^- and CH_3COO^-

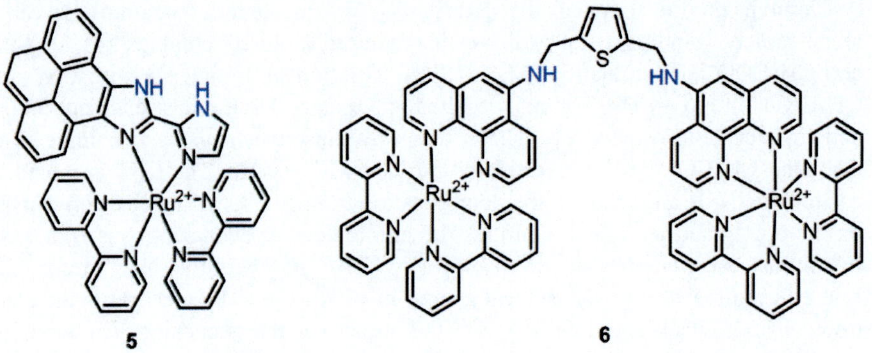

Fig. 5 Molecular structures of Ru(II) complexes **5** and **6** as the chemosensors for CN^- and SCN^-, respectively

N atom and the Ru(II) complex. The increase of luminescence intensity was thus recorded for SCN^- detection in CH_3CN-HEPES buffer solution (1:1, v/v, pH = 7.2).

In an early study, Lin et al. reported a Ru(II) complex (**7**) for highly selective detection of F^- in water by naked eye and luminescence (Fig. 6) [96]. Complex **7** was developed by incorporating a Schiff-base ligand with two bpy ligands. In the presence of F^-, the conversion of quinonehydrazone moiety to azophenol could occur, resulting in a remarkable red shift of absorption spectra from 475 to 580 nm and a solution color change from orange to blue-violet. The binding of F^- also led

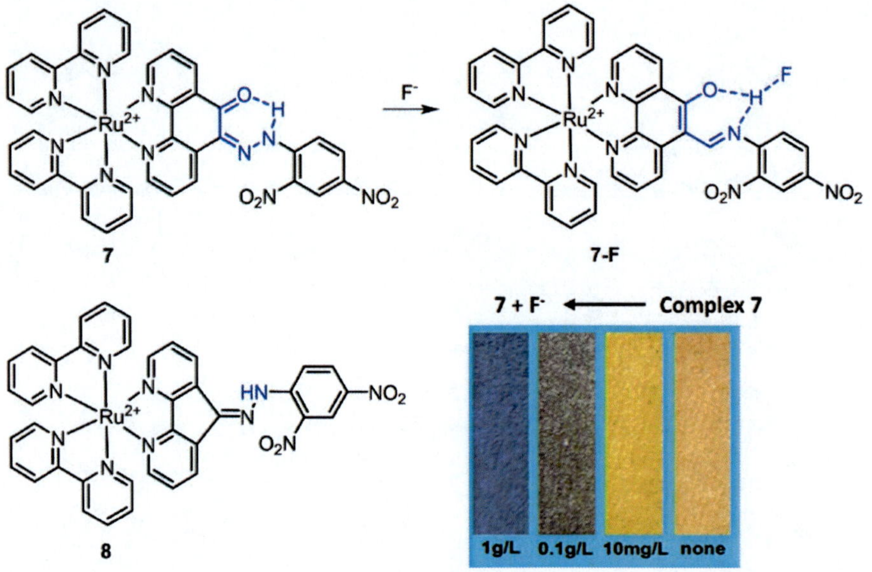

Fig. 6 Molecular structure of Ru(II) complexes **7** and **8** as chemosensors for F^-. The test paper prepared by complex **7** was then used for F^- detection in aqueous solution. Adapted with permission from Ref. [96]. Copyright 2006 Royal Society of Chemistry

to the increase of luminescence intensity at 630 nm. Although the spectrometric responses were measured in CH_3CN, the test paper prepared by staining of complex **7** also showed color changes in aqueous solution. In a later study, the same group modified the 1,10-phenanthroline-5,6-dione ligand to produce complex **8** for F^- detection (Fig. 6) [97]. In the presence of F^-, a similar red shift of absorption spectra (from 467 to 580 nm), color change (from yellow to magenta) and luminescence enhancement were obtained, which were attributed to the F^--mediated hydrogen bonding and deprotonation of N–H. The test papers were also prepared for F^- detection in aqueous solution with ten times higher sensitivity (LoD = 1 ppm) than complex **7**.

2.2 Response Based on Specific Reactions

Compared with the above-discussed response mechanism of hydrogen bonding and Lewis acid-base interactions, the chemosensors developed by the mechanism of chemical reaction have high sensitivity and selectivity [78, 98, 99]. The interference from water is also minimized because specific chemical reactions are involved in the sensing mechanism of these chemosenosors. Aldehyde is a strong electron-withdrawing group that can quench the MLCT emission of Ru(II) complex. In a previous study, a Ru(II) complex (**9**, $[Ru(CHO\text{-}bpy)_3]^{2+}$) with three aldehyde functionalized bpy ligands was designed and synthesized by Zhang et al. as the chemosensor for biothiol (cysteine-Cys and homocysteine-Hcy) detection in DMSO-HEPES buffer [100]. In a later study, Zhang et al. found that the nucleophilic addition of aldehyde can also be triggered by hydrogen sulfite (HSO_3^-) in acidic buffer [24],

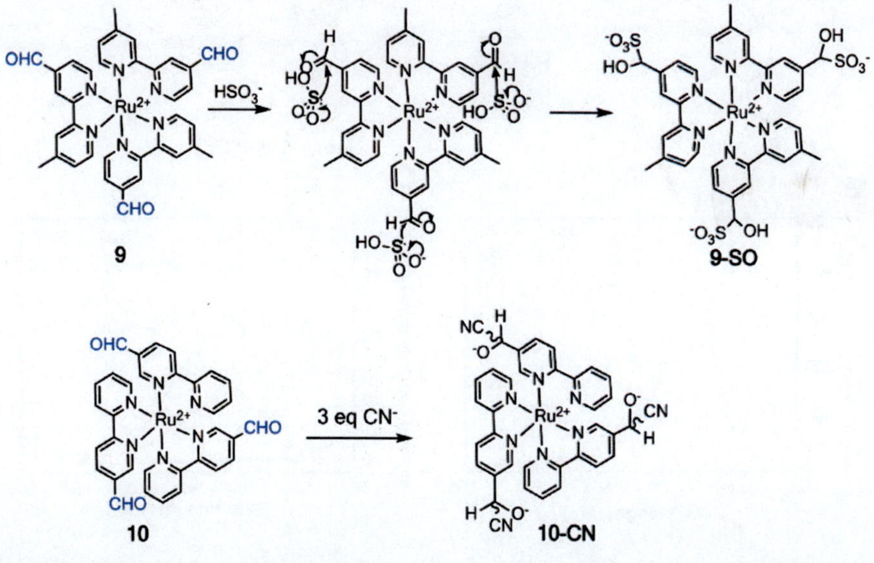

Fig. 7 Molecular structures of Ru(II) complexes **9** and **10** and their reactions with HSO_3^- and CN^-, respectively

thus allowing complex **9** to be used as a chemosensor for HSO_3^- detection in phosphate-buffered saline (PBS) buffer (50 mM, pH=5) (Fig. 7) [101]. The reaction of HSO_3^- and complex **9** led to the formation of **9-SO**, accompanied by an increase of luminescence at 620 nm. The result of HSO_3^- detection in wine and sugar samples showed that complex **9** has good precision and accuracy for HSO_3^- in food samples. Using 5-formyl-2,2′-bipyridine as the ligand, Zhu et al. prepared a Ru(II) complex (**10**) as the luminescence chemosensor for CN^- detection (Fig. 7) [102]. Similar to $[Ru(CHO\text{-}bpy)_3]^{2+}$, complex **10** in CH_3CN-H_2O (6:4) showed weak emission at 700 nm. The nucleophilic attack by CN^- led to the formation of **10-CN**, accompanied by a blue shift and increase of emission at 618 nm. Complex **10** with a detection limit of 0.75 μM was then used for the preparation of a test paper for naked eye CN^- detection.

In addition to aldehyde, nucleophilic addition between the azo (N=N) group and HSO_3^- has recently been exploited for the development of chemosensors for HSO_3^- detection (Fig. 8A) [24, 85]. Owing to the PeT from the Ru(II) center to the attached azo-2,4-dinitrobenzene (DNB), Ru(II) complex **11** (Ru-azo) exhibited weak emission in 25 mM PBS buffer of pH 7.4. The HSO_3^- triggered reaction with the azo group led to the formation of **11**-SO3 (Ru-SO3), accompanied by an enhancement in luminescence at 635 nm. More interestingly, complex **11** has a long emission lifetime of 258 ns, which enabled its use for background-free luminescence detection

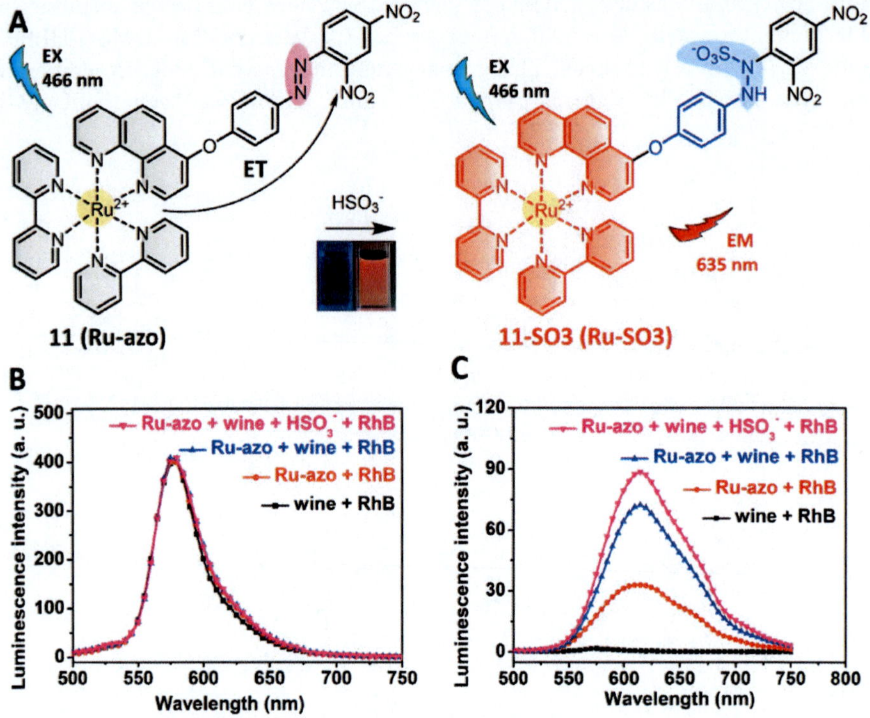

Fig. 8 Molecular structure of Ru(II) complex **11** and its reaction with HSO_3^- (**A**). Detection of HSO_3^- in rhodamine (RhB)-contaminated wine samples by steady-state luminescence (**B**) and TGL (**C**) analyses. Adapted with permission from Ref. [85]. Copyright 2020 Royal Society of Chemistry

of HSO_3^- through a TGL mode. In a wine sample containing rhodamine (spiked as the artificial background signal), steady-state luminescence analysis of HSO_3^- failed (Fig. 8B). TGL analysis eliminated the background signals and allowed for HSO_3^- detection with high accuracy and precision (Fig. 8C).

2.3 Response Based on Displacement of Metal Ions

By modifying the ligands with additional coordination sites, the produced Ru(II) complexes are capable of binding to other metal ions, such as Cu^{2+}, Co^{2+}, Zn^{2+} and Hg^{2+}, to form heterobimetallic complexes. The heterobimetallic Ru(II) complexes thus can be used as chemosensors for anion sensing through a displacement approach. Different from the hydrogen bonding-based anion sensing approach, the displacement-based response mechanism also allowed the chemosensors to be used for anion detection in water because the water molecules are not involved in the displacement processes. Among various heterobimetallic Ru(II) complexes, the one with Cu^{2+} (Ru(II)–Cu(II)) has been widely studied because Ru(II) complex luminescence can be quenched by Cu^{2+} binding through electron and energy transfers. Then, Cu^{2+} can be displaced in the presence of several anions, including sulfide (S^{2-}), CN^- and pyrophosphate (PPi) (Fig. 9A).

By coupling di(2-picolyl)amine (DPA) to one bpy ligand, Zhang et al. synthesized a Ru(II) complex (**12**) and then demonstrated the corresponding heterobimetallic Ru(II)–Cu(II) complex as the chemosensor for S^{2-} detection (Fig. 9B) [86]. Complex **12** showed high luminescence ($\phi = 3.07\%$) in HEPES buffer (pH 7.2), while the luminescence was completely quenched upon binding to Cu^{2+} through a 1:1 coordination stoichiometry. In the presence of S^{2-}, Cu^{2+} was then displaced to form the original complex **12**, which was accompanied by the recovery of luminescence. This heterobimetallic Ru(II)–Cu(II) complex showed high sensitivity to S^{2-} (LoD = 20.7 nM), allowing its use for S^{2-} detection in three wastewater samples. In a similar work, Li et al. modified the bpy ligand with a Cu^{2+} receptor, 1,4,7,10-tetraazacyclododecane (cyclen), and then demonstrated the Ru(II) complex (**13**) for sequential Cu^{2+} and S^{2-} detection (Fig. 9B) [103]. Compared with complex **12**, complex **13** has better selectivity to Cu^{2+} binding, while the formed **13-Cu** is not as sensitive as **12-Cu** for S^{2-} sensing.

The displacement approach has also been employed for the development of Ru(II) complex chemosensors for CN^-. In 2017, two Ru(II) complexes (**14** and **15**) were synthesized by Zheng and co-workers, and the corresponding heterobimetallic Ru(II)–Cu(II) complexes were used for CN^- detection in 20 mM HEPES buffer (pH 7.2) (Fig. 9B) [104]. A recovery of luminescence was observed after the displacement of Cu^{2+} from **14-Cu** to **15-Cu** to form $[Cu(CN^-)X]^{n-}$ and complexes **14** and **15**. The LoD for CN^- was then determined to be 0.36 and 0.87 μM using **14-Cu** and **15-Cu** as the chemosensors, respectively. Similarly, Zhang et al. reported a Ru(II) complex (**16**) for PPi detection in 2018 (Fig. 9B) [105], in which one Cu^{2+} from the **16-Cu** was displaced by the addition of two PPi. The chemosensor **16-Cu** showed high sensitivity (LoD = 0.58 nM) for PPi detection in 10 mM HEPES buffer (pH 7.4).

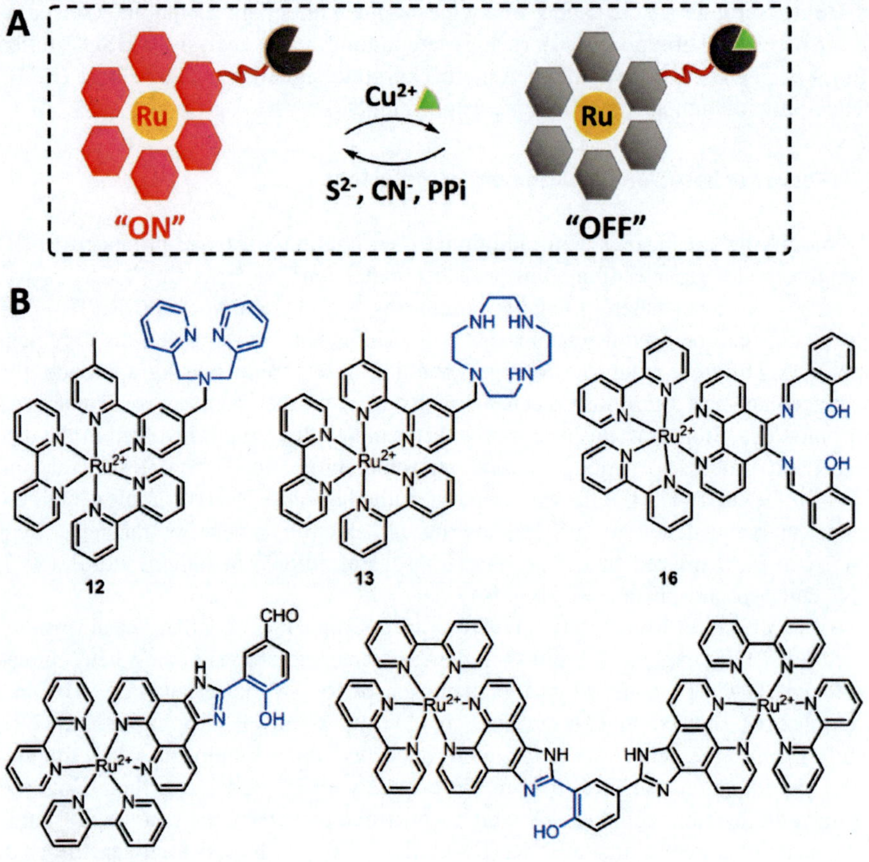

Fig. 9 The principle of displacement-based Ru(II) complex chemosensors for anion detection (**A**) and the molecular structures of Ru(II) complexes **12–16**

3 Ru(II) Complex Chemosensors for pH

Similar to most pH sensors, Ru(II) complex chemosensors for pH have mainly been developed through the mechanism of protonation and deprotonation of several functional groups, such as imidazole [106, 107], hydroxyl (–OH) [108–110], carboxyl (–COOH) [111], pyridine [112, 113] and others [114]. As a result of the protonation-deprotonation process, the molecular structures and the electron density distribution of Ru(II) complex are changed, leading to the variations of absorption and emission intensity/wavelength. In this section, the progress in the development of Ru(II) complex chemosensors for pH will be briefly discussed.

As described above, deprotonation of N–H of Ru(II) complex **5**'s 2,2′-biimidazole ligand occurs under basic conditions or binding with CN^- [94]. In 2020, Tormo et al. also reported the use of imidazole-based ligand for the development of Ru(II) complex chemosensors for pH [115]. In later research, deprotonation of imidazole N–H under neutral and basic conditions was exploited by Yu et al. for the development of Ru(II) complex **17** ($[Ru(bim)^2(pip)]^{2+}$) for pH sensing and imaging (Fig. 10A) [116]. The increase of pH led to the deprotonation of all imidazole N–H from both bim and pip ligands, resulting in red shift of absorption spectra and decrease of MLCT emission. Complex **17** has a pKa of 4.49 and low cytotoxicity, enabling lysosome imaging in U251 cells. Luminescence images of U251 cells with complex **17** and LysoTracker Red showed good co-localization (Fig. 10B), and then the application of complex **17** in monitoring of intracellular pH changes was demonstrated by treating the U251 cells with lysosomal acidification inhibitor (bafilomycin A1).

In 2015, Zheng and coworkers synthesized two Ru(II) complexes (**18** and **19**) and investigated their absorption and luminescence responses to pH (Fig. 11A) [112]. Protonation-deprotonation process occurs on the imidazole N–H and both imidazole N–H and pyridine for complexes **18** and **19**, respectively. The two-step protonation-deprotonation processes resulted in complex **18** with $pKa_1 = 0.98 \pm 0.04$ and $pKa_2 = 8.34 \pm 0.03$, while the three-step protonation-deprotonation processes of complex **19** exhibited $pKa_1 = 1.86 \pm 0.02$, $pKa_2 = 3.43 \pm 0.04$ and $pKa_3 = 9.07 \pm 0.08$. Coordination of the terpyridine (tpy) ligand of complex **19** with Re(I), a heterobimetallic Ru(II)–Re(I) complex **20**, was developed by Zheng et al. in 2014 (Fig. 11A) [113]. The coordination of the tpy ligand blocked the protonation-deprotonation of one pyridine; thus, a two-step protonation-deprotonation process was obtained. The pKa_1 and pKa2 value of complex **20** was 1.38 ± 0.03 and 6.84 ± 0.04, respectively. Interestingly, the coordination with Re(I) significantly improved the biocompatibility, allowing complex **20** to be used for luminescence imaging in HeLa cells.

Dinuclear Ru(II) complex **21** was then synthesized by Meng and co-workers in 2017 (Fig. 11A) [117]. Ru(II) complex **21** showed NIR emission at 760 nm with a large Stokes shift of 254 nm and lifetime (τ) 108.3 ± 0.4 ns. Different from complex **20**, a three-step protonation-deprotonation process was observed on imidazole N–H at the second ligand. pKa values of complex **21** changed to $pKa_1 = 1.36 \pm 0.02$, $pKa_2 = 5.76 \pm 0.05$ and $pKa_3 = 9.01 \pm 0.14$. Through modification of the second ligand, the same group recently reported a Ru(II) complex (**22**) as a chemosensor for pH imaging and cancer cell discrimination (Fig. 11A) [118]. Compared with complex **20**, similar photophysical properties, including intense NIR emission (~700 nm) and large Stokes shift (~200 nm) were obtained for complex **22**. More

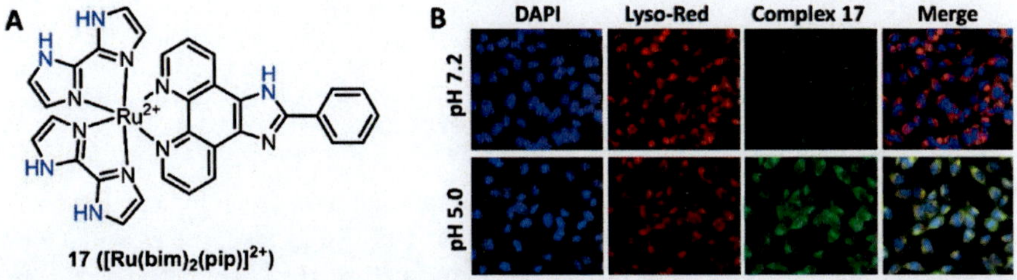

Fig. 10 Molecular structure of pH-sensitive Ru(II) complex **17** (**A**). Luminescence co-localization imaging of U251 cells stained with complex **17** and DYPI and LysoTracker Red (**B**). Adapted with permission from Ref. [116]. Copyright 2017 Elsevier

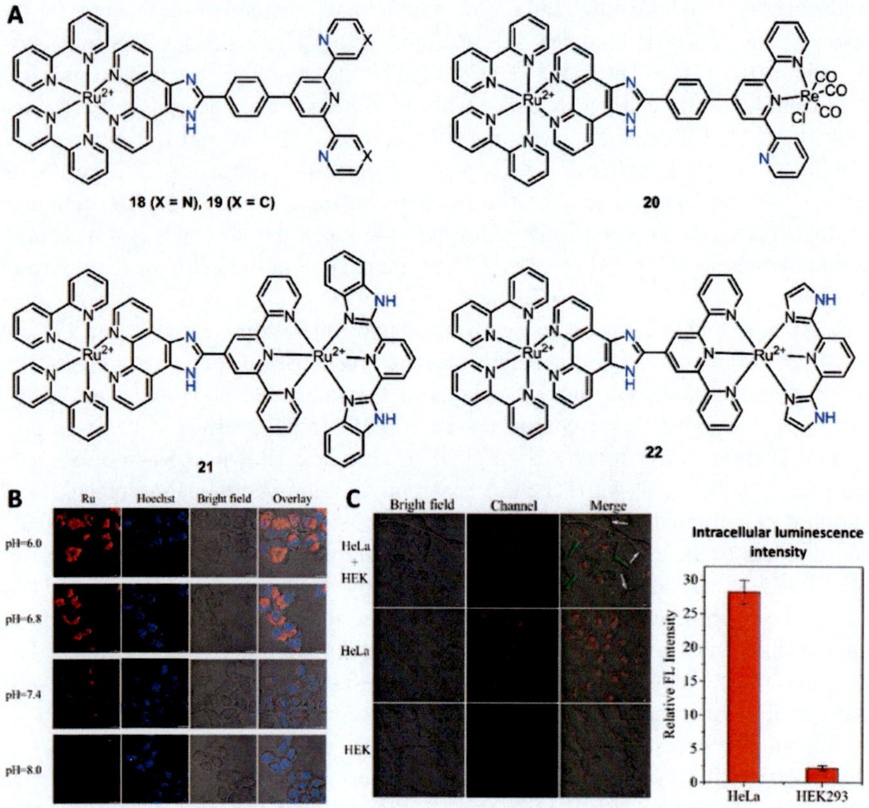

Fig. 11 Molecular structures of pH-sensitive Ru(II) complexes **18–22** (**A**) and the application of complex **22** for HeLa cancer cell pH imaging (**B**) and discrimination from healthy HEK293 cells (**C**). Adapted with permission from Ref. [118]. Copyright 2020 American Chemical Society

importantly, the pKa_2 value of complex **22** was determined to be 7.87, which is closer to the physiological value (i.e., 7.0–7.4). This allowed complex **22** to be used for luminescence imaging of intracellular pH in lysosomes (Fig. 11B). Moreover, imaging of HeLa cells showed about 13-fold higher intensity than HEK293 cells (Fig. 11C), demonstrating the "distinguishing" ability of complex **22** to identify the tumor and healthy cells.

4 Ru(II) Complex Chemosensors for Metal Ions

In addition to the anions, metal ions also play important roles in biological and environmental systems. Some metal ions, such as Cu^{2+}, Fe^{3+} and Zn^{2+}, are essential elements in the human body, while the metal ions, such as Hg^{2+}, Cd^{2+} and Cr^{3+}, are highly toxic, causing several problems for biological and environmental systems [119]. To detect these metal ions in biological and environmental systems, a number of Ru(II) complex chemosensors have been developed in the past few decades. In this section, the progress in the development of chemosensors for Cu^{2+}, Hg^{2+} [120] and others will be discussed according to the types of ions.

4.1 Ru(II) Complex Chemosensors for Cu^{2+}

As described above, the binding of Ru(II) complexes with Cu^{2+} could lead to the quenching of their luminescence through an excited-state electron transfer or energy transfer mechanism [86, 104, 105]. By virtue of this mechanism, a series of Ru(II) complexes have been developed as luminescence "ON–OFF" chemosensors for Cu^{2+} determination and imaging (Fig. 12). Complex **17** with biimidazole ligands showed good performance in pH sensing with the protonation-deprotonation mechanism [116]. In a recent study, the biimidazole-coupled phen ligand was employed as the binding site for the development of Ru(II) complex chemosensor (**23**) for Cu^{2+} detection by Li and co-workers [121]. The coordination (1:1 bonding ratio) of Cu^{2+} with complex **23**'s biimidazole led to the formation of a stable cyclic structure. The quenched emission of complex **23** showed a linearity with Cu^{2+} concentration in the range of 0.25–12 μM, and the LoD was 83.3 nM. The application of complex **23** was then demonstrated by Cu^{2+} detection in tap and lake water samples and imaging in A549 cells. For simple modification of imidazole to pyrazol, Cu^{2+} "ON–OFF" Ru(II) complex **24** was then reported by Xia and colleagues [122]. Different from complex **23**, complex **24** coordinated with Cu^{2+} through a 1:2 stoichiometry. This complex showed higher sensitivity to Cu^{2+} (LoD = 17.8 nM) compared with complex **23**. The test paper was then prepared for Cu^{2+} detection in river water samples. Interestingly, complex **25** with quinoline substitution showed a 1:1 stoichiometry when binding to Cu^{2+} in HEPES buffer (LoD = 50.67 nM) [123]. The test paper was also prepared using complex **25** as the chemosensor for Cu^{2+} detection. In another study, Zhang et al. reported a Ru(II) complex **26** for Cu^{2+} detection in aqueous solution and imaging in live pea aphids (LoD = 244 nM) [124]. Complex **26** was developed through coordination with two phen ligands and one 2-(2-hydroxyphenyl) imidazo[4,5-f][1,10]phenanthroline. The Cu^{2+} binding with 2-hydroxyphenyl imidazo through a 1:1 stoichiometry quenched complex **26**'s emission. Similar to Cu^{2+}-coordinated complexes **23** and **25**, the Cu^{2+} can be displaced by the addition of

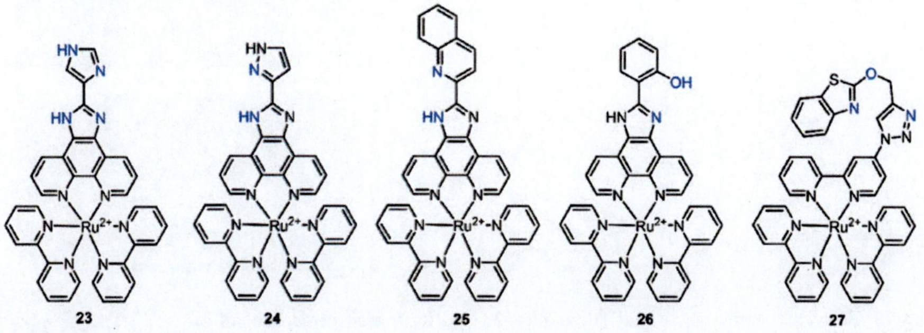

Fig. 12 Molecular structure of Ru(II) complexes **23–27** as the chemosensors for Cu^{2+}

EDTA (ethylene diamine tetraacetic acid), which allows the reset of the chemosensors for further detection of Cu^{2+}.

In addition to imidazole, some other groups, such as DPA [86, 125–128], 1,8-naphthyridine [129], 1,3-benzothiazole [130], carboxyl [131] and others [132], have also been utilized as the response units for the development of Ru(II) complex chemosensors for Cu^{2+}. For example, complex **27** reported by Ramachandran and colleagues was capable of detecting phosphate anions through C–H-anion interaction and Cu^{2+} through coordination with triazole, benzothiazole and the "O" linker, respectively [130] (Fig. 12). The application of this chemosensor (**27**) for Cu^{2+} detection (LoD = 700 nM) and imaging was also demonstrated by luminescence Cu^{2+} imaging in MCF-7 cells.

Similar to complex **15**, the pyridine "linker" of dinuclear Ru(II) complex **28** has also been developed as the chemosensor for Cu^{2+} detection in H_2O/CH_3CN (1:1, v/v) [133], in HEPES buffer solution (10 mM, pH 7.4) (Fig. 13A) [134]. Complex **28** showed high luminescence ($\phi = 0.06$) at 600 nm, while the Ru(II) complex's MLCT emission was quenched after coordination of Cu^{2+} with imidazole and pyridine. This chemosensor showed high sensitivity (LoD = 33.3 nM) and selectivity, reversibility (in the presence of EDTA) and good biocompatibility and cell membrane permeability, enabling it to be used for luminescence imaging. In addition, the Ru(II) complex's large two-photon absorption (TPA) cross-section enabled complex **28** to be used for TP imaging of Cu^{2+} in HeLa cells and zebrafish (Fig. 13B).

Although "OFF–ON" luminescence response chemosensors feature high sensitivity and selectivity, and excellent performance in luminescent bioimaging, the development of turn "ON" response chemosensors remains a challenge due to the intrinsic luminescence quenching property of Cu^{2+}. In 2009, a phenothiazine-coupled Ru(II) complex **29** was developed by Ajayakumar as the luminescence turn

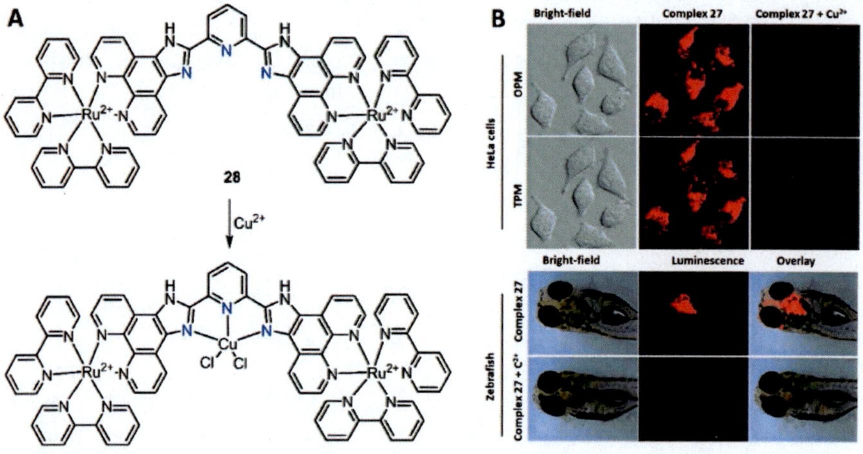

Fig. 13 Molecular structure of Ru(II) complex **28** and its response mechanism to Cu^{2+} (**A**). One-photon microscopy (OPM) and two-photon microscopy (TPM) imaging of Cu^{2+} in cells and zebrafish (**B**). Adapted with permission from Ref. [134]. Copyright 2013 Wiley

"ON" chemosensor for Cu^{2+} detection (Fig. 14A) [135]. Complex **29** was almost non-luminescent ($\phi = 0.0035$ in CH_3CN) because of the PeT from electron-rich phenothiazine to the Ru(II) center. In the presence of Cu^{2+}, the oxidation of phenothiazine to phenothiazine-5-oxide inhibited the PeT process, and thus the emission of complex **29** was switched "ON." In another research, Zhang et al. reported a Ru(II) complex "OFF–ON" luminescence chemosensor **30** for Cu^{2+} detection and imaging (Fig. 14A) [136]. Complex **30** with an *o*-(phenylazo)aniline structure showed weak luminescence in HEPES buffer solution (20 mM, pH 7.4). Cu^{2+}-mediated oxidative cyclization led to > 80-fold enhancement in luminescence at 599 nm. The large enhancement in luminescence allowed complex **30** for highly sensitive (LoD = 4.42 nM) and selective detection of Cu^{2+} in buffer and imaging of Cu^{2+} in pea aphids (Fig. 14B).

Despite the quenching of most Ru(II) complexes' emission by Cu^{2+} binding, the heterobimetallic Ru(II)–Cu(II) complexes provided an excellent platform for further development of "OFF–ON" response chemosensors for the detection of various analytes, such as anions [137], adenosine triphosphate (ATP) [138], amino acids [139, 140], redox biology [128] and other metal ions [141]. For example, in 2012, Wang et al. reported a Ru(II) complex for sequential detection of Cu^{2+} and Cr^{3+} in aqueous solution [141]. The coordination of a complex with Cu^{2+} produced the non-luminescent Ru(II)–Cu(II) complex, and this heterobimetallic complex showed high selectivity to Cr^{3+} in NaOAc-HOAc buffer (pH 5.6). A high sensitivity of this chemosensor to Cr^{3+} was also obtained with a LoD 66 nM.

4.2 Ru(II) Complex Chemosensors for Hg^{2+}

Because of the high binding affinity of S atom with Hg^{2+}, a series of S atom-bearing Ru(II) complexes have been developed as the chemosensors for Hg^{2+} detection. Previous research has revealed the colorimetric response of N-719 to Hg^{2+}, in which the absorption spectra of N-719 were blue-shifted after binding of Hg^{2+} [142, 143]. The N-719 functionalized upconversion nanoparticles (UCNPs) were then prepared,

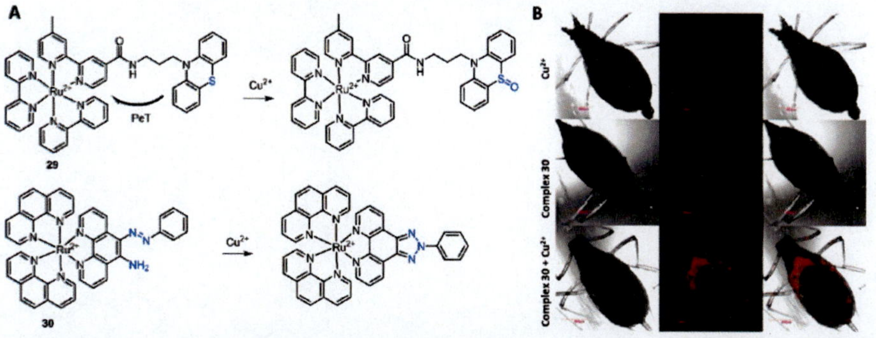

Fig. 14 Molecular structure of Ru(II) complexes **29**, **30** and their response reaction with Cu^{2+} (**A**). The application of Ru(II) complex **30** for Cu^{2+} imaging in pea aphids. Adapted with permission from Ref. [136]. Copyright 2015 Springer Nature

and the application of the ratiometric upconversion luminescence (UCL) nanosensor for Hg^{2+} detection and imaging was also demonstrated [142]. In 2015, the N-719 derivative Ru(II) complex **31** was prepared by Fan and co-workers for colorimetric and luminescent determination of Hg^{2+} [144]. Similar to N-719, the response of complex **31** to Hg^{2+} was ascribed to the binding of electron-deficient Hg^{2+} to the electron-rich sulfur atom of NCS (thiocyante) groups. A 40-nm blue shift (from 525 to 485 nm) of absorption and a remarkable increase of luminescence at 720 nm were observed upon binding of complex **31** to Hg^{2+}. In another study, Li et al. reported that a cyclometallated Ru(II) complex functionalized UCNPs for Hg^{2+} detection in water [145]. By modifying the cyclometallated Ru(II) complex with a propylsulfonate-coupled hemi-cyanine, Ru(II) complex **32** was then produced as the chemosensor for Hg^{2+} colorimetric analysis [146]. The Hg^{2+}-initialized conversion of coordination from C to S atom led to a remarkable absorption and solution color change from dark red to light yellow. In a subsequent study, the same group modified sulfonate to carboxyl to prepare a polymer membrane for colorimetric detection of Hg^{2+} [147].

The binding of S atom to Hg^{2+} has also been exploited for the development of Ru(II) complexes **33** and **34** for Hg^{2+} detection. For complex **33**, oxathiacrown ether is able to bind Hg^{2+}, resulting in a 30 nm blue shift of absorption spectra and color change from red-orange to yellow [148]. For complex **34**, Hg^{2+} can bind to four benzothiazole S atoms from two complex **34**s through a 1:2 binding stoichiometry [149]. The response of Hg^{2+} was accompanied by a significant increase of complex **34**'s emission at 656 nm. Nevertheless, complex **34** showed poor selectivity to Hg^{2+}, as luminescence enhancement at 630 nm was observed in the presence of Ag^{+}. The response of complex **34** to Ag^{+} was due to the binding of Ag^{+} to benzothiazoles' S atoms through a 1:1 stoichiometry (Fig. 15).

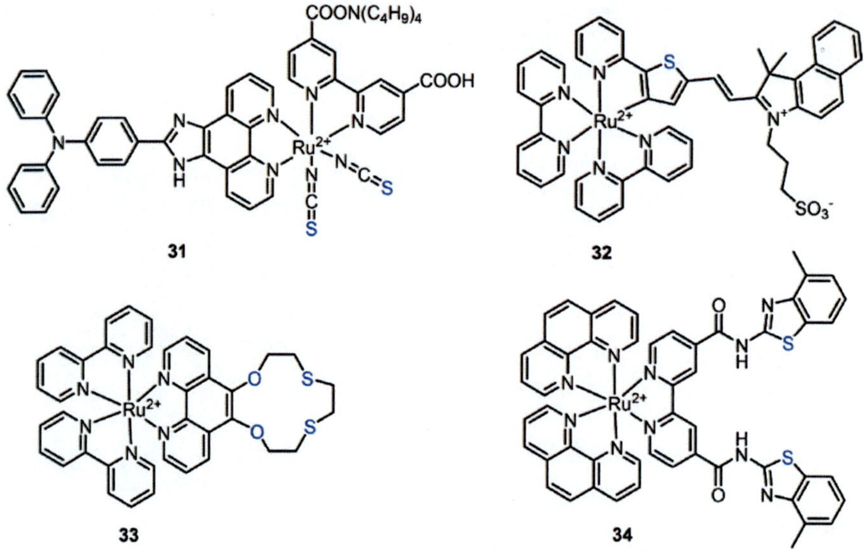

Fig. 15 Molecular structure of Ru(II) complexes **31–34** for Hg^{2+} detection

Based on the Hg^{2+}-mediated desulfation and intramolecular cyclic guanylation of thiourea reaction, Ru et al. developed Ru(II) complex **35** as an "OFF–ON" luminescence response chemosensor for Hg^{2+} (Fig. 16A) [150]. Complex **35** showed weak luminescence ($\phi = 0.4\%$), while the emission significantly increased upon reacting with Hg^{2+} $\phi = 2.2\%$). This reaction-based chemosensor (**35**) showed high selectivity and sensitivity (LoD = 8 nM) for Hg^{2+} detection. More importantly, the long lifetime emission of Ru(II) complex ($\tau_{35} = 215$ ns, $\tau_{35\text{-}Hg^{2+}} = 785$ ns) enabled the background-free TGL analysis of Hg^{2+}. With this Ru(II) complex **35**, luminescence imaging of Hg^{2+} in SMMC-7721 cells was then demonstrated (Fig. 16B). Using the same response reaction, the same research group then developed a phen ligand-based Ru(II) complex **36** for Hg^{2+} detection (Fig. 16A) [151]. Compared with complex **35**, complex **36** has higher sensitivity with a detection limit down to 5.4 nM. Similarly, the applications of this chemosensor for TGL Hg^{2+} detection and luminescence imaging of Hg^{2+} in SMMC-7721 cells were also demonstrated.

4.3 Ru(II) Complex Chemosensors for Other Metal Ions

Ru(II) complex-based chemosensors have also been developed for the detection of other metal ions [152], such as Fe^{3+} [153], Fe^{2+} [77], Na^{+} [154], Ag^{+} [149], Co^{2+} [155], Ba^{2+} [156] and Zn^{2+} [157]. For example, the 1,8-naphthyridine linked dinuclear Ru(II) complex has been reported for both Cu^{2+} and Fe^{3+} detection [153], and the azacrown ether coupled Ru(II) complex was found to respond to Ba^{2+} ion

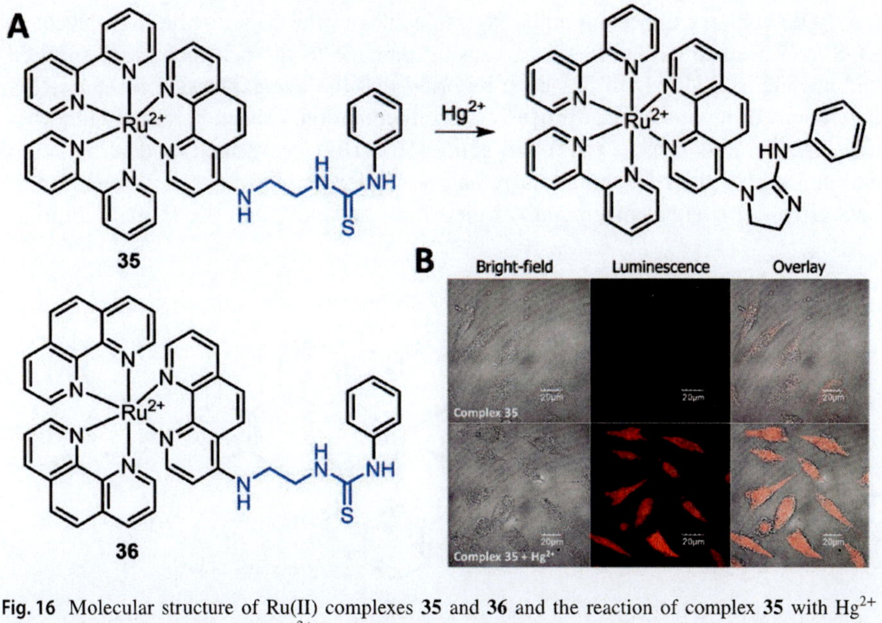

Fig. 16 Molecular structure of Ru(II) complexes **35** and **36** and the reaction of complex **35** with Hg^{2+} (**A**). Luminescence imaging of Hg^{2+} in SMMC-7721 cells using complex **35** as the chemosensor (**B**). Adapted with permission from Ref. [150]. Copyright 2015 American Chemical Society

[156]. The high binding affinity of the tpy ligand with Fe^{2+} allowed Zheng et al. to use Ru(II) complex **18** for Fe^{2+} detection [158]. The coordination of Fe^{2+} with complex **18** through a 1:2 binding stoichiometry was accompanied by the quenching of Ru(II) complex's emission at 608 nm. In 2014, Kumar and colleagues developed Ru(II) complex **37** for the detection of Pd^{2+} (Fig. 17) [159]. In the presence of Pd^{2+}, a 16-nm blue shift of the MLCT absorption band and the emerging of a new absorption at 565 nm were observed, which was accompanied by a solution color change from orange to dark red. The emission at 670 nm of complex **37** ($\phi = 3.5\%$) was about 12-fold decreased ($\phi = 0.12\%$) because of the paramagnetic properties of Pd^{2+} ions. Recently, Xie and colleagues reported a Ru(II) complex **38** as the chemosensor for gold(III) (Au^{3+}) detection (Fig. 17) [160]. Complex **38** showed high luminescence in PBS buffer (pH 7.2), while its emission was remarkably quenched after binding with Au^{3+}. Complex **38** showed high sensitivity (LoD = 135 nM) and selectivity to Au^{3+}, and low cytotoxicity, allowing its use for Au^{3+} imaging in living cells and zebrafish. The application of this chemosensor for Au^{+} drug release in anti-inflammation drugs was then demonstrated in inflamed microphage cells and LPS-induced zebrafish.

5 Ru(II) Complex Chemosensors for Reactive Biomolecules

Ru(II) complexes have also been designed and synthesized as the chemosensors for the determination of small reactive biomolecules, including reactive nitrogen/oxygen/sulfur/carbonyl species (RNS/ROS/RSS/RCS) and amino acids. These biomolecules play very important roles in biological systems [25, 27, 161]. For example, ROS/RNS are important signaling molecules in the body, while high levels of ROS/RNS lead to oxidative stress, causing damage to the cell membrane, protein and nucleic acids [162]. It has been reported that the overexpression of ROS/RNS is implicated in pathological processes in inflammation, cancer, cardiovascular disease, Alzheimer's disease (AD) and aging [162, 163]. Nevertheless, determination and monitoring their levels in situ remain a challenge due to (1) the limited numbers of robust chemosensors and (2) their high reactivity and the short lifetime of

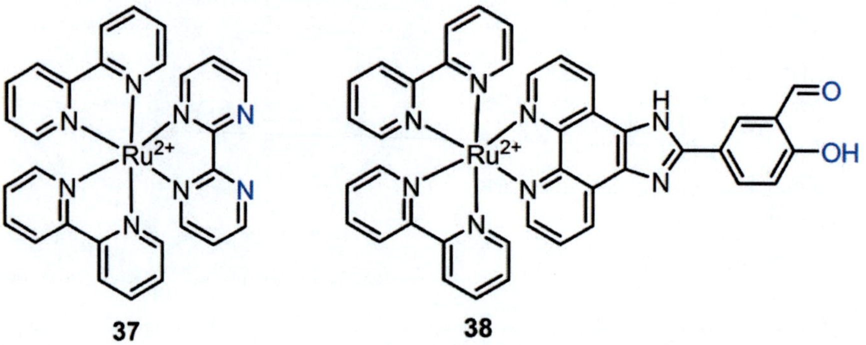

37 **38**

Fig. 17 Molecular structures of Ru(II) complexes **37** and **38** for Pd^{2+} and Au^{3+} detection, respectively most species, particularly the ROS/RNS [99, 164–166]. This section will outline the contributions of Ru(II) complex chemosensors to the determination and imaging of these small biomolecules.

5.1 Ru(II) Complex Chemosensors for RNS

RNS, mainly nitric oxide (NO), peroxynitrite ($ONOO^-$) and nitroxyl (HNO), are endogenous biomolecules that play essential roles in various biological processes [167]. Of these RNS, NO is produced by conversion of arginine through nitric oxide synthase-mediated oxidation [167]. NO has been found to be associated with blood vessel health and signaling pathways [168]. The reaction of NO with superoxide ($O_2^{\bullet -}$) forms $ONOO^-$, a highly reactive biomolecule, contributing to signaling transduction in living systems [169]. In 2010, Zhang et al. pioneered the use of Ru(II) complexes for reactive biomolecule detection and demonstrated the first Ru(II) complex chemosensor for NO detection [170]. This Ru(II) complex (**39**) was simply designed by coupling of NO response unit, 3,4-diaminophenoxy, to Ru(II)-bpy luminophore (Fig. 18A). After reacting with NO, the electron density of the triazolephenoxy product was reduced, thus inhibiting the PeT process and turning "ON" the MLCT emission. Ru(II) complex **39** has high sensitivity and selectivity, and fast response to NO, allowing its further use for NO determination and imaging in animal and plant cells [170, 171]. Later research showed that complex **39** has better performance in endothelial NO detection compared with the DAF-FM (Fig. 18B), demonstrating the usefulness of complex **39** as a promising chemosensor

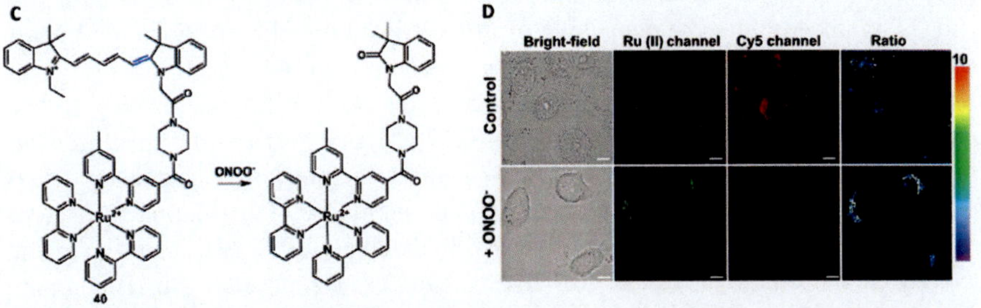

Fig. 18 Molecular structure of Ru(II) complexes **39** and **40** and their response reaction with NO (**A**) and $ONOO^-$ (**C**), respectively. Comparison of sensitivity of complex **39** with DAF-FM in cell-free media (**B**). Adapted with permission from Ref. [172]. Copyright 2019 Springer Nature. Luminescence ratiometric imaging of ONOO− in HeLa cells by complex 40 (D). Adapted with permission from Ref. [177]. Copyright 2018 Royal Society of Chemistry

for NO clinical investigations [172]. Further efforts have also been devoted to modifying complex **39** for the development of other chemosensors for NO detection [173–175].

To determine the levels of $ONOO^-$ in living cells [176], Zhang et al. developed a Ru(II) complex **40** by coupling of Ru(II) complex with a cyanine 5 (Cy5) dyes (Fig. 18C) [177]. In this FRET-based chemosensor, the Ru(II) complex served as the energy donor, and Cy5 served as the energy acceptor. In the presence of $ONOO^-$, the oxidation-cleavage led to the corruption of the electron transfer (ET) process [178], accompanied by the decrease of Cy5's emission and increase of Ru(II) complex's luminescence. Complex **40** also showed fast response, high sensitivity and selectivity to $ONOO^-$, and low cytotoxicity, facilitating its application in ratiometric detection and imaging of mitochondrial $ONOO^-$ in living HeLa cells (Fig. 18D).

5.2 Ru(II) Complex Chemosensors for ROS

ROS is a group of reactive biomolecules and free radicals derived from molecular oxygen [162, 164, 179]. Generally, ROS include $O_2^{\bullet -}$, hydroxyl radicals (•OH), hydrogen peroxide (H_2O_2), singlet oxygen ($^1O_2$) and hypochlorous acid (HOCl) [164]. These short-lived species display high reactivity, causing oxidation of other biomolecules, such as DNA, lipids and proteins [162]. In the past years, a number of Ru(II) complex chemosensors have been developed for the detection and imaging of $^1O_2$ and HOCl [180], which will be discussed in this section.

Coupling with anthracene [181], pyrene [182] and 1,8-naphthalimide [183], early research revealed that Ru(II) complexes' emissions can be quenched by the energy transfer mechanism. The triplet states of pyrene and 1,8-naphthalimide have similar energy levels to ^{3}MLCT of Ru(II) complexes [46]; the energy transfer between these triplet states enables a longer emission lifetime with lower quantum yields compared with $[Ru(bpy)_3]^{2+}$ prototype complex [182–184]. Different from pyrene and 1,8-naphthalimide, the energy of anthracene's triplet state lies 1460 cm^{-1} below the Ru(II) complexes' ^{3}MLCT [185], which leads to the quenching of ^{3}MLCT states as the energy transfer from Ru(II) complexes' ^{3}MLCT to anthracene's triplet state [181]. Upon specific oxidation of anthracene with $^1O_2$, the corresponding endoperoxide is produced; thus, the quenching of ^{3}MLCT is inhibited [186]. Inspired by this mechanism, Zhang et al. designed three Ru(II) complexes (**41a, b, c**) for $^1O_2$ detection in buffer (Fig. 19) [187]. All three complexes showed high selectivity to $^1O_2$ over other ROS and RNS, and the LoD was 0.17 μM, using **41a** as the chemosensor.

For the determination of HOCl, Ru(II) complex chemosensors have been developed by exploiting different response reactions, including (1) oxidation of S atom [188, 189], (2) amines (including dibenzoylhydrazine) [190, 191] and (3) oxime derivatives and others [192, 193]. In 2013, Zhang et al. reported a Ru(II) complex chemosensor **42** for HOCl detection and imaging [188]. Complex **42** was developed by conjugating a Ru(II) complex luminophore and 2,4-dinitrophenyl (DNP) quencher through an S linker (Fig. 20). The specific oxidation of the S linker with HOCl led to the cleavage of electron transfer acceptor (DNP quencher), accompanied by an increase of MLCT emission at 626 nm. The applications of complex **42**

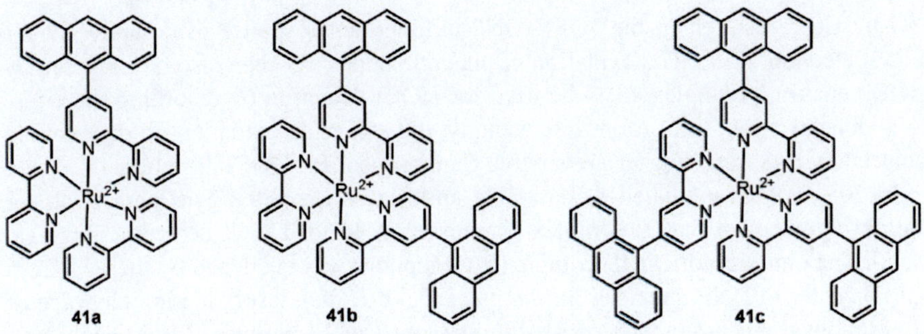

Fig. 19 Molecular structures of Ru(II) complexes **41a**, **b**, **c** for $^1O_2$ detection

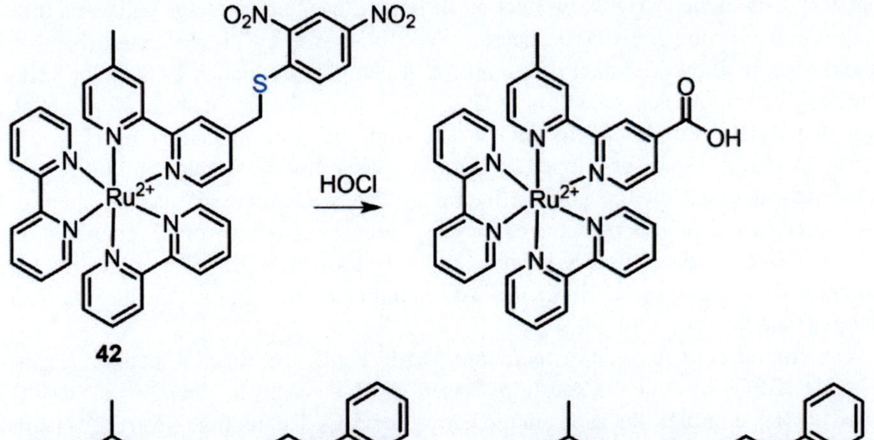

Fig. 20 Molecular structure of Ru(II) complexes **42** and **43** and the oxidations with HOCl

for imaging of exogenous HOCl in HeLa and endogenous HOCl in RAW 264.7 cells were then demonstrated. HOCl-mediated oxidation of phenothiazine's "S" has also been exploited for developing Ru(II) complexes for HOCl detection in biological [194] and environmental samples [189]. For example, in 2014, Liu et al. reported a Ru(II) complex **43** as the reversible chemosensor for HOCl determination and imaging (Fig. 20) [194]. An early study found that the Ru(II) complex emissions can be quenched by phenothiazine through a PeT mechanism, and the oxidation of

"S" by Cu^{2+} resulted in the "OFF–ON" luminescent response of complex **29** for Cu^{2+} detection [135]. The oxidation of phenothiazine's "S" can also be triggered by HOCl, enabling complex **43** to be used for HOCl detection by recording the emission changes at 605 nm. More interestingly, the sulfoxide oxidation product can be reduced by H_2S, resulting in a reversible chemosensor for HOCl detection.

In 2013, HOCl-mediated oxidation of amide was exploited by Zhang et al. for the development of Ru(II) complex chemosensor **44** for HOCl detection (Fig. 21) [195]. The emission of Ru(II) complex luminophore was quenched by the PeT from Ru(II) center to DNP quencher. In the presence of HOCl, fast oxidation-cleavage of DNP allowed the production of the luminescent Ru(II) complex. More than 1100-fold enhancement in luminescence at 626 nm was obtained immediately after the reaction of complex **44** with HOCl. The application of complex **44** for imaging of phagocytosis-induced HOCl production in RAW 264.7 macrophage cells was then successfully demonstrated. The decrease and blue shift of an absorption band at about 450 nm allowed further development of complex **44**-loaded UCNPs for background-free luminescent detection of HOCl [53]. In a subsequent study, Zhang et al. reported Ru(II) complex **45** for HOCl detection and lysosomal imaging (Fig. 21) [191]. Complex **45** was developed by coupling of the Ru(II) complex and ferrocenyl moiety through a HOCl-responsive hydrazine linker. Complex **45** can be internalized through a caveolae-mediated endocytosis process. The lysosomal accumulation of complex **45** allowed its use for imaging of lysosomal HOCl. The application of complex **45** in imaging of HOCl in vivo, including in flea and zebrafish, was then demonstrated.

O-nitroaniline can react specifically with HOCl to form a benzofurazan-1-oxide (BFO) in aqueous solution. Based on this reaction, Zhang et al. developed Ru(II) complex **46** as a chemosensor for HOCl detection (Fig. 22) [196]. The MLCT emission was quenched because of the PeT mechanism, while the

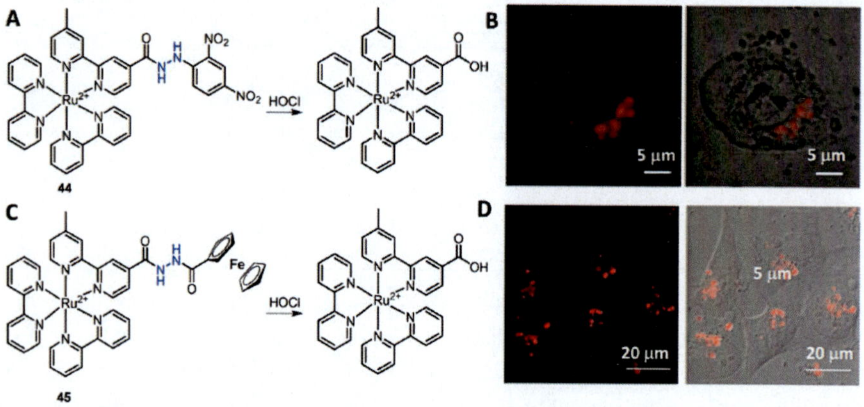

Fig. 21 Molecular structure of Ru(II) complexes **44** (**A**) and **45** (**C**) and oxidations with HOCl. Luminescence imaging of phagocytosis-induced HOCl production in RAW 264.7 macrophage cells using Ru(II) complex **44** (**B**) and lysosomal HOCl in MDA-MB-231 cells using Ru(II) complex **45** (**D**). Adapted with permission from Ref. [191, 195]. Copyright 2013, 2015 Elsevier

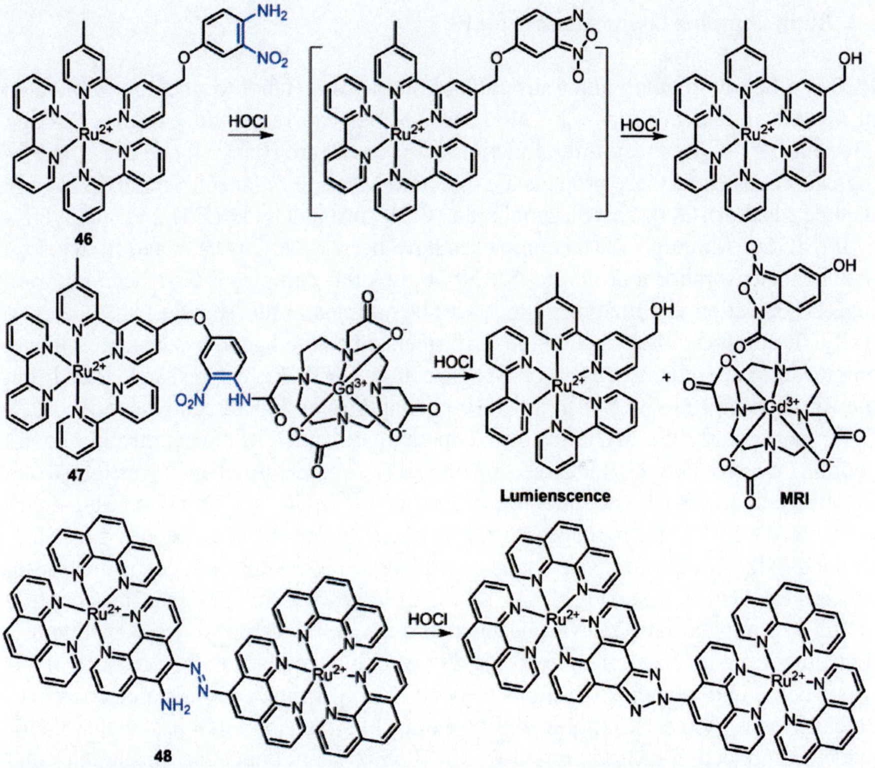

Fig. 22 Molecular structure of Ru(II) complexes **46**–**48** and their oxidations with HOCl

products of complex **46** reacting with HOCl showed intense emission in borate buffer. Complex **46** was then applied to imaging of exogenous HOCl in HeLa and endogenous HOCl in neutrophils. In a subsequent study, Shi et al. modified complex **46** with a Gd-DOTA contrast agent to produce the hetrobimetallic Ru(II)–Gd(III) complex **47** for bimodal (luminescence and magnetic resonance imaging (MRI)) determination and imaging of HOCl (Fig. 22) [197]. Upon reacting with HOCl, both luminescence and MR signals were increased. The increase of luminescence was attributed to the corruption of PeT between *o*-nitroaniline and Ru(II) complex, and the increase of MR signal was ascribed to the increased number of inner-water molecules. Then, complex **47** was applied to luminescence and MRI detection of HOCl in drug-induced acute liver and kidney injury in a mouse. Following the development of complex **30** for Cu^{2+} detection [136], the same group found that the dinuclear Ru(II) complex **48** can also be used for HOCl detection (Fig. 22) [198]. The oxidation-cyclization of the azo and amino group in complex **48** produced a triazole derivative, accompanied by > 50-fold enhancement of MLCT emission at 600 nm. Moreover, this oxidation-cyclization showed high selectivity to HOCl over other ROS and Cu^{2+}. Complex **48** was then applied to luminescence imaging of HOCl in mouse.

5.3 Ru(II) Complex Chemosensors for RCS

RCS is a family of small and transient carbon-based metabolites that are implicated in several biological process [199]. These species can react with proteins through covalent bonds. As a result, the functions of proteins are changed, and thus the biological processes of these proteins are affected. In living organisms, the RCS mainly include glyoxal (GO), carbon monoxide (CO), formaldehyde (FA) and methylglyoxal (MGO). Although Ru(II) complexes have been widely investigated in developing chemosensors for ions and ROS/RNS, only a few complexes have been reported for RCS detection and imaging, which will be outlined in this section.

Ru(II) complex with 4,5-diamino-1,10-phenanthroline ligand has been previously employed as the chemosensor for NO detection [173]. Recent research found that the Ru(II) complexes' 4,5-diamino-1,10-phenanthroline ligand can respond to RCS [200], particularly the MGO to form 2-methylpyrazino-1,10-phenanthroline ligand coordinated products [201]. Based on this reaction, Zhang et al. investigated the capability of the Ru(II) complex chemosensor for MGO determination and imaging in RAW 264.7 macrophages and flea. Recently, Zhang et al. reported a "dual-key-and-lock" Ru(II) complex chemosensor **49** for lysosomal FA determination in cancer cells and tumors (Fig. 23A) [202]. Complex **49** was designed by coupling of Ru(II) complex with a DNP quencher through an FA-responsive linker. Interestingly, the response reaction of complex **49** can only take place in the presence of FA (first "key") under acidic conditions (second "key"), which allow FA detection specifically in lysosomes. Complex **49** has a long lifetime ($\tau = 330.4$ ns), which facilitates the application of background-free TGL analysis in human serum samples and mouse organs. Luminescence imaging results clearly showed that complex **49** could be used for lysosomal FA detection in HeLa cells (Fig. 23B). With this FA-responsive Ru(II) complex, in vivo and ex vivo imaging results confirmed the much higher FA levels in tumor cells and tissues (Fig. 23C).

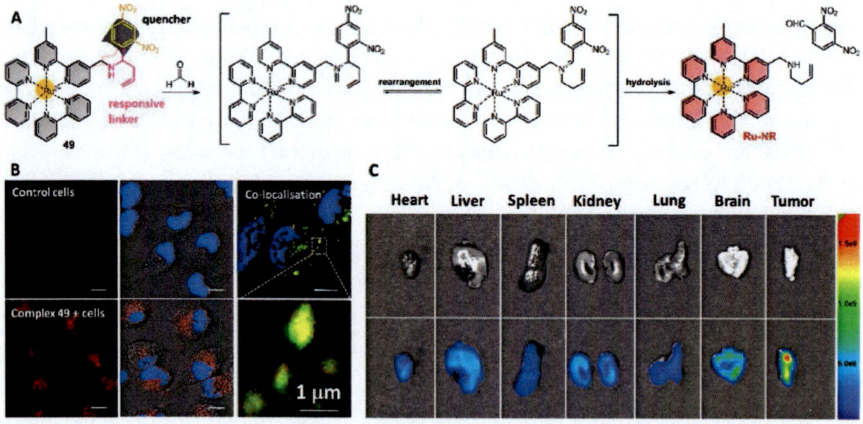

Fig. 23 Molecular structure of Ru(II) complex **49** and its response reaction with FA (**A**). The luminescence imaging of intracellular FA in lysosomes (**B**) and ex vivo imaging of FA in different mouse organs (**C**). Adapted with permission from Ref. [202]. Copyright 2019 American Chemical Society

5.4 Ru(II) Complex Chemosensors for RSS

Hydrogen sulfide (H_2S) is one of the major RSS in the human body and is involved in various biological processes. This endogenous gaseous molecule is produced by CBS (cystathionine β-synthase) and CSE (cystathionine γ-lyase) catalyzed reaction with thiol-containing biomolecules [203]. Recent research has also revealed that the H_2S is a gasotransmitter and a regulator of critical biological processes [204, 205]. The metabolites of H_2S, such as polysulfides and persulfides, are also important RSS that may have similar or divergent regulatory roles in living systems [206, 207]. Recently, by exploiting the response mechanisms of sulfoxide reduction [194], displacement of Cu^{2+} [137], cleavage of 7-nitro-2,1,3-benzoxadiazoles (NBD) [208] and DNP [209], a few Ru(II) complexes have been developed as chemosensors for RSS detection [194], and these RSS-responsive Ru(II) complexes will be discussed in this section.

In 2018, coupling of Ru(II) luminophore and DNP quencher with a H_2S-responsive linker, Du et al. reported Ru(II) complex **50** as the chemosensor for H_2S (Fig. 24A) [209]. Complex **50** was weakly luminescent because of the PeT from the Ru(II) center to DNP. In the presence of H_2S, the cleavage of the linker led to more than 86-fold enhancement in emission intensity at 612 nm. The long lifetime (> 300 ns) of both complex **50** and the products facilitated the TGL assay of H_2S in human blood serum samples and mouse organs without any background signals. The H_2S concentration was 47.70 ± 4.50 μM and < 1.2 μmol/g in sera and mouse organs, respectively. Luminescent imaging results showed that complex **50** accumulated in lysosomes, thus allowing lysosomal H_2S to be detected. By exploiting the H_2S-triggered thiolysis of NBD, Ru(II) complex **51** was recently developed by Liu et al. for the detection of H_2S in PBS buffer and in living organisms (Fig. 24B) [208]. Similar to complex **50**, the emission quenching of complex **51** was attributed to the PeT from the Ru(II) center to DNB. The thiolysis of complex **51** showed high

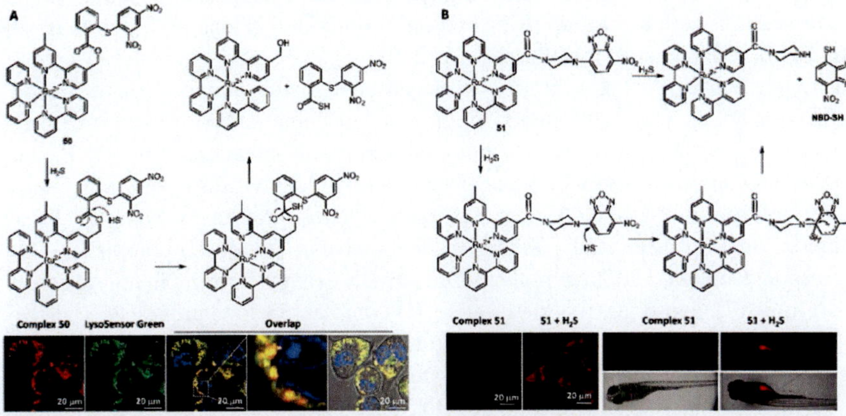

Fig. 24 Molecular structures of complex **50** (**A**) and **51** (**B**) and the application of complex **50**, **51** for H_2S imaging in cells and zebrafish. Adapted with permission from Ref. [208, 209]. Copyright 2018 Wiley and 2021 Elsevier

sensitivity (LoD = 177.3 nM) and selectivity to H_2S over other biothiols, including Cys, Hcy and glutathione (GSH). Complex **51** was then applied to visualize the H_2S in living HeLa cells and zebrafish.

6 Ru(II) Complex Chemosensors for Amino Acids

Amino acids are building blocks of proteins and play critical roles in living organisms [210]. In living organisms, the amino acids are involved in synthesis of proteins and other nitrogen-containing species, such as hormones, enzymes, creatine and some neurotransmitters. In the past few years, luminescent Ru(II) complexes have contributed significantly to the development of chemosensors for various amino acids, such as biothiols (Cys, Hcy, and GSH), histidine (His) and others, which will be discussed in this section.

6.1 Ru(II) Complex Chemosensors for Biothiols

GSH is one of the most abundant biothiols in the human body with intracellular concentrations ranging from 1 to 10 mM and about 1 mM in blood [161]. The intracellular Cys concentration is in the range of 30–200 μM and about 250 μM in blood [161]. Different from GSH and Cys, Hcy has a much lower concentration in the body, i.e., 5–15 μM in cells and about 10 μM in blood [211]. Recent studies have also revealed that the concentration of biothiols is implicated in different conditions, such as inflammation, cardiovascular diseases, HIV infection and cancers [161]. To detect biothiols in the body, a series of Ru(II) complex chemosensors have been developed through the following response mechanisms, (1) nucleophilic substitution and cleavage of the sulfonamide or sulfonate ester bond [212–214], (2) cyclization of aldehyde group with amino and thiol groups [100, 215, 216] and (3) others [217], such as reaction with azo group [218] and α,β-unsaturated ketone [219], cleavage of NBD and displacement of metal ions [139, 220].

Through coupling of 2,4-dinitrobenzenesulfonyl to the amine of phen ligand, Ru(II) complex **52** was developed for biothiols determination by Ji and co-workers in 2010 (Fig. 25) [221]. Upon the reaction with biothiols, cleavage of the DNP electron acceptor led to a 90-fold enhancement of MLCT emission at 598 nm. Imaging of biothiols in NCI-H446 cells was then demonstrated using complex **52** as chemosensor. In a prior study, Zhang et al. found that the Ru(II) complexes with two response units could have higher "OFF-to-ON" ratios for thiophenol detection [222]. In 2017, Gao et al. reported on Ru(II) complex **53** for luminescence detection of biothiols (Fig. 25) [223]. In this Ru(II) complex, two DNP quenchers were linked to two bpy ligands through a sulfonate ester bond, which enabled the quenching of MLCT emission and the "OFF–ON" response to biothiols. A morpholine moiety was conjugated to the third bpy ligand, allowing complex **53** with lysosome

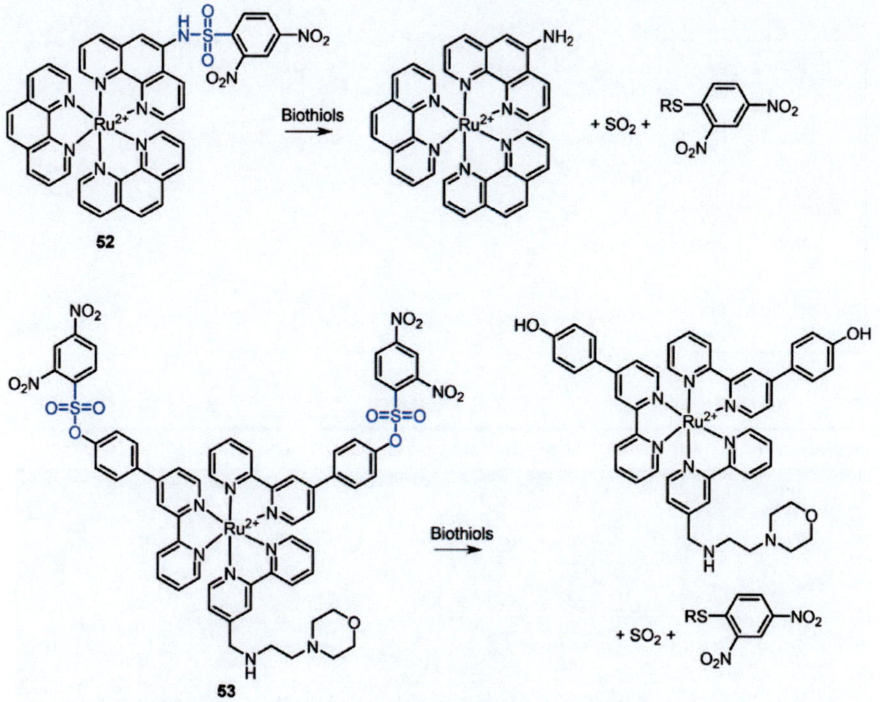

Fig. 25 Molecular structures of Ru(II) complex **52** and **53** and their response reactions with biothiols

targeting ability. The capability of complex **53** for background-free TGL detection of biothiols was also demonstrated.

Although a number of biothiol-sensitive chemosensors have been reported, the measurement of total biothiols and determination the level of each one remains a challenge. In 2020, Liu et al. reported a "Two Birds with One Stone" Ru(II) complex **54** for the detection and discrimination of biothiols in vitro and in vivo (Fig. 26A) [224]. Complex **54** was developed through coupling of two different signaling units (Ru(II) complex and NBD) through a "luminophore-responsive linker-luminophore" approach. In the presence of GSH, the cleavage of "O" ether bond led to the formation of a luminescent Ru(II) complex and non-fluorescent NBD-SR1. In contrast, the reaction of complex **54** with Cys and Hcy led to the formation of a red-emitting Ru(II) complex and NBD-SR2 that can further undergo a five- or six-member cyclic intermediate-associated rearrangement to form corresponding green-emitting NBD-NR. This allowed for discrimination of GSH from Cys and Hcy under steady-state luminescence measurements. Moreover, under the TGL measurement model, the total biothiol concentration was obtained as elimination of the emission from NBD-NR. The GSH and Cys/Hcy concentrations were thus determined by measuring the same sample with both steady-state and TGL models. The time-gated luminescence imaging of

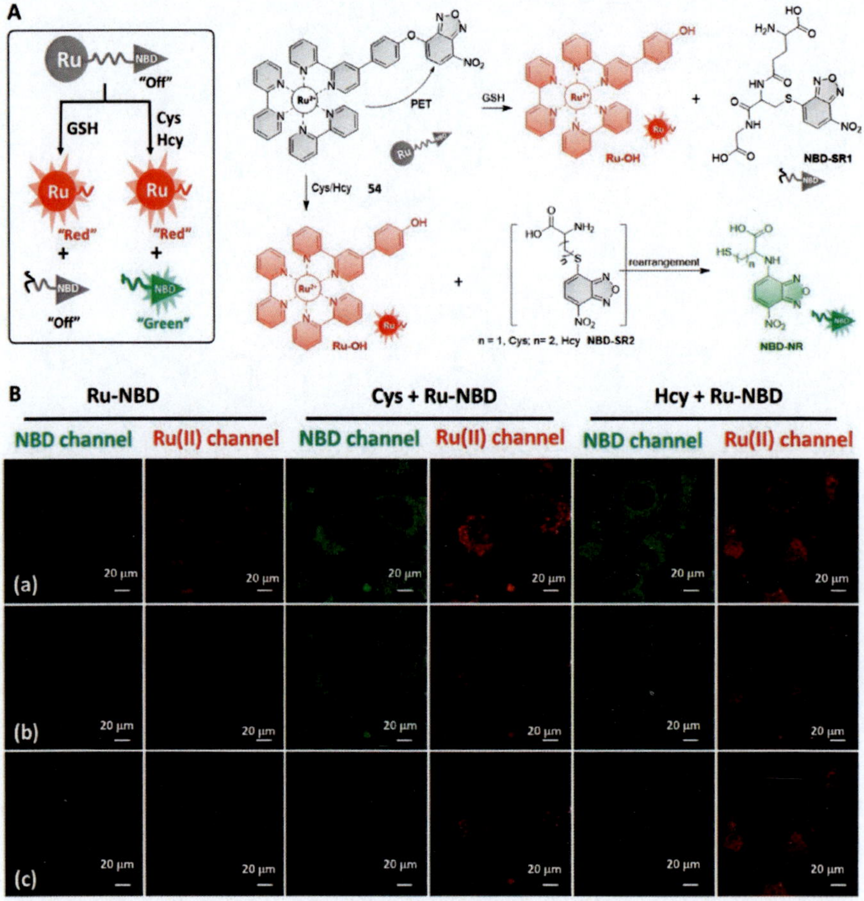

Fig. 26 Strategy for the development of Ru(II) complex **54** for biothiol detection and discrimination (**A**). Luminescence and TGL imaging (a, 0–12 ns; b, 0–4 ns; c, 4–12 ns) of HeLa cells incubated with complex **54** (**B**). Adapted with permission from Ref. [224]. Copyright 2020 Wiley

intracellular biothiols was then demonstrated, showing that the NBD emission was eliminated after a 4-ns delay ($\tau_{NBD\text{-}NR} = 0.8$ ns) (Fig. 26B).

6.2 Ru(II) Complex Chemosensors for Other Amino Acids

Ru(II) complexes have also been developed for the detection of other amino acids, such as methionine (Met) and histidine (His), through the response mechanism of amino acid-dominated binding of metal ions (e.g., Cu^{2+} and Ni^{2+}) [140, 220, 225]. For example, Gao et al. developed Ru(II) complex **55** in 2015 and then used the heterobimetallic Ru(II)–Ni(II) complex as the chemosensor for His detection (Fig. 27) [226]. Complex **55** showed intense luminescence in EtOH/HEPES buffer (50 mM, pH 7.2, 2:3, v/v), and its emission was quenched upon binding to Ni^{2+}.

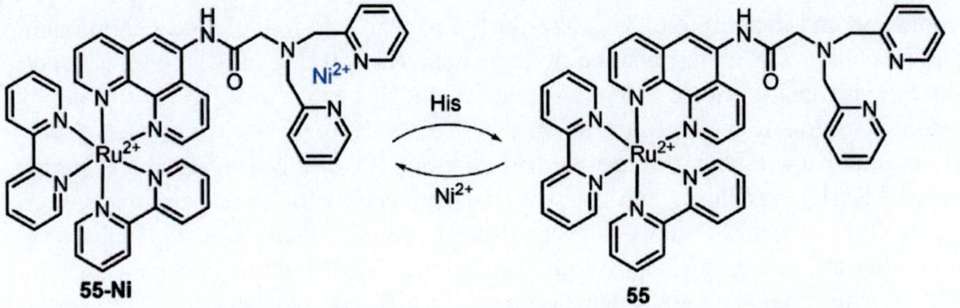

Fig. 27 Molecular structure of Ru(II) complex **55** and heterobimetallic Ru(II)–Ni(II) complex for His detection

In the presence of His, the displacement of Ni^{2+} led to the recovery of Ru(II) complex's emission at 603 nm. The LoD was 265 nM at this test condition. Imaging of His in HeLa cells, zebrafish and flea was then performed using complex **55** as the chemosensor.

7 Conclusions

The past few decades have witnessed rapid progress in the development of chemosensors for biological investigations and environmental sample determinations. Of various chemosensors, the ones using Ru(II) complex-based luminophores are particularly interesting because of their abundant photo-physical/-chemical properties. Consequently, a series of Ru(II) complex chemosensors have been developed for the detection of ions and small biomolecules in recent years. In this review, the progress in the development of Ru(II) complex chemosensors for the detection of anions, metal ions, reactive biomolecules (ROS, RNS, RSS, and RCS) and amino acids was summarized, particularly focused on those chemosensors that can be used for target analyte detection in aqueous solution and imaging in living systems.

By carefully surveying the reported Ru(II) complex chemosensors, it is clear that the Ru(II) complex has been a useful platform for the determination of various analyte levels in vitro and in vivo. Nevertheless, there is still some room for further development of Ru(II) complex chemosensors in future studies. As it has been confirmed that the photo-physical/-chemical properties of Ru(II) complexes are varied by changes of the coordination ligands [227–230], the toxicity, particularly the photo-toxicity, could be one of the key considerations in developing Ru(II) complex chemosensors in the future. Generally, low cytotoxicity is one of the essential criteria for a chemosensor to be applied in biological studies. Moreover, the triplet nature of the emission state of Ru(II) complexes with long phosphorescence lifetime is easily quenched by oxygen [56]. Therefore, development of the Ru(II) complex chemosensors with minimal quenching from the surrounding environments, particularly the levels of oxygen, is also demanded. With intense background autofluorescence in biological systems limiting the use of other probes, recent research has confirmed that indeed Ru(II) complexes with prolonged emission lifetime can be used for TGL bioassays and imaging [28, 31, 202]. Such a background-free bioassay and imaging approach allow the determination of target analytes in living cells with higher sensitivity and signal-to-noise (S/N) ratio, which can be further investigated for biomolecule detection in vivo in future studies.

In summary, taking together with the unique photo-physical/-chemical properties of Ru(II) complexes and the potential applications of chemosensors, ongoing research is expected to develop robust Ru(II) complex chemosensors for the determination and imaging of ions and biomolecules in the future. We hope that this review will provide a knowledge base for the Ru(II) complex chemosensor area and inspire the readers to contribute to this promising research field in the future.

Acknowledgements The authors gratefully acknowledge the financial support from Australian Research Council (DE170100092) and National Health and Medical Research Council (APP1175808).

Funding Open Access funding enabled and organized by CAUL and its Member Institutions.

Declarations

Conflict of interest The authors declare that they have no conflict of interest.

References

1. Schreiber Stuart L, Kotz Joanne D, Li M, Aubé J, Austin Christopher P, Reed John C, Rosen H, White EL, Sklar Larry A, Lindsley Craig W, Alexander Benjamin R, Bittker Joshua A, Clemons Paul A, de Souza A, Foley Michael A, Palmer M, Shamji Alykhan F, Wawer Mathias J, McManus O, Wu M, Zou B, Yu H, Golden Jennifer E, Schoenen Frank J, Simeonov A, Jadhav A, Jackson Michael R, Pinkerton Anthony B, Chung Thomas DY, Griffin Patrick R, Cravatt Benjamin F, Hodder Peter S, Roush William R, Roberts E, Chung D-H, Jonsson Colleen B, Noah James W, Severson William E, Ananthan S, Edwards B, Oprea Tudor I, Conn PJ, Hopkins Corey R, Wood Michael R, Stauffer Shaun R, Emmitte Kyle A, Brady Linda S, Driscoll J, Li Ingrid Y, Loomis Carson R, Margolis Ronald N, Michelotti E, Perry Mary E, Pillai A, Yao Y (2015) Advancing biological understanding and therapeutics discovery with small-molecule probes. Cell 161:1252–1265
2. Park S-H, Kwon N, Lee J-H, Yoon J, Shin I (2020) Synthetic ratiometric fluorescent probes for detection of ions. Chem Soc Rev 49:143–179
3. Singh H, Tiwari K, Tiwari R, Pramanik SK, Das A (2019) Small molecule as fluorescent probes for monitoring intracellular enzymatic transformations. Chem Rev 119:11718–11760
4. Jiang X, Wang L, Carroll SL, Chen J, Wang MC, Wang J (2018) Challenges and opportunities for small-molecule fluorescent probes in redox biology applications. Antioxid Redox Signal 29:518–540

5. Ma S, Chen G, Xu J, Liu Y, Li G, Chen T, Li Y, James TD (2021) Current strategies for the development of fluorescence-based molecular probes for visualizing the enzymes and proteins associated with Alzheimer's disease. Coord Chem Rev 427:213553
6. Rahimi F, Chatzimichail S, Saifuddin A, Surman AJ, Taylor-Robinson SD, Salehi-Reyhani A (2020) A review of portable high-performance liquid chromatography: the future of the field? Chromatographia 83:1165–1195
7. Balcaen L, Bolea-Fernandez E, Resano M, Vanhaecke F (2015) Inductively coupled plasma—Tandem mass spectrometry (ICP-MS/MS): a powerful and universal tool for the interference-free determination of (ultra)trace elements—a tutorial review. Anal Chim Acta 894:7–19
8. Theiner S, Loehr K, Koellensperger G, Mueller L, Jakubowski N (2020) Single-cell analysis by use of ICP-MS. J Anal At Spectrom 35:1784–1813
9. Park SY, Yoon SA, Cha Y, Lee MH (2021) Recent advances in fluorescent probes for cellular antioxidants: detection of NADH, hNQO1, H2S, and other redox biomolecules. Coord Chem Rev 428:213613
10. Withers PJ, Bouman C, Carmignato S, Cnudde V, Grimaldi D, Hagen CK, Maire E, Manley M, Du Plessis A, Stock SR (2021) X-ray computed tomography. Nat Rev Methods Primers 1:18
11. Avasthi A, Caro C, Pozo-Torres E, Leal MP, García-Martín ML (2020) Magnetic nanoparticles as MRI contrast agents. Top Curr Chem 378:40
12. Vaquero JJ, Kinahan P (2015) Positron emission tomography: current challenges and opportunities for technological advances in clinical and preclinical imaging systems. Annu Rev Biomed Eng 17:385–414
13. Willmann JK, van Bruggen N, Dinkelborg LM, Gambhir SS (2008) Molecular imaging in drug development. Nat Rev Drug Discov 7:591–607
14. Ametamey SM, Honer M, Schubiger PA (2008) Molecular imaging with PET. Chem Rev 108:1501–1516
15. Meng Q, Wu M, Shang Z, Zhang Z, Zhang R (2022) Responsive gadolinium(III) complex-based small molecule magnetic resonance imaging probes: design, mechanism and application. Coord Chem Rev 457:214398
16. Kobayashi H, Ogawa M, Alford R, Choyke PL, Urano Y (2010) New strategies for fluorescent probe design in medical diagnostic imaging. Chem Rev 110:2620–2640
17. Feng H, Meng Q, Ta HT, Zhang R (2020) Development of "dual-key-and-lock" responsive probes for biosensing and imaging. New J Chem 44:12890–12896
18. Wu L, Huang J, Pu K, James TD (2021) Dual-locked spectroscopic probes for sensing and therapy. Nat Rev Chem 5:406–421
19. Fernández-Suárez M, Ting AY (2008) Fluorescent probes for super-resolution imaging in living cells. Nat Rev Mol Cell Biol 9:929–943
20. Liu C, Gao X, Yuan J, Zhang R (2020) Advances in the development of fluorescence probes for cell plasma membrane imaging. Trends Analyt Chem 133:116092
21. Zhou X, Lee S, Xu Z, Yoon J (2015) Recent progress on the development of chemosensors for gases. Chem Rev 115:7944–8000
22. Wongkongkatep J, Ojida A, Hamachi I (2017) Fluorescence sensing of inorganic phosphate and pyrophosphate using small molecular sensors and their applications. Top Curr Chem 375:30
23. Gori M, Thakur A, Sharma A, Flora SJS (2021) Organic-molecule-based fluorescent chemosensor for nerve agents and organophosphorus pesticides. Top Curr Chem 379:33
24. Feng H, Liu J, Qaitoon A, Meng Q, Sultanbawa Y, Zhang Z, Xu ZP, Zhang R (2021) Responsive small-molecule luminescence probes for sulfite/bisulfite detection in food samples. Trends Anal Chem 136:116199
25. Zhang R, Song B, Yuan J (2018) Bioanalytical methods for hypochlorous acid detection: recent advances and challenges. Trends Anal Chem 99:1–33
26. Sedgwick AC, Brewster JT, Wu T, Feng X, Bull SD, Qian X, Sessler JL, James TD, Anslyn EV, Sun X (2021) Indicator displacement assays (IDAs): the past, present and future. Chem Soc Rev 50:9–38
27. Wu L, Sedgwick AC, Sun X, Bull SD, He X-P, James TD (2019) Reaction-based fluorescent probes for the detection and imaging of reactive oxygen, nitrogen, and sulfur species. Acc Chem Res 52:2582–2597
28. Zhang KY, Yu Q, Wei H, Liu S, Zhao Q, Huang W (2018) Long-Lived Emissive Probes for Time-Resolved Photoluminescence Bioimaging and Biosensing. Chem Rev 118:1770–1839

29. Aulsebrook ML, Graham B, Grace MR, Tuck KL (2018) Lanthanide complexes for luminescence-based sensing of low molecular weight analytes. Coord Chem Rev 375:191–220
30. Heffern MC, Matosziuk LM, Meade TJ (2014) Lanthanide probes for bioresponsive imaging. Chem Rev 114:4496–4539
31. Zhang R, Yuan J (2020) Responsive metal complex probes for time-gated luminescence biosensing and imaging. Acc Chem Res 53:1316–1329
32. Connell TU, Donnelly PS (2018) Labelling proteins and peptides with phosphorescent d6 transition metal complexes. Coord Chem Rev 375:267–284
33. Caporale C, Massi M (2018) Cyclometalated iridium(III) complexes for life science. Coord Chem Rev 363:71–91
34. Wing-Wah Yam V, Kam-Wing Lo K, Kit-Mai Fung W, Wang C-R (1998) Design of luminescent polynuclear copper(I) and silver(I) complexes with chalcogenides and acetylides as the bridging ligands. Coord Chem Rev 171:17–41
35. Lo KK-W (2015) Luminescent rhenium(I) and iridium(III) polypyridine complexes as biological probes, imaging reagents, and photocytotoxic agents. Acc Chem Res 48:2985–2995
36. Lo KK-W, Tsang KH-K, Sze K-S, Chung C-K, Lee TK-M, Zhang KY, Hui W-K, Li C-K, Lau JS-Y, Ng DC-M, Zhu N (2007) Non-covalent binding of luminescent transition metal polypyridine complexes to avidin, indole-binding proteins and estrogen receptors. Coord Chem Rev 251:2292–2310
37. Lo KK-W, Louie M-W, Zhang KY (2010) Design of luminescent iridium(III) and rhenium(I) polypyridine complexes as in vitro and in vivo ion, molecular and biological probes. Coord Chem Rev 254:2603–2622
38. Coogan MP, Fernández-Moreira V (2014) Progress with, and prospects for, metal complexes in cell imaging. Chem Commun 50:384–399
39. Liu Z, He W, Guo Z (2013) Metal coordination in photoluminescent sensing. Chem Soc Rev 42:1568–1600
40. Zhao Q, Huang C, Li F (2011) Phosphorescent heavy-metal complexes for bioimaging. Chem Soc Rev 40:2508–2524
41. Campagna S, Puntoriero F, Nastasi F, Bergamini G, Balzani V (2007) In: Balzani V, Campagna S (eds) Photochemistry and photophysics of coordination compounds I. Springer Berlin Heidelberg, Berlin, Heidelberg, pp 117–214
42. Ma D-L, Ma VP-Y, Chan DS-H, Leung K-H, He H-Z, Leung C-H (2012) Recent advances in luminescent heavy metal complexes for sensing. Coord Chem Rev 256:3087–3113
43. Zhao Q, Li F, Huang C (2010) Phosphorescent chemosensors based on heavy-metal complexes. Chem Soc Rev 39:3007–3030
44. Singh A, Barman P (2021) Recent advances in schiff base ruthenium metal complexes: synthesis and applications. Top Curr Chem 379:29
45. Balzani V, Ceroni P, Credi A, Venturi M (2021) Ruthenium tris(bipyridine) complexes: Interchange between photons and electrons in molecular-scale devices and machines. Coord Chem Rev 433:213758
46. McClenaghan ND, Leydet Y, Maubert B, Indelli MT, Campagna S (2005) Excited-state equilibration: a process leading to long-lived metal-to-ligand charge transfer luminescence in supramolecular systems. Coord Chem Rev 249:1336–1350
47. Lo KK-W (2020) Molecular design of bioorthogonal probes and imaging reagents derived from photofunctional transition metal complexes. Acc Chem Res 53:32–44
48. Demas JN, DeGraff BA (2001) Applications of luminescent transition platinum group metal complexes to sensor technology and molecular probes. Coord Chem Rev 211:317–351
49. Keefe MH, Benkstein KD, Hupp JT (2000) Luminescent sensor molecules based on coordinated metals: a review of recent developments. Coord Chem Rev 205:201–228
50. Tyson DS, Castellano FN (1999) Intramolecular singlet and triplet energy transfer in a ruthenium(II) diimine complex containing multiple pyrenyl chromophores. J Phys Chem A 103:10955–10960
51. Li M-J, Wong KM-C, Yi C, Yam VW-W (2012) New ruthenium(II) complexes functionalized with coumarin derivatives: synthesis, energy-transfer-based sensing of esterase, cytotoxicity, and imaging studies. Chem Eur J 18:8724–8730
52. Yang D, Guan S, Niu Y, Xie Z, Zhou S, Qu X (2018) Construction of a hypoxia responsive upconversion nanosensor for tumor imaging by fluorescence resonance energy transfer from carbon dots to ruthenium complex. J Mater Chem B 6:2315–2322

53. Zhang R, Liang L, Meng Q, Zhao J, Ta HT, Li L, Zhang Z, Sultanbawa Y, Xu ZP (2019) Responsive upconversion nanoprobe for background-free hypochlorous acid detection and bioimaging. Small 15:1803712
54. Yip AM-H, Lo KK-W (2018) Luminescent rhenium(I), ruthenium(II), and iridium(III) polypyridine complexes containing a poly(ethylene glycol) pendant or bioorthogonal reaction group as biological probes and photocytotoxic agents. Coord Chem Rev 361:138–163
55. Lo KK-W, Choi AW-T, Law WH-T (2012) Applications of luminescent inorganic and organometallic transition metal complexes as biomolecular and cellular probes. Dalton Trans 41:6021–6047
56. Ji S, Wu W, Wu W, Song P, Han K, Wang Z, Liu S, Guo H, Zhao J (2010) Tuning the luminescence lifetimes of ruthenium(II) polypyridine complexes and its application in luminescent oxygen sensing. J Mater Chem 20:1953–1963
57. Payne SJ, Fiore GL, Fraser CL, Demas JN (2010) Luminescence oxygen sensor based on a ruthenium(II) star polymer complex. Anal Chem 82:917–921
58. Zhou C, Zhao W-x, You F-t, Geng Z-x, Peng H-s (2019) Highly stable and luminescent oxygen nanosensor based on ruthenium-containing metallopolymer for real-time imaging of intracellular oxygenation. ACS Sens 4:984–991
59. Ghosh A, Das P, Gill MR, Kar P, Walker MG, Thomas JA, Das A (2011) Photoactive RuII–polypyridyl complexes that display sequence selectivity and high-affinity binding to duplex DNA through groove binding. Chem Eur J 17:2089–2098
60. Nonat AM, Quinn SJ, Gunnlaugsson T (2009) Mixed f−d coordination complexes as dual visible- and near-infrared-emitting probes for targeting DNA. Inorg Chem 48:4646–4648
61. Olson EJC, Hu D, Hörmann A, Jonkman AM, Arkin MR, Stemp EDA, Barton JK, Barbara PF (1997) First observation of the key intermediate in the "light-switch" mechanism of [Ru(phen)$_2$d ppz]^2+. J Am Chem Soc 119:11458–11467
62. Gill MR, Garcia-Lara J, Foster SJ, Smythe C, Battaglia G, Thomas JA (2009) A ruthenium(II) polypyridyl complex for direct imaging of DNA structure in living cells. Nat Chem 1:662–667
63. Baggaley E, Gill MR, Green NH, Turton D, Sazanovich IV, Botchway SW, Smythe C, Haycock JW, Weinstein JA, Thomas JA (2014) Dinuclear ruthenium(II) complexes as two-photon, time-resolved emission microscopy probes for cellular DNA. Angew Chem Int Ed 53:3367–3371
64. Saadallah D, Bellakhal M, Amor S, Lefebvre J-F, Chavarot-Kerlidou M, Baussanne I, Moucheron C, Demeunynck M, Monchaud D (2017) Selective luminescent labeling of DNA and RNA quadruplexes by π-extended ruthenium light-up probes. Chem Eur J 23:4967–4972
65. Sheet SK, Sen B, Thounaojam R, Aguan K, Khatua S (2017) Ruthenium(II) complex-based luminescent bifunctional probe for Ag^+ and phosphate ions: Ag^+-assisted detection and imaging of rRNA. Inorg Chem 56:1249–1263
66. Wragg A, Gill MR, Turton D, Adams H, Roseveare TM, Smythe C, Su X, Thomas JA (2014) Tuning the cellular uptake properties of luminescent heterobimetallic iridium(III)–ruthenium(II) DNA imaging probes. Chem Eur J 20:14004–14011
67. Boynton AN, Marcélis L, Barton JK (2016) [Ru(Me$_4$phen)$_2$d ppz]^2+, a light switch for DNA mismatches. J Am Chem Soc 138:5020–5023
68. Puckett CA, Barton JK (2008) Mechanism of cellular uptake of a ruthenium polypyridyl complex. Biochemistry 47:11711–11716
69. Puckett CA, Barton JK (2007) Methods to explore cellular uptake of ruthenium complexes. J Am Chem Soc 129:46–47
70. Berni E, Gosse I, Badocco D, Pastore P, Sojic N, Pinet S (2009) Differential photoluminescent and electrochemiluminescent detection of anions with a modified ruthenium(II)–bipyridyl complex. Chem Eur J 15:5145–5152
71. Aletti AB, Gillen DM, Gunnlaugsson T (2018) Luminescent/colorimetric probes and (chemo-) sensors for detecting anions based on transition and lanthanide ion receptor/binding complexes. Coord Chem Rev 354:98–120
72. Busschaert N, Caltagirone C, Van Rossom W, Gale PA (2015) Applications of supramolecular anion recognition. Chem Rev 115:8038–8155
73. Jose DA, Kar P, Koley D, Ganguly B, Thiel W, Ghosh HN, Das A (2007) Phenol- and catechol-based ruthenium(II) polypyridyl complexes as colorimetric sensors for fluoride ions. Inorg Chem 46:5576–5584
74. Mo H-J, Niu Y-L, Zhang M, Qiao Z-P, Ye B-H (2011) Photophysical, electrochemical and anion sensing properties of Ru(ii) bipyridine complexes with 2,2′-biimidazole-like ligand. Dalton Trans 40:8218–8225

75. Baggi G, Boiocchi M, Ciarrocchi C, Fabbrizzi L (2013) Enhancing the anion affinity of urea-based receptors with a $Ru(terpy)_2^{2+}$ chromophore. Inorg Chem 52:5273–5283
76. Bhaumik C, Saha D, Das S, Baitalik S (2011) Synthesis, structural characterization, photophysical, electrochemical, and anion-sensing studies of luminescent homo- and heteroleptic ruthenium(II) and osmium(II) complexes based on terpyridyl-imidazole ligand. Inorg Chem 50:12586–12600
77. Das S, Karmakar S, Mardanya S, Baitalik S (2014) Synthesis, structural characterization, and multichannel anion and cation sensing studies of a bifunctional Ru(ii) polypyridyl–imidazole based receptor. Dalton Trans 43:3767–3782
78. Khatua S, Samanta D, Bats JW, Schmittel M (2012) Rapid and highly sensitive dual-channel detection of cyanide by bis-heteroleptic ruthenium(II) complexes. Inorg Chem 51:7075–7086
79. Chowdhury B, Sinha S, Ghosh P (2016) Selective sensing of phosphates by a new bis-heteroleptic RuII complex through halogen bonding: a superior sensor over its hydrogen-bonding analogue. Chem Eur J 22:18051–18059
80. Chowdhury B, Khatua S, Dutta R, Chakraborty S, Ghosh P (2014) Bis-heteroleptic ruthenium(II) complex of a triazole ligand as a selective probe for phosphates. Inorg Chem 53:8061–8070
81. Hao Y, Yang P, Li S, Huang X, Yang X-J, Wu B (2012) Selective anion sensing by a ruthenium(ii)–bipyridyl-functionalized tripodal tris(urea) receptor. Dalton Trans 41:7689–7694
82. Gale PA, Caltagirone C (2015) Anion sensing by small molecules and molecular ensembles. Chem Soc Rev 44:4212–4227
83. Saha D, Das S, Mardanya S, Baitalik S (2012) Structural characterization and spectroelectrochemical, anion sensing and solvent dependence photophysical studies of a bimetallic Ru(ii) complex derived from 1,3-di(1H-imidazo[4,5-f][1,10]phenanthroline-2-yl)benzene. Dalton Trans 41:8886–8898
84. Sun Y, Hudson ZM, Rao Y, Wang S (2011) Tuning and switching MLCT phosphorescence of $[Ru(bpy)_3]^{2+}$ complexes with triarylboranes and anions. Inorg Chem 50:3373–3378
85. Zhang W, Xi X, Wang Y-L, Du Z, Liu C, Liu J, Song B, Yuan J, Zhang R (2020) Responsive ruthenium complex probe for phosphorescence and time-gated luminescence detection of bisulfite. Dalton Trans 49:5531–5538
86. Zhang R, Yu X, Yin Y, Ye Z, Wang G, Yuan J (2011) Development of a heterobimetallic Ru(II)–Cu(II) complex for highly selective and sensitive luminescence sensing of sulfide anions. Anal Chim Acta 691:83–88
87. Ghosh A, Ganguly B, Das A (2007) Urea-based ruthenium(II)—polypyridyl complex as an optical sensor for anions: synthesis, characterization, and binding studies. Inorg Chem 46:9912–9918
88. Bhaumik C, Das S, Saha D, Dutta S, Baitalik S (2010) Synthesis, characterization, photophysical, and anion-binding studies of luminescent heteroleptic bis-tridentate ruthenium(II) complexes based on 2,6-bis(benzimidazole-2-yl)pyridine and 4′-substituted 2,2′:6′,2″ Terpyridine derivatives. Inorg Chem 49:5049–5062
89. Gale PA, Howe ENW, Wu X, Spooner MJ (2018) Anion receptor chemistry: highlights from 2016. Coord Chem Rev 375:333–372
90. Karmakar S, Maity D, Mardanya S, Baitalik S (2015) Pyrene and imidazole functionalized luminescent bimetallic Ru(ii) terpyridine complexes as efficient optical chemosensors for cyanide in aqueous, organic and solid media. Dalton Trans 44:18607–18623
91. Hancock LM, Marchi E, Ceroni P, Beer PD (2012) Anion sensing in aqueous media by photoactive transition-metal bipyridyl rotaxanes. Chem Eur J 18:11277–11283
92. Mondal D, Bar M, Mukherjee S, Baitalik S (2016) Design of Ru(II) complexes based on anthraimidazoledione-functionalized terpyridine ligand for improvement of room-temperature luminescence characteristics and recognition of selective anions: experimental and DFT/TD-DFT study. Inorg Chem 55:9707–9724
93. Maity D, Vyas G, Bhatt M, Paul P (2015) Detection of NaCN in aqueous media using a calixarene-based fluoroionophore containing ruthenium(II)-bipyridine as the fluorogenic unit. RSC Adv 5:6151–6159
94. Mardanya S, Karmakar S, Bar M, Baitalik S (2015) Pyrene-biimidazole based Ru(II) and Os(II) complexes as highly efficient probes for the visible and near-infrared detection of cyanide in aqueous media. Dalton Trans 44:21053–21072
95. Cheng F, He C, Yu S, Yin H (2017) Highly selective SCN—fluorescent sensing by a Ru(II) complex containing functionalized polypyridine. Inorganica Chim Acta 462:43–49
96. Lin Z-h, Ou S-j, Duan C-y, Zhang B-g, Bai Z-p (2006) Naked-eye detection of fluoride ion in water: a remarkably selective easy-to-prepare test paper. Chem Commun 624–626

97. Lin Z-h, Zhao Y-g, Duan C-y, Zhang B-g, Bai Z-p (2006) A highly selective chromo- and fluorogenic dual responding fluoride sensor: naked-eye detection of F^- ion in natural water via a test paper. Dalton Trans 3678–3684
98. Su X, Hu R, Li X, Zhu J, Luo F, Niu X, Li M, Zhao Q (2016) Hydrophilic indolium cycloruthenated complex system for visual detection of bisulfite with a large red shift in absorption. Inorg Chem 55:745–754
99. Chan J, Dodani SC, Chang CJ (2012) Reaction-based small-molecule fluorescent probes for chemoselective bioimaging. Nat Chem 4:973–984
100. Zhang R, Yu X, Ye Z, Wang G, Zhang W, Yuan J (2010) Turn-on luminescent probe for cysteine/homocysteine based on a ruthenium(II) complex. Inorg Chem 49:7898–7903
101. Zhang W, Liu H, Zhang F, Wang Y-L, Song B, Zhang R, Yuan J (2018) Development of a ruthenium(II) complex-based luminescence probe for detection of hydrogen sulfite in food samples. Microchem J 141:181–187
102. Zhu J-W, Ou H-D, Xu N, Deng W, Yao Z-J (2020) Ruthenium-based phosphorescent probe for selective and naked-eye detection of cyanide in aqueous media. Dyes Pigm 176:108196
103. Li M, Liang Q, Zheng M, Fang C, Peng S, Zhao M (2013) An efficient ruthenium tris(bipyridine)-based luminescent chemosensor for recognition of Cu(ii) and sulfide anion in water. Dalton Trans 42:13509–13515
104. Zheng Z-B, Huang Q-Y, Han Y-F, Zuo J, Ma Y-N (2017) Ruthenium(II) complex-based chemosensors for highly sensitive and selective sequential recognition of copper ion and cyanide. Sens Actuators B Chem 253:203–212
105. Zhang S-T, Li P, Kou X, Xiao D (2018) Highly selective and sensitive luminescent turn-on probe for pyrophosphate detection in aqueous solution. ChemistrySelect 3:10057–10063
106. Cheng F, Chen J, Wang F, Tang N, Hou N, Chen G (2012) Synthesis, characterization, and pH-induced luminescence switching of two trinuclear Ru(II) polypyridyl complexes containing imidazole. Transit Met Chem 37:415–421
107. Yin H, Liu Z, Yu S, Yang Y, Dong J, Yang X, Wang F, He C, Cheng F (2021) A binuclear ruthenium polypyridyl complex: synthesis, characterization, pH luminescence sensor and electrochemical properties. Transit Met Chem 46:49–56
108. Jing B, Wu T, Tian C, Zhang M, Shen T (2000) pH-dependent luminescence of ruthenium(II) polypyridine complexes. Bull Chem Soc Jpn 73:1749–1755
109. Bai G-Y, Wang K-Z, Duan Z-M, Gao L-H (2004) Luminescent pH sensing and DNA binding properties of a novel ruthenium(II) complex. J Inorg Biochem 98:1017–1022
110. Yu F, Shen C, Zheng T, Chu W-K, Xiang J, Luo Y, Ko C-C, Guo Z-Q, Lau T-C (2016) Acid-base behaviour in the absorption and emission spectra of ruthenium(II) complexes with hydroxy-substituted bipyridine and phenanthroline ligands. Eur J Inorg Chem 2016:3641–3648
111. Gonçalves HMR, Maule CD, Jorge PAS, Esteves da Silva JCG (2008) Fiber optic lifetime pH sensing based on ruthenium(II) complexes with dicarboxybipyridine. Anal Chim Acta 626:62–70
112. Zheng Z-B, Kang S-Y, Zhao Y, Zhang N, Yi X, Wang K-Z (2015) pH and copper ion luminescence on/off sensing by a dipyrazinylpyridine-appended ruthenium complex. Sens Actuators B Chem 221:614–624
113. Zheng Z-B, Wu Y-Q, Wang K-Z, Li F (2014) pH luminescence switching, dihydrogen phosphate sensing, and cellular uptake of a heterobimetallic ruthenium(ii)–rhenium(i) complex. Dalton Trans 43:3273–3284
114. San Tan S, Yanagisawa S, Inagaki K, Morikawa Y, Kassim MB (2017) Augmented pH-sensitivity absorbance of a ruthenium(II) bis(bipyridine) complex with elongation of the conjugated ligands: an experimental and theoretical investigation. Phys Chem Chem Phys 19:25734–25745
115. Tormo L, Bustamante N, Colmenarejo G, Orellana G (2010) Can luminescent Ru(II) polypyridyl dyes measure pH directly? Anal Chem 82:5195–5204
116. Yu H-j, Hao Z-f, Peng H-l, Rao R-h, Sun M, Alana WR, Ran C, Chao H, Yu L (2017) Near-infrared lysosome pH tracker and naked-eye colorimetric nucleic acids sensor based on ruthenium complexes $[Ru(bim)_2(dppz)]^{2+}$ and $[Ru(bim)_2(pip)]^{2+}$. Sens Actuators B Chem 252:313–321
117. Meng T-T, Wang H, Zheng Z-B, Wang K-Z (2017) pH-switchable "Off–On–Off" near-infrared luminescence based on a dinuclear ruthenium(II) complex. Inorg Chem 56:4775–4779
118. Liu H-Y, Zhang S-Q, Cui M-C, Gao L-H, Zhao H, Wang K-Z (2020) pH-sensitive near-IR emitting dinuclear ruthenium complex for recognition, two-photon luminescent imaging, and subcellular localization of cancer cells. ACS Appl Bio Mater 3:5420–5427

119. Carter KP, Young AM, Palmer AE (2014) Fluorescent sensors for measuring metal ions in living systems. Chem Rev 114:4564–4601
120. Cheng F, Ren M, He C, Yin H (2016) Luminescent chemosensor for Hg^{2+} ion based on a dinuclear Ru(II) complex containing open chain azacrown ether. Inorg Chim Acta 450:170–175
121. Li M, Sheth S, Xu Y, Song Q (2020) Ru(II)-bipyridine complex as a highly sensitive luminescent probe for Cu^{2+} detection and cell imaging. Microchem J 156:104848
122. Xia XL, Zhang DB, Zhang JL, Pu SZ (2020) Highly sensitive ruthenium complex-based fluorescent probe for copper ion detection. Tetrahedron 76:131526
123. Kumar P, Kumar S (2020) Copper ion luminescence on/off sensing by a quinoline-appended ruthenium(II)-polypyridyl complex in aqueous media. J Mol Struct 1202:127242
124. Zhang Y, Liu Z, Zhang Y, Xu Y, Li H, Wang C, Lu A, Sun S (2015) A reversible and selective luminescent probe for Cu^{2+} detection based on a ruthenium(II) complex in aqueous solution. Sens Actuators B Chem 211:449–455
125. Lin Q-T, Pei L-M, Xu W-C, Chao H, Ji L-N (2012) $[Ru(bpy)_2(pipdpa)]^{2+}$ as a highly sensitive and selective luminescent chemosensor for Cu^{2+} in aqueous solution. Inorg Chem Commun 16:104–106
126. Li X, Chen Y, Meng J, Lü K, Zhang A, Zhang B (2011) A new luminescent ruthenium(II) polypyridine-derived dipicolylamine complex as a sensor for Cu^{2+} ions. Chin J Chem 29:1947–1950
127. Liu X-W, Xiao Y, Zhang S-B, Lu J-L (2017) A selective luminescent sensor for the detection of copper (II) ions based on a ruthenium complex containing DPA moiety. Inorg Chem Commun 84:56–58
128. Geißer B, Alsfasser R (2003) Probing the aqueous copper(II) coordination chemistry of bifunctional chelating amino acid ligands with a luminescent ruthenium chromophore. Dalton Trans 612–618
129. He C, Yu S, Ma S, Liu Z, Yao L, Cheng F, Liu P (2019) A novel ruthenium(II) polypyridyl complex bearing 1,8-naphthyridine as a high selectivity and sensitivity fluorescent chemosensor for Cu^{2+} and Fe3+ ions. Molecules 24
130. Ramachandran M, Anandan S (2020) Triazole appending ruthenium(II) polypyridine complex for selective sensing of phosphate anions through C–H–anion interaction and copper(II) ions via cancer cells. New J Chem 44:6186–6196
131. He C-L, Ren F-L, Zhang X-B, Dong Y-Y, Zhao Y (2006) A fluorescent chemosensor for copper(II) based on a carboxylic acid-functionalized tris(2,2′-bipyridine)-ruthenium(II) complex. Anal Sci 22:1547–1551
132. Comba P, Krämer R, Mokhir A, Naing K, Schatz E (2006) Synthesis of new phenanthroline-based heteroditopic ligands—highly efficient and selective fluorescence sensors for copper(II) ions. Eur J Inorg Chem 2006:4442–4448
133. Cheng F, Tang N, Miao K, Wang F (2014) A dinuclear ruthenium(II) polypyridyl complex containing a terpy-like fragment for Cu^{2+} probing. Z Anorg Allg Chem 640:1816–1821
134. Zhang P, Pei L, Chen Y, Xu W, Lin Q, Wang J, Wu J, Shen Y, Ji L, Chao H (2013) A dinuclear ruthenium(II) complex as a one- and two-photon luminescent probe for biological Cu^{2+} detection. Chem Eur J 19:15494–15503
135. Ajayakumar G, Sreenath K, Gopidas KR (2009) Phenothiazine attached $Ru(bpy)_3{}^{2+}$ derivative as highly selective "turn-ON" luminescence chemodosimeter for Cu^{2+}. Dalton Trans 1180–1186
136. Zhang Y, Liu Z, Yang K, Zhang Y, Xu Y, Li H, Wang C, Lu A, Sun S (2015) A ruthenium(II) complex as turn-on Cu(II) luminescent sensor based on oxidative cyclization mechanism and its application in vivo. Sci Rep 5:8172
137. Ye Z, An X, Song B, Zhang W, Dai Z, Yuan J (2014) A novel dinuclear ruthenium(ii)–copper(ii) complex-based luminescent probe for hydrogen sulfide. Dalton Trans 43:13055–13060
138. Padilla-Tosta Miguel E, Lloris José M, Martínez-Máñez R, Pardo T, Sancenón F, Soto J, Marcos MD (2001) ATP recognition through a fluorescence change in a multicomponent dinuclear system containing a $Ru(Tpy)_2{}^{2+}$ fluorescent core and a cyclam–Cu^{2+} complex. Eur J Inorg Chem 2001:1221–1226
139. Zhang P, Wang Y, Huang W, Zhao Z, Li H, Wang H, He C, Liu J, Zhang Q (2018) "Turn off-on" phosphorescent sensor for biothiols based on a Ru–Cu ensemble. Sens Actuators B Chem 255:283–289

140. Zhang S-T, Li P, Liao C, Luo T, Kou X, Xiao D (2018) A highly sensitive luminescent probe based on Ru(II)-bipyridine complex for Cu^{2+}, l-Histidine detection and cellular imaging. Spectrochim Acta A 201:161–169
141. Wang J-N, Qi Q, Zhang L, Li S-H (2012) Turn-On luminescent sensing of metal cations via quencher displacement: rational design of a highly selective chemosensor for chromium(III). Inorg Chem 51:13103–13107
142. Liu Q, Peng J, Sun L, Li F (2011) High-efficiency upconversion luminescent sensing and bioimaging of Hg(II) by chromophoric ruthenium complex-assembled nanophosphors. ACS Nano 5:8040–8048
143. Duan C, Liang L, Li L, Zhang R, Xu ZP (2018) Recent progress in upconversion luminescence nanomaterials for biomedical applications. J Mater Chem B 6:192–209
144. Fan S-H, Shen J, Wu H, Wang K-Z, Zhang A-G (2015) A highly selective turn-on colorimetric and luminescence sensor based on a triphenylamine-appended ruthenium(II) dye for detecting mercury ion. Chin Chem Lett 26:580–584
145. Li X, Wu Y, Liu Y, Zou X, Yao L, Li F, Feng W (2014) Cyclometallated ruthenium complex-modified upconversion nanophosphors for selective detection of Hg^{2+} ions in water. Nanoscale 6:1020–1028
146. Li X, Du K, Xie C, Wu Y, Zhang B, Tang D (2020) A highly sensitive and selective colorimetric probe based on a cycloruthenated complex: an Hg^{2+}-promoted switch of thiophene coordination. Dalton Trans 49:2024–2032
147. Wu Y, Cheng X, Xie C, Du K, Li X, Tang D (2020) A polymer membrane tethered with a cycloruthenated complex for colorimetric detection of Hg^{2+} ions. Spectrochim Acta A 228:117541
148. Odhiambo RA, Aluoch AO, Njenga LW, Kagwanja SM, Wandiga SO, Wendt OF (2018) Synthesis, characterisation and ion-binding properties of oxathiacrown ethers appended to $[Ru(bpy)_2]^{2+}$. Selectivity towards Hg^{2+}, Cd^{2+} and Pb^{2+}. RSC Adv 8:3663–3672
149. Khatua S, Schmittel M (2013) A single molecular light-up sensor for quantification of Hg^{2+} and Ag+ in aqueous medium: high selectivity toward Hg^{2+} over Ag^{+} in a mixture. Org Lett 15:4422–4425
150. Ru J, Tang X, Ju Z, Zhang G, Dou W, Mi X, Wang C, Liu W (2015) Exploitation and application of a highly sensitive Ru(II) complex-based phosphorescent chemodosimeter for Hg^{2+} in aqueous solutions and living cells. ACS Appl Mater Interfaces 7:4247–4256
151. Ru J, Mi X, Guan L, Tang X, Ju Z, Zhang G, Wang C, Liu W (2015) Design and application of a water-soluble phosphorescent Ru(ii) complex as turn-on sensing material for Hg^{2+}. J Mater Chem B 3:6205–6212
152. Cheng F, Tang N, Yue X (2009) A new family of Ru(II) polypyridyl complexes containing open-chain crown ether for Mg^{2+} and Ca^{2+} probing. Spectrochim Acta A 71:1944–1951
153. He C, Yu S, Ma S, Liu Z, Yao L, Cheng F, Liu P (2019) A novel ruthenium(II) polypyridyl complex bearing 1,8-naphthyridine as a high selectivity and sensitivity fluorescent chemosensor for Cu^{2+} and Fe^{3+} ions. Molecules 24:4032
154. Cheng F, Tang N (2008) A novel family of Ru(II) polypyridyl complexes containing benzocrown ether for Na^{+} and Li^{+} probing. Inorg Chem Commun 11:400–403
155. Li C-Y, Zhang X-B, Jin Z, Han R, Shen G-L, Yu R-Q (2006) A fluorescent chemosensor for cobalt ions based on a multi-substituted phenol-ruthenium(II) tris(bipyridine) complex. Anal Chim Acta 580:143–148
156. Lin H, Cinar ME, Schmittel M (2010) Comparison of ruthenium(II) and cyclometalated iridium(III) azacrown ether phenanthroline hybrids for the detection of metal cations by electrochemiluminescence. Dalton Trans 39:5130–5138
157. Barigelletti F, Flamigni L, Calogero G, Hammarström L, Sauvage J-P, Collin J-P (1998) A functionalized ruthenium(II)-bis-terpyridine complex as a rod-like luminescent sensor of zinc(II). Chem Commun 2333–2334
158. Zheng Z-B, Duan Z-M, Ma Y-Y, Wang K-Z (2013) Highly sensitive and selective difunctional ruthenium(II) complex-based chemosensor for dihydrogen phosphate anion and ferrous cation. Inorg Chem 52:2306–2316
159. Kumar A, Chhatwal M, Singh AK, Singh V, Trivedi M (2014) A fast and selective probe for monitoring Pd^{2+} in aqueous medium via the dual-optical readout. Chem Commun 50:8488–8490
160. Xie Z, Wen J, Sun S, Zhang J, Deng X, Han S, Wang L, Zhang B, Hong C, Sun S (2021) In-vitro and in-vivo monitoring of gold(III) ions from intermediate metabolite of sodium aurothiomalate through water-soluble ruthenium (II) complex-based luminescent probe. Bioorg Chem 110:104749

161. Zhang R, Yong J, Yuan J, Ping XuZ (2020) Recent advances in the development of responsive probes for selective detection of cysteine. Coord Chem Rev 408:213182
162. Liu J, Wu M, Zhang R, Xu ZP (2021) Oxygen-derived free radicals: Production, biological importance, bioimaging, and analytical detection with responsive luminescent nanoprobes. View 20200139
163. Wu Y, Vazquez-Prada KX, Liu Y, Whittaker AK, Zhang R, Ta HT (2021) Recent advances in the development of theranostic nanoparticles for cardiovascular diseases. Nanotheranostics 5:499–514
164. Winterbourn CC (2008) Reconciling the chemistry and biology of reactive oxygen species. Nat Chem Biol 4:278–286
165. Kwon N, Kim D, Swamy KMK, Yoon J (2021) Metal-coordinated fluorescent and luminescent probes for reactive oxygen species (ROS) and reactive nitrogen species (RNS). Coord Chem Rev 427:213581
166. Chen X, Wang F, Hyun JY, Wei T, Qiang J, Ren X, Shin I, Yoon J (2016) Recent progress in the development of fluorescent, luminescent and colorimetric probes for detection of reactive oxygen and nitrogen species. Chem Soc Rev 45:2976–3016
167. Patel RP, McAndrew J, Sellak H, White CR, Jo H, Freeman BA, Darley-Usmar VM (1999) Biological aspects of reactive nitrogen species. Biochim Biophys Acta Bioenerg 1411:385–400
168. Vidanapathirana AK, Psaltis PJ, Bursill CA, Abell AD, Nicholls SJ (2021) Cardiovascular bioimaging of nitric oxide: achievements, challenges, and the future. Med Res Rev 41:435–463
169. Wang S, Chen L, Jangili P, Sharma A, Li W, Hou J-T, Qin C, Yoon J, Kim JS (2018) Design and applications of fluorescent detectors for peroxynitrite. Coord Chem Rev 374:36–54
170. Zhang R, Ye Z, Wang G, Zhang W, Yuan J (2010) Development of a ruthenium(II) complex based luminescent probe for imaging nitric oxide production in living cells. Chem Eur J 16:6884–6891
171. Zhang W, Zhao D, Zhang R, Ye Z, Wang G, Yuan J, Yang M (2011) A ruthenium(ii) complex based turn-on electrochemiluminescence probe for the detection of nitric oxide. Analyst 136:1867–1872
172. Vidanapathirana AK, Pullen BJ, Zhang R, Duong M, Goyne JM, Zhang X, Bonder CS, Abell AD, Bursill CA, Nicholls SJ, Psaltis PJ (2019) A novel ruthenium-based molecular sensor to detect endothelial nitric oxide. Sci Rep 9:1720
173. Zhang W, Zhang J, Zhang H, Cao L, Zhang R, Ye Z, Yuan J (2013) Development and application of a ruthenium(II) complex-based photoluminescent and electrochemiluminescent dual-signaling probe for nitric oxide. Talanta 116:354–360
174. Yu X, Zhang R, Ye Z, Song B, Yuan J (2013) Design and synthesis of a ruthenium(II) complex-based luminescent probe for highly selective and sensitive luminescence detection of nitric oxide. J Fluoresc 23:1113–1120
175. Ghosh A, Das P, Saha S, Banerjee T, Bhatt HB, Das A (2011) Diamine derivative of a ruthenium(II)-polypyridyl complex for chemodosimetric detection of nitrite ion in aqueous solution. Inorg Chim Acta 372:115–119
176. Ma J, Wu J, Liu W, Wang P, Fan Z (2012) Ruthenium(II) complex-based fluorescent sensor for peroxynitrite. Spectrochim Acta A 94:340–345
177. Zhang W, Liu Y, Gao Q, Liu C, Song B, Zhang R, Yuan J (2018) A ruthenium(II) complex–cyanine energy transfer scaffold based luminescence probe for ratiometric detection and imaging of mitochondrial peroxynitrite. Chem Commun 54:13698–13701
178. Jia X, Chen Q, Yang Y, Tang Y, Wang R, Xu Y, Zhu W, Qian X (2016) FRET-based mito-specific fluorescent probe for ratiometric detection and imaging of endogenous peroxynitrite: Dyad of Cy3 and Cy5. J Am Chem Soc 138:10778–10781
179. Gomes A, Fernándes E, Lima JLFC (2006) Use of fluorescence probes for detection of reactive nitrogen species: a review. J Fluoresc 16:119–139
180. Wang H, Zhang R, Bridle KR, Jayachandran A, Thomas JA, Zhang W, Yuan J, Xu ZP, Crawford DHG, Liang X, Liu X, Roberts MS (2017) Two-photon dual imaging platform for in vivo monitoring cellular oxidative stress in liver injury. Sci Rep 7:45374
181. Wilson GJ, Launikonis A, Sasse WHF, Mau AWH (1997) Excited-state processes in ruthenium(II) bipyridine complexes containing covalently bound arenes. J Phys Chem A 101:4860–4866
182. Tyson DS, Bialecki J, Castellano FN (2000) Ruthenium(II) complex with a notably long excited state lifetime. Chem Commun 2355–2356
183. Tyson DS, Luman CR, Zhou X, Castellano FN (2001) New Ru(II) chromophores with extended excited-state lifetimes. Inorg Chem 40:4063–4071

184. Tyson DS, Henbest KB, Bialecki J, Castellano FN (2001) Excited state processes in ruthenium(II)/pyrenyl complexes displaying extended lifetimes. J Phys Chem A 105:8154–8161
185. Wilson GJ, Sasse WHF, Mau AWH (1996) Singlet and triplet energy transfer processes in ruthenium(II) bipyridine complexes containing covalently bound arenes. Chem Phys Lett 250:583–588
186. Yin H-J, Liu Y-J, Gao J, Wang K-Z (2017) A highly sensitive and selective visible-light excitable luminescent probe for singlet oxygen based on a dinuclear ruthenium complex. Dalton Trans 46:3325–3331
187. Ye Z, Song B, Yin Y, Zhang R, Yuan J (2013) Development of singlet oxygen-responsive phosphorescent ruthenium(II) complexes. Dalton Trans 42:14380–14383
188. Zhang R, Ye Z, Song B, Dai Z, An X, Yuan J (2013) Development of a ruthenium(II) complex-based luminescent probe for hypochlorous acid in living cells. Inorg Chem 52:10325–10331
189. Zhang J, Zhang D, Xiao L, Pu S (2021) Development of an ultrasensitive Ru(II) complex-based fluorescent probe with phenothiazine unit for selective detection HOCl and its application in water samples. Dyes Pigm 188:109179
190. Yu X, Zhang W, Ye Z, Song B, Yuan J (2015) Development of a functional ruthenium(II) complex that can act as a photoluminescent and electrochemiluminescent dual-signaling probe for hypochlorous acid. J Fluoresc 25:997–1004
191. Cao L, Zhang R, Zhang W, Du Z, Liu C, Ye Z, Song B, Yuan J (2015) A ruthenium(II) complex-based lysosome-targetable multisignal chemosensor for in vivo detection of hypochlorous acid. Biomaterials 68:21–31
192. Sen B, Sheet SK, Patra SK, Koner D, Saha N, Khatua S (2019) Highly selective detection of hypochlorous acid by a bis-heteroleptic Ru(II) complex of pyridyl-1,2,3-triazole ligand via C(sp2)–H hydroxylation. Inorg Chem 58:9982–9991
193. Li M, Wang Y, Yang Y, Gao Y, Zhao M, Zheng M, Peng S (2015) Oximated ruthenium tris-bipyridyl complex: synthesis and luminescent response specifically for ClO^- in water containing multiple ions. Dalton Trans 44:14071–14076
194. Liu F, Gao Y, Wang J, Sun S (2014) Reversible and selective luminescent determination of ClO−/H2S redox cycle in vitro and in vivo based on a ruthenium trisbipyridyl probe. Analyst 139:3324–3329
195. Zhang R, Song B, Dai Z, Ye Z, Xiao Y, Liu Y, Yuan J (2013) Highly sensitive and selective phosphorescent chemosensors for hypochlorous acid based on ruthenium(II) complexes. Biosens Bioelectron 50:1–7
196. Ye Z, Zhang R, Song B, Dai Z, Jin D, Goldys EM, Yuan J (2014) Development of a functional ruthenium(II) complex for probing hypochlorous acid in living cells. Dalton Trans 43:8414–8420
197. Shi W, Song B, Liu Z, Zhang W, Tan M, Song F, Yuan J (2020) Smart bimodal imaging of hypochlorous acid in vivo using a heterobimetallic ruthenium(II)–gadolinium(III) complex probe. Anal Chem 92:11145–11154
198. Liu Z, Gao K, Wang B, Yan H, Xing P, Zhong C, Xu Y, Li H, Chen J, Wang W, Sun S (2016) A dinuclear ruthenium(II) complex as turn-on luminescent probe for hypochlorous acid and its application for in vivo imaging. Sci Rep 6:29065
199. Ohata J, Bruemmer KJ, Chang CJ (2019) Activity-based sensing methods for monitoring the reactive carbon species carbon monoxide and formaldehyde in living systems. Acc Chem Res 52:2841–2848
200. Kele P, Li X, Duerkop A (2013) New luminescent ruthenium probes for detection of diacetyl. Microchem J 108:156–160
201. Zhang W, Zhang F, Wang Y-L, Song B, Zhang R, Yuan J (2017) Red-emitting ruthenium(II) and iridium(III) complexes as phosphorescent probes for methylglyoxal in vitro and in vivo. Inorg Chem 56:1309–1318
202. Liu C, Zhang R, Zhang W, Liu J, Wang Y-L, Du Z, Song B, Xu ZP, Yuan J (2019) "Dual-key-and-lock" ruthenium complex probe for lysosomal formaldehyde in cancer cells and tumors. J Am Chem Soc 141:8462–8472
203. Kolluru GK, Shen X, Kevil CG (2020) Reactive sulfur species. Arterioscler Thromb Vasc Biol 40:874–884
204. Gruhlke MCH, Slusarenko AJ (2012) The biology of reactive sulfur species (RSS). Plant Physiol Biochem 59:98–107
205. Pluth MD, Zhao Y, Cerda MM (2020) In: Chenoweth DM (ed) methods in enzymology. Academic Press, pp 149–164

206. Yuan S, Shen X, Kevil CG (2017) Beyond a gasotransmitter: hydrogen sulfide and polysulfide in cardiovascular health and immune response. Antioxid Redox Signal 27:634–653
207. Jiang C, Huang H, Kang X, Yang L, Xi Z, Sun H, Pluth MD, Yi L (2021) NBD-based synthetic probes for sensing small molecules and proteins: design, sensing mechanisms and biological applications. Chem Soc Rev 50:7436–7495
208. Liu C, Liu J, Zhang W, Wang Y-L, Gao X, Song B, Yuan J, Zhang R (2021) A Ruthenium(II) complex-based probe for colorimetric and luminescent detection and imaging of hydrogen sulfide in living cells and organisms. Anal Chim Acta 1145:114–123
209. Du Z, Song B, Zhang W, Duan C, Wang Y-L, Liu C, Zhang R, Yuan J (2018) Quantitative monitoring and visualization of hydrogen sulfide in vivo using a luminescent probe based on a ruthenium(II) complex. Angew Chem Int Ed 57:3999–4004
210. Cheng Z, Kuru E, Sachdeva A, Vendrell M (2020) Fluorescent amino acids as versatile building blocks for chemical biology. Nat Rev Chem 4:275–290
211. Niu L-Y, Chen Y-Z, Zheng H-R, Wu L-Z, Tung C-H, Yang Q-Z (2015) Design strategies of fluorescent probes for selective detection among biothiols. Chem Soc Rev 44:6143–6160
212. Zheng Z-B, Cui J-C, Han Y-F, Ge Y-Q, Zuo J, Hao W-X (2019) Development of fast-response turn-on phosphorescent probes for biothiols based on ruthenium(II) complexes. Anal Methods 11:2341–2350
213. Zheng Z-B, Han Y-F, Ge Y-Q, Cui J-C, Zuo J, Nie K (2019) Rapid and selective detection of biothiols by novel ruthenium(II) complex-based phosphorescence probes. Spectrochim Acta A 216:328–334
214. Zhang W, Zhang R, Zhang J, Ye Z, Jin D, Yuan J (2012) Photoluminescent and electrochemiluminescent dual-signaling probe for bio-thiols based on a ruthenium(II) complex. Anal Chim Acta 740:80–87
215. Li M-J, Zhan C-Q, Nie M-J, Chen G-N, Chen X (2011) Selective recognition of homocysteine and cysteine based on new ruthenium(II) complexes. J Inorg Biochem 105:420–425
216. Chen H, Li X, Wu Y, Gao W, Bai R (2012) A ruthenium(II) complex with environment-responsive dual emission and its application in the detection of cysteine/homocysteine. Dalton Trans 41:13292–13297
217. Ye Z, Gao Q, An X, Song B, Yuan J (2015) A functional ruthenium(II) complex for imaging biothiols in living bodies. Dalton Trans 44:8278–8283
218. Li G-Y, Liu J-P, Huang H-Y, Wen Y, Chao H, Ji L-N (2013) Colorimetric and luminescent dual-signaling responsive probing of thiols by a ruthenium(II)-azo complex. J Inorg Biochem 121:108–113
219. Li Y, Shi N, Li M (2019) An efficient ruthenium(II) tris(bipyridyl)-based chemosensor for the specific detection of cysteine and its luminescence imaging in living zebrafish. New J Chem 43:18517–18524
220. Chow C-F, Chiu BKW, Lam MHW, Wong W-Y (2003) A trinuclear heterobimetallic Ru(II)/Pt(II) complex as a chemodosimeter selective for sulfhydryl-containing amino acids and peptides. J Am Chem Soc 125:7802–7803
221. Ji S, Guo H, Yuan X, Li X, Ding H, Gao P, Zhao C, Wu W, Wu W, Zhao J (2010) A highly selective OFF-ON red-emitting phosphorescent thiol probe with large stokes shift and long luminescent lifetime. Org Lett 12:2876–2879
222. Zhang R, Ye Z, Yin Y, Wang G, Jin D, Yuan J, Piper JA (2012) Developing red-emissive ruthenium(II) complex-based luminescent probes for cellular imaging. Bioconjug Chem 23:725–733
223. Gao Q, Zhang W, Song B, Zhang R, Guo W, Yuan J (2017) Development of a novel lysosome-targeted ruthenium(II) complex for phosphorescence/time-gated luminescence assay of biothiols. Anal Chem 89:4517–4524
224. Liu C, Liu J, Zhang W, Wang Y-L, Liu Q, Song B, Yuan J, Zhang R (2020) "Two birds with one stone" ruthenium(II) complex probe for biothiols discrimination and detection in vitro and in vivo. Adv Sci 7:2000458
225. Yang Y, Li Y, Zhi X, Xu Y, Li M (2020) A red-emitting luminescent probe for sequentially detecting Cu^{2+} and cysteine/histidine in aqueous solution and its imaging application in living zebrafish. Dyes Pigm 183:108690
226. Gao Q, Song B, Ye Z, Yang L, Liu R, Yuan J (2015) A highly selective phosphorescence probe for histidine in living bodies. Dalton Trans 44:18671–18676

227. Li X, Gorle AK, Sundaraneedi MK, Keene FR, Collins JG (2018) Kinetically-inert polypyridylruthenium(II) complexes as therapeutic agents. Coord Chem Rev 375:134–147
228. Ko C-N, Li G, Leung C-H, Ma D-L (2019) Dual function luminescent transition metal complexes for cancer theranostics: the combination of diagnosis and therapy. Coord Chem Rev 381:79–103
229. Liu J, Zhang C, Rees TW, Ke L, Ji L, Chao H (2018) Harnessing ruthenium(II) as photodynamic agents: encouraging advances in cancer therapy. Coord Chem Rev 363:17–28
230. Knoll JD, Turro C (2015) Control and utilization of ruthenium and rhodium metal complex excited states for photoactivated cancer therapy. Coord Chem Rev 282–283:110–126